349.1 afw: van't middagront AC
130.9 afw: van't middagront CB
400.0 afwijk: van't midd: ront AB
240.0 afw: van't midd: ront
349.1 afw: van't middag ront
109.1 afw: van't middagront HC

om HG te vinden

gelijk Rad: ∠H tot AH alzoo tang: ∠GAH tot HG
                              14 gr: 15 m ∠GAH
100000 :   240.9 ::   25397
                          24
                       ─────
                       101588
                        50794
                       ─────
                       60.952 voor HG in tiende deelen

In heeft de driehoek HGC ook drie bekende palen
te weten HG en HC ende ∠H regt, kies dooh zoek ik
de hoek HCG

gelijk HC: tot Rad ∠H, alzoo HG tot tang: ∠HCG
   HC   HG
  109.1 :   100000 :: 60.9½   60954000/55066 tang van
                       100000   109.1   (29 gr: 15½ m:
                       ──────   voor ∠HCG
                       6090000
                         50000
                       ──────
                       6095000

~~55 gr tang van 29 gr: voor ∠HCG~~

Om GC te vinden

gelijk Rad: ∠H tot HC alzoo secans: ∠HCG tot GC
   HC   29 gr: 15½ ∠HCG
100000 : 109.1 :: 114649
                     109.1
                  ─────
                  114649
                 1030800
                 1145490
                 ─────
                 124979969 voor GC in tiende deelen dat is 124.9⁷⁰

Voorts kan men ook cf vinden
van de □ Dc. cf = AC. CB door de 35: 3 Euclid.
en daer om is

D.c AC :: CB : cf
     349.1   130.9
     349.1   ...
     940.0   ...

Sailing School

SAILING

INFORMATION CULTURES
*Ann Blair, Anthony Grafton,
and Earle A. Havens, Series Editors*

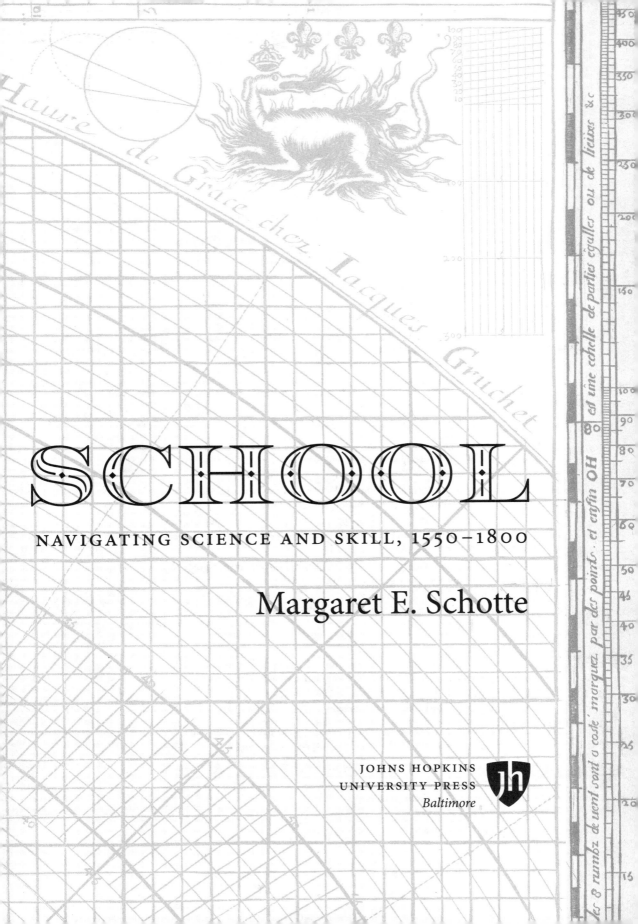

SCHOOL

NAVIGATING SCIENCE AND SKILL, 1550–1800

Margaret E. Schotte

JOHNS HOPKINS
UNIVERSITY PRESS
Baltimore

This publication was made possible in part by the Barr Ferree Foundation Fund for Publications, Department of Art and Archaeology, Princeton University.

© 2019 Margaret E. Schotte
All rights reserved. Published 2019
Printed in the United States of America on acid-free paper
9 8 7 6 5 4 3 2 1

Johns Hopkins University Press
2715 North Charles Street
Baltimore, Maryland 21218-4363
www.press.jhu.edu

Library of Congress Cataloging-in-Publication Data

Names: Schotte, Margaret E., 1976– author.
Title: Sailing school : navigating science and skill, 1550–1800 / Margaret E. Schotte.
Description: Baltimore : Johns Hopkins University Press, 2019. | Series: Information cultures | Includes bibliographical references and index.
Identifiers: LCCN 2018041206 | ISBN 9781421429533 (hardcover : alk. paper) | ISBN 9781421429540 (electronic) | ISBN 1421429535 (hardcover : alk. paper) | ISBN 1421429543 (electronic)
Subjects: LCSH: Navigation — Research — Europe. | Navigation — Study and teaching — Europe. | Navigation — Europe — History.
Classification: LCC VK455 .S36 2019 | DDC 623.8907/04 — dc23
LC record available at https://lccn.loc.gov/2018041206

A catalog record for this book is available from the British Library.

Title-page art: Sinical quadrant. S. Le Cordier, "Quartier de Proportion" (Havre de Grâce: J. Gruchet, 1682). Bibliothèque nationale de France.
Endpapers: A. van Asson, Schatkamer ([Maassluis?, Netherlands], ca. 1705), ff. 83r–v, [134v]–135. York University Libraries, CTASC, Toronto, ON.

Special discounts are available for bulk purchases of this book. For more information, please contact Special Sales at 410-516-6936 or specialsales@press.jhu.edu.

Johns Hopkins University Press uses environmentally friendly book materials, including recycled text paper that is composed of at least 30 percent post-consumer waste, whenever possible.

For Heather

How many Schools are Founded to instruct
The Youth design'd proud Vessels to Conduct?
What Volumes writ, to teach the safest way
To guide these floating Castles on the Sea?

                    W. M., IN DANIEL NEWHOUSE,
                    The Whole Art of Navigation
                    (London, 1685)

Contents

- ix   Acknowledgments
- xii   Editorial Practices

- 3   INTRODUCTION

- 17   PROLOGUE. A Model Education
  — Seville, ca. 1552

- 27   CHAPTER ONE. From the Water to the Writing Book
  — Amsterdam, ca. 1600

- 63   CHAPTER TWO. "By the Shortest Path": Developing Mathematical Rules
  — Dieppe, 1675

- 93   CHAPTER THREE. Hands-On Theory along the Thames
  — London, 1683

- 115   CHAPTER FOUR. Paper Sailors, Classroom Lessons
  — The Netherlands, ca. 1710

- 149   CHAPTER FIVE. Lieutenant Riou Is Put to the Test
  — The Southern Indian Ocean, 1789

- 173   EPILOGUE. Sailing by the Book, ca. 1800

- 185   Glossary
- 191   Notes
- 247   Bibliography
- 287   Index

Color plates appear following page 116

Acknowledgments

This book has taken me on a long journey, tacking between the shores of maritime history and the history of science, across oceans of old books. Thanks go to Anthony Grafton for his unflagging enthusiasm for this wide-ranging project and for his scrupulous feedback on how best to revive the voices of these sailors. I owe a special thanks to Eileen Reeves, who discerned the best organization for the project long before I did. Michael Gordin offered astute guidance; and María Portuondo pushed me to clarify the argument and exposition. It has been many years since Ann Blair first offered bracing critiques of my writing. In the time since she has given wise and warm advice on how to cope with information overload and other occupational hazards—merci bien.

My work has been shaped by insightful comments from participants in the History of Science Program Seminar at Princeton University. I am fortunate to have had the opportunity to discuss my research with Alexi Baker, Richard Blakemore, Peter Burke, H. Floris Cohen, Paul Cohen, Karel Davids, Richard Dunn, Danielle Fauque, Robert D. Hicks, Stephen Johnston, Matthew L. Jones, Alston Kennerley, Pamela O. Long, Bertie Mandelblatt, Mauricio Nieto Olarte, Nicholas Rodger, Neil Safier, Pamela H. Smith, Jacob Soll, Dirk Tang, and participants in the 2016 conference dedicated to Waldseemueller's Carta Marina at the Library of Congress. I owe particular thanks to Willem Mörzer Bruyns for his introductions to valuable maritime sources and people and to Diederick Wildeman for helping me track down a significant number of *schatkamers*. Margarette

Lincoln shared an opportune insight about the stage productions at Sadler's Wells, and Dániel Margócsy provided images of an ideal set of documents. For inspiration and sustaining tea and treats along the way, I am indebted to Natalie Zemon Davis, Joan Keenan, Germaine Warkentin, and Abby Zanger.

Aspects of this research have been presented at the Toronto French History Seminar, the Max Planck Institute for the History of Science, the History of Science colloquium at Johns Hopkins University, the National Maritime Museum (Greenwich), the "Global Maritime History" EMSI 2016 conference (San Marino, CA), the Herzog August Bibliothek (Wolfenbüttel), and meetings of the Canadian Nautical Research Society, History of Science Society, Society for the History of Technology, and Scientiae. Comments from panelists and audience members at these venues enriched my analysis. I would like to thank the International Maritime Economic History Association for recognizing my work with the Frank Broeze Prize for Outstanding Doctoral Thesis in Maritime History.

Research for this book was supported by the Princeton Institute for International and Regional Studies and the John Carter Brown Library. The Social Science and Humanities Research Council of Canada funded the early stages of this project as well as the production of the book index. The Liberal Arts and Professional Studies faculty at York University also contributed research funding and a publication subvention. For their generous support of the illustrations in this volume, I wish to express my gratitude to the Barr Ferree Foundation Fund for Publications, Department of Art and Archaeology, Princeton University.

I would like to gratefully acknowledge the institutions that granted me access to their collections and permission to reproduce images, and to thank the librarians, archivists, and other staff members who assisted me in person and from a distance. These collections are listed by country. The Netherlands: Het Scheepvaartmuseum (Amsterdam); Maritiem Museum Rotterdam; Nationaal Archief, Koninklijke Bibliotheek (Den Haag); Museum Boerhaave, Universiteitsbibliotheek Leiden. United Kingdom: Cambridge University Library, Pepys Library (Magdalene College, Cambridge); Bodleian Library (Oxford); Guildhall records at the London Metropolitan Archives, Caird Library of the National Maritime Museum; National Archives (Kew); Plymouth and West Devon Record Office. France: Archives nationales de France, Bibliothèque nationale—Arsenal, Richelieu, Tolbiac, Musée national de la Marine, Bibliothèque Sainte-Géneviève (Paris); Societé historique de la défense (Vincennes). United States: Columbia University, Morgan Library, New-York Historical Society, New York Public Library (New York); John Carter Brown Library, John Hay Library (Rhode Island); Houghton Library (Massachusetts). Canada: Bibliothèque et Archives nationales du Québec, Musée de la civilisation—bibliothèque du Séminaire de Québec (Québec); Thomas Fisher Library (Toronto). At collections I was unable to visit, staff provided kind assistance with images from the Albertina Museum (Vienna), Biblioteca Nacional España, Christ's Hospital, Library of New South Wales, the Rijksmuseum, Universitätsbibliothek Marburg, and Universiteitsbibliotheek Utrecht; details about manuscript holdings were helpfully confirmed by staff at Medway Archives and the Middle Temple Library (UK). Closest to home, it has been a delight to work with all of the staff at the Clara Thomas Archives and Special Collections.

Revising and expanding this project has borne some resemblance to the dry-land portion of an early modern navigator's training process. After the hands-on phase in the archives, it involved considerable time seated at a desk attempting to generate a coherent manuscript from numerous books. I would like to thank Matthew R. McAdam and William Krause for their assistance with this undertaking, as well as the production and marketing teams at Johns Hopkins University Press for their efforts to create a volume that effectively showcases these fascinating sources. I am glad to have had help from my meticulous and patient copy editor Brian MacDonald and incisive comments and suggestions from the two anonymous reviewers.

My colleagues at York University have been unfailingly supportive, in particular Tom Cohen and Rachel Koopmans, who each read the manuscript and improved it immeasurably, and Michael Moir, who abetted my interest in all things nautical. Joel Silverberg constructively reviewed my discussion of the mathematical and technical aspects of celestial navigation. My appreciation extends to family and friends who hosted me during the research process: Rodolphe Chamonal, Anya Cowan, Joanna Gröning, the Sarner family, Amy Ryan, Willy van Doorn, Sjaan van de Waerdt, and Ekkie Vermeer. Heartfelt thanks to Hannah-Louise Clark, Elaine Coburn, Franziska Exeler, Yulia Frumer, Sarah Milov, Renée Raphael, and Aviva Rothman for ongoing camaraderie and conversation.

Finally, I am forever grateful to my family. Diane and Leo Schotte, my biggest supporters, gave me maritime roots and a love of exploring. My grandmother, Peggy Thompson, passed on to me her love of books and history. Warm thanks as well to Jan Schotte, Connie Hanson, and Gillian M. S. Weindling. Most of all I thank Corinna, Juliet, and Spencer for the joy they bring to my life, and Heather Hanson, my lodestar and beloved companion on this adventure.

Editorial Practices

For reasons of space, the originals of quotations in foreign languages have been omitted here and are available online, keyed to the footnote numbers. These can be accessed at the following site: http://hdl.handle.net/10315/35728.

Where no published English edition is available, translations are my own.

Minor changes have been made to the spelling of titles (i/j, u/v, vv/w), accents, and punctuation (/ to ,) to facilitate readability. The titles of English and Dutch primary sources are capitalized according to early modern conventions rather than current standards.

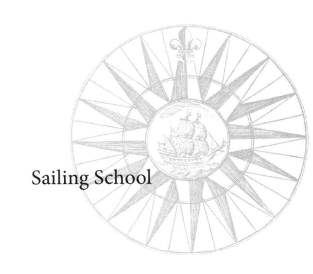

Sailing School

I.1.
"The pilot takes his ease, the sailor carries tobacco." J. van Breen, *Stiermans Gemack, ofte een korte Beschrijvinge vande Konst der Stierlieden* ('s Gravenhage, 1662), frontispiece.
Collection Maritiem Museum Rotterdam

Introduction

The frontispiece of a 1662 textbook by Joost van Breen, a Dutch navigation instructor and poet, offers an evocative central vignette (figure I.1). High in the ship's transom, as if painted there, four men cross a rocky beach; bowed under heavy bales of tobacco, they are the common seamen who manned and loaded early modern trading vessels. These rough sailors, possibly illiterate and thought to be prone to drunkenness, enjoyed little respect. On the right-hand side of the image, by contrast, another pair of mariners stand relaxed next to their boats, waiting for the crewmen to complete their backbreaking work. These are the navigators, responsible for the safe transit over rough oceans of not just the tobacco and other trade goods but the crew, munitions, and the ship itself. For this daunting task, they were paid twice the wage of common sailors and spared the monotony of hauling on ropes and loading cargo. In the pithy motto found underneath the vignette, Van Breen explains that "the pilot takes his ease" while "the sailor carries tobacco."[1]

The instruments that encircle the scene, and especially the items held by the figure standing to the right of the vignette's frame, illustrate the navigator's task, tools, and training. In his right hand, that navigator grasps a coil of rope tied to a wooden chip — a "log and line" that he would pay out in his vessel's wake to help him assess her speed. Van Breen was not impressed by this tool, deeming it an "imperfect help," although he admitted that he himself had used it at sea and remained none the worse for it.[2] In his left hand, the man holds aloft a book. This appears too small to be a volume of sea charts and too large to be an al-

manac of the Sun's position or the time of the tides, but it could be the navigator's personal logbook in which he recorded details of the voyage. Van Breen may have intended it to stand in for his own textbook — *The Navigator's Ease* (*Stiermans Gemack*), a manual that explained how to use all of this equipment as well as how to calculate a vessel's course and position. These two tools capture the double technical demands of early modern European navigational practice: the log and line required skilled observation and deep familiarity with waves and currents, whereas the book presumed literacy. Navigation was not only a traditional art, which men learned through hands-on practice, but also a theoretical science codified in published works.

Within this panorama — where mariners ride the waves in the shadow of the hulking merchantman and wield all manner of instruments — we find the multivalent themes that are woven together in this book. It is, first and foremost, the story of how those navigators learned their skills and the processes by which common seamen were formed into competent navigators. Because the navigator's training and expertise are deeply intertwined with questions of education and technology, this study also entails a history of information in the early modern period — how it was developed, codified, and transmitted. Navigational training provides an excellent yet unexplored means to examine the process by which a traditional craft was transformed into an applied science over the course of the sixteenth, seventeenth, and eighteenth centuries. This transition, it turns out, depended in large part on the printing press. The modest book held aloft by the man at the right of Van Breen's image had at least as great an effect on the maritime world as did the imposing cannons upon which he stands.

Navigation was integral to early modern imperial expansion. No body of theory and its related practices were more crucial to fostering international exchange and competition. Seventeenth-century commentators deemed the art of navigation one of "the greatest Benefits in this World" for its role in supporting trade and invention and enriching nations.[3] While other historians have assessed the training of navigators, this study aims to do something quite different.[4]

First, it is transnational. Most maritime histories are narrowly national, and yet maritime communities across Europe had a shared heritage.[5] Thus they can — and should — be analyzed across political and linguistic boundaries. Authorities in England, France, and the Netherlands set out to educate their mariners in remarkably similar ways. They took inspiration from the prevailing model of higher education — university textbooks, lectures, and oral examinations — as well as from Iberian books that carried the new techniques north. When local differences do appear within this common model, they matter all the more, for they show startlingly concrete connections among economics, politics, and applied science.

Second, this study defines print culture broadly, aiming to capture all textual materials produced by and for navigators. During this period mariners relied more and more on charts, almanacs, and logbooks to carry out their days' work. They also sought out published nautical manuals for lessons about the theoretical underpinnings of their profession. Although this turn to textual authority signals a profound shift in how maritime knowledge was conceived and transmitted, the authors and readers of this welter of salt-tinged texts have never been critically analyzed. The chapters that follow will read "top-down" administrative documents against the grain, mine popular textbooks for signs of use, and closely analyze manuscript workbooks created by sailors themselves. By means of such close and careful analysis, it is possible to trace the multiple ways in which the processes of print and writing changed the practice of navigation.

Third, this study seeks to recover actual practices from two locales that seem very separate: classroom and ship's deck.[6] Over more than two centuries, contemporaries debated the balance between theory and practice, engaging with larger questions about the nature of expertise. Whereas other studies of premodern education focus on more elite students, the present volume draws upon a range of evidence — from idealized curricula and legislation to lists of classroom instruments and telling lacunae in textbooks — to determine how such training was actually implemented for disparate groups of early modern European navigators. For instance, in chapter three we see that Royal Navy examiners asked hopeful young gentlemen to demonstrate their ability to tie knots during their lieutenants' exams. By contrast, naval masters, men who had already spent years at sea, faced no similar test of dexterity. As no seventeenth-century textbook taught knot tying, the future lieutenants were expected to have learned this skill from other sailors in their brief time at sea before their exams — a strategy designed to ensure that aspiring officers had been conscientious trainees.

The overall picture that emerges from this new analysis is one of navigators as highly skilled technicians — but also as analytical, mathematical, and intelligent. Sailors and navigators have rarely been included in accounts of the Scientific Revolution because they have been perceived as insufficiently mathematical, conservative rather than innovative.[7] However, historians of science have steadily expanded our notion of the participants in this transformative period by considering tacit and bodily knowledge, both material and technical contributions and theoretical and textual ones.[8] Navigators, for their part, are overdue for inclusion in the annals of early modern natural philosophy precisely because of their intermediary position between "high" and "low" science.[9] Mariners did far more than simply accept the directives of land-based theoreticians; as this volume demonstrates, they adapted technologies to suit their daily tasks, disseminated these improvements within their communities, and provided critical feedback to their superiors. Collectively, seamen were better educated and more technically adroit than their reputation leads us to expect. Once we see how comfortable mariners were with innovation, numbers, and theory, the Scientific Revolution shows itself to be larger and more complex: intellectual influences flowed not only down from the elites but also up from practitioners, stretching much further into society than has previously been recognized.

THE SCIENCE OF CREATING SKILLED NAVIGATORS

In the project of European discovery and colonialism, navigators were crucial players. However, it took years to develop the necessary skills, causing critical shortages in an age of naval and mercantile expansion. Some type of training program was essential for navies and merchant fleets to replace those men lost to shipwrecks, wars, and even offers from better-paying vessels.[10] J. B. P. Willaumez, an officer in the post–Revolutionary French navy, articulated the crux of the problem that plagued every marine administrator throughout the period of this study: compared to building new ships, "sailors are more difficult to create."[11] In thinking about how best to increase the numbers of these valuable mariners, experts sought to identify which characteristics made men especially suited to the more challenging tasks of the navigator. Could one develop the necessary intuitive skills only if one was born to a maritime family or, in the case of officers, an elite one? On the other hand, if such professionals could be trained — not born but made — was there a certain amount of time on the water that would allow a mariner to cross a threshold to competence? Where could maritime nations find suitable candidates for these mercantile and naval roles, and how could they expedite their training?

Many administrators followed a pragmatic policy: after seeking out men who were already familiar with the sea, they introduced them to the new theories required by celestial navigation. Coastal cities with strong trading and fishing traditions were obvious targets for such searches, but even regions "with few boats" could develop good reputations for able men. The French viewed the colonial wilds as particularly promising: the fishery on the Grand Banks of Newfoundland was deemed a "most useful novitiate," while New France had the potential to become a nursery of "navigators, fishermen, sailors and artisans."[12] In 1634, Nathaniel Boteler, a privateer *cum* colonial administrator, brainstormed suggestions for finding more suitable navigators for England's navy. In his *Dialogues* he suggested half a dozen possible sources of "seminaries of good Saylers," ranging from the Thames, where "fresh-water" ferrymen had just begun to get their "Sea Leggs and Sea Stomachs," up through the merchant companies, whose veteran crewmen would have significant technical skills as well as a knack for surviving the unhealthy climates they would encounter on long-distance voyages.[13]

However, we shall see that, as the seventeenth century gave way to the eighteenth, daily shipboard tasks grew more mathematical. Consequently, the ideal navigator shifted from a hoary old salt — who might emerge from one of Boteler's nurseries capable of wielding a handful of traditional tools — to a radically different type of practitioner, one who could calculate times and angles, read precise observations off instruments, and keep records of all this in his logbook.[14] As the navigators' tasks evolved, so too did the plans for training them. Entrepreneurial instructors opened a variety of small private schools, which were soon followed by larger, often state-funded institutions. Very few studies have explored the student experience in these classrooms or how the curricula were implemented.[15] Over more than two hundred years, the pendulum swung between theoretical and practical approaches, inside and outside the schoolroom walls. In the case studies that follow, some experts lobby on the side of theory, convinced that a solid foundation in mathematics would improve the quality of the navigators' observations and reduce their outright errors, while other authorities recommend eschewing theory entirely, concerned that humble fishermen could neither learn complicated mathematics nor be shaped into suitable candidates for naval service. Entrepreneurial authors and instrument makers often contend that it would be better to teach techniques and tools that avoid calculation.[16]

These perspectives — about the relative value of theory and practice, book learning and sailing experience, foundations and shortcuts — reflect the land-based educators' deep-seated biases about their maritime students. At the same time, they connect to wider premodern debates about technical education and the relationship between mind and hand. As Pamela Long has shown, the classical origins of the juxtaposition between *ars* and *scientia* led to many centuries of privilege for the thinking man over the one who worked with his hands.[17] And yet, as Long has also shown, from the late Middle Ages onward workers of many different kinds documented their expertise in writing. These treatises by (and for) men who worked with their hands — from artists to engineers — played an important role in gradually opening new arenas in which these experienced men could apply their skills.

Early modern conversations about the role of experience therefore extended to the question of who was best suited to write about and teach the crucial subject of navigation. Did one need to have experience at sea? Or did the complex theories demand someone more learned? Joost van Breen, for his part, loudly advertised his experience on the title page of *Stiermans Gemack*: he was a "lover of the liberal arts, having formerly been to sea, and now serving as equipment master for the Admiralty of Zeeland." And that was not all: he was also a teacher, author, inventor, and an examiner for the Dutch East India Company.[18] It is feasible to interrogate each of these groups for clues about maritime education, for they all left more records than the sailors themselves. Similarly accessible are the administrators who supervised practitioners of all kinds—monarchs, their ministers, the bureaucratic arm of the navy or merchant companies. Although few of these men had personal seagoing experience, they nonetheless played important roles in these communities. They wrote legislation, assembled funding, and recruited manpower. But while their guidelines and preferences shaped the local course of study, formally and informally, their idealistic lesson plans and condescending descriptions of mariners must be analyzed with care. The views of these "experts" remind us of the apt contemporary motto, "The best navigators stand on land." It was all too easy to obtain good-quality observations from a stable place on shore or in a quiet study, but quite another matter to do so on a boat's deck in the midst of a roiling sea.[19]

These early debates—about the balance between theory and practice and about the nature of expertise—extended well beyond the maritime sphere. New research into various fields, from commerce to medicine, has brought to light similarly contentious negotiations over the appropriate ratio of theory to practice.[20] Recent studies on the role of experts have explored the connections between politics and technology, the environment, and science. Eric Ash reminds us that in those realms, just as in the field of navigation, "the relationship between theory and practice was not always a purely binary or adversarial one."[21] However, the long history of disdain for manual workers influenced modern historiography as well. For much of the twentieth century, artisans and craftsmen were overlooked by historians of education.[22] Because of archival or societal bias, maritime historians were similarly inclined to examine the careers and training of more elite figures.[23] Thus, both maritime and social historians have considered the social aspects of life on board but not the intellectual culture that did in fact exist. Where an earlier generation of economic historians emphasized sailors' role as a labor force, a newer body of work considers them as cultural forces—pirates, mutineers, political actors, or the first global citizens.[24]

Scholars have avidly documented technological developments in navigation and shipbuilding, but the sailors themselves are often presented as timeless. For instance, maritime historian Olaf Janzen asserted that "a mariner of 1700 would have had no difficulty working a ship of 1800."[25] Of course, such stasis is impossible. Over the early modern period, navigation saw major changes—in technology and technique, but also in scope and purpose. In the same period, its practitioners lived through shifts in print culture, literacy, politics, and economics. There was no question that the art—or science—of navigation was profoundly altered. And yet, despite these changes, sailors would nonetheless retain some of their traditional devices. On their quest for safe and efficient routes across the open oceans, navigators drew upon a wide range of strategies, the old alongside the new.

Navigators are thus an ideal lens through which to consider the intersection of print, practice, and technical education. These diverse men who developed navigational expertise were a self-made elite: they were rewarded with higher wages not due to their social rank but in recognition of the technical skills they mastered. In the merchant fleet at least, this professional class was viewed by contemporaries as open to all; with sufficient effort, an ordinary sailor could enter its ranks. A literate young mariner with a head for figures might work his way up to navigator and, from there, after a decade or more, to captain. (This was less true in the navies, particularly in the eighteenth century, where social status played a large role in one's career prospects.) In both the merchant marine and the navy, men faced examinations to certify their competence at a series of milestones.[26] Many manuals written for general practitioners — like Joost van Breen's — focused on the competencies that could transform them into navigators. And these books could indeed have a transformative effect — not only on a mariner's career but on how he viewed the world around him. These lessons could discipline a man into a sharp-eyed observer, a keen reader of charts and waves, comfortable with math as well as memory. To understand how print shaped both sailors and their nautical practice in the early modern period, we must explore the emergence, circulation, and use of these texts.

WHEN PRINT MEETS THE WAVES

What impact did the printing press have on navigation? At first this question appears determinist and somewhat nonsensical. Early modern sailors seem far removed from the world of print. Their field has been roundly considered one of intuition and memory. Yet ships and sailors produced far more paper than one might imagine. In addition to the bureaucratic records that swelled during this period, and the charts (on vellum or paper) that were an increasingly prominent tool for wayfinding, mariners personally collected and created numerous documents on every voyage. Chief among these were logbooks, but sailors also acquired almanacs, licensing paperwork, financial documents, and miscellaneous personal ephemera.[27] Over the course of the seventeenth century, virtually all these documents made the transition into print. There were two other important categories of textual production: student notebooks — manuscript "treasure chests" (*schatkamers*) — in which members of the maritime community puzzled through lessons related to their profession; and hundreds, if not thousands, of publications that centered on navigation and the sea. When analyzed together, this torrent of maritime textbooks and their manuscript corollaries form an ideal corpus to illuminate the complicated interactions among authors, teachers, and readers, each with their claims to varied expertise.[28]

In light of this wealth of textual material, we might expect more engagement between historians of the book and maritime history. The sheer volume of material published on the subject of navigation indicates the presence of a sizable and diverse readership — but this has yet to be extensively analyzed.[29] Although most studies of maritime books are limited to national bibliographies, scholars are beginning to explore the oeuvre of important publishers or regions, particularly in the Netherlands.[30] In her provocative analysis of the early modern period, Elizabeth Eisenstein made grand claims about the impact of the printing press, asserting that its technological and social effects brought about revolutionary changes. Critics have reacted against such fervor, suggesting that there was instead continuity with the age of manuscript; more individual agency for the printers, authors, and readers who interacted with the powerful press; and considerable variation inherent to the early modern printed book.[31] With these caveats in mind, I take up the question of how printing changed the science and practice of navigation from the sixteenth to the eighteenth century. Some of the effects are quickly evident. The press offered improvements that would prove particularly far-reaching for precisely the types of book used by mariners and aspiring mariners. The two most obvious are the ability to reprint large quantities of data — the astronomical tables that became ubiquitous in the age of celestial navigation — and the ease of reproducing images, both cartographic and diagrammatic. These and other technical modifications would in turn lead to significant educational and practical changes in the maritime world.

The corpus of nautical manuals upon which this study draws is impressive in scope and variety. Across Europe, this period of nearly three centuries (1509–1800) saw published more than six hundred distinct navigational titles. Some of these works were reprinted in upward of twenty editions. This tally includes general introductory texts (among them cartographic texts that furnish substantial explanatory text within the prefatory material) as well as specialized titles that focus on more specific topics or instruments.[32] These flowed from presses across Europe — the rate of production was initially highest in Spain but then took off dramatically in the Netherlands, England, and, to a lesser extent, France. The content of these nautical texts differed by region. This difference reflected both the size and the financial means of their local audiences, as well as the difficulty of the journeys to be made. Navigators who needed to return safely from Asia required different theoretical and practical training from those who repeatedly sailed familiar coastal trajectories around the Mediterranean or even the far northern whaling and fishing grounds.[33] However, more significant to this story than such differences are the parallels between texts and the ease with which information was transposed from one market to another. Despite infelicitous translations and other perceived shortcomings of print, from a surprisingly early date the maritime community warmed to the view that books were an effective means of transmitting knowledge, even when it came to traditional tacit practices.

Scholars disagree about the audience for these published manuals. David Waters, author of the major history of the development of navigation in Elizabethan England, argues that the spate of maritime books that appeared in late sixteenth-century London was primarily produced for and read by gentlemen.[34] However, abundant surviving evidence indicates that these texts were in fact read widely. Although certain treatises may well have remained in the hands of elite men who never set foot on a boat, most of these materials were enthusiastically taken up by members of diverse maritime communities. The market for maritime books expanded steadily, particularly in England and the Netherlands. With the spread of printing, it became far harder to prevent information from flowing across political and linguistic borders.[35] By the turn of the seventeenth century, maritime communities were thus characterized by the unexpectedly ready exchange of knowledge, both on board among multinational crews and on shore, in person and in print.[36] The political importance of a skilled navy and an imposing merchant fleet seemed to encourage rather than obstruct the liberal borrowing of models and techniques from rivals. Thus, in schoolrooms across maritime Europe we find similar pedagogical expectations, practices, and ideas.

The idea that a navigator's tasks could be described and transmitted on paper was a relatively new one. Artisans in other non-learned fields had begun to embrace the written word in the fourteenth and fifteenth centuries as a way of promoting or profiting from their skills. Sailors were slower to see a need for textual records.[37] Their worldview was based on observations and tempo — the shape of coastal landmarks, the rate at which waves broke across the boat's bow, or the speed at which the stars rotated. However, by the sixteenth century in Iberia, and the seventeenth century in more northern polities, mariners began making efforts to translate these impressions of the world around them into rules that could be written down. (Which way did the stars move?) Once these were recorded, they needed to be explained. (How could the stars help one tell time?) Thus was born the genre of the nautical manual.[38]

The growing field of the history of reading and the more specific study of early modern reading practices both offer additional strategies to help us uncover early mariners' cognitive practices. In particular, what demands did these books place on the reader's memory? In what ways did these books function as external memories? How could they help the mariner cope with information overload?[39] These textual accounts naturally privilege the perspective of the author over that of the practitioner, although there are valuable examples of practitioners who chose to publish.[40] Close attention to bibliographic statistics concerning the production and reception of editions can capture salient shifts in popularity across social strata. Similarly, the stereotypes encapsulated in print about the "self-conceitedness and obstinacy" of seamen and "ignorant" authors can help us better understand the fluctuating social hierarchy of each of these groups.[41] And the physical marks left on bindings and margins open tantalizing glimpses into how these books were read, referenced, and used across early modern Europe. All these issues help us to develop a more complete picture of how information circulated within the intellectual world of mariners, importantly cutting across the perceived barrier between manuscript and print. This story also underscores the weight of the act of publication, as well as the various responses to a text in print and manuscript form. Publishers would update editions, classroom teachers would creatively excerpt questions, and readers would dispute with the author.

By tracking shifting content and conventions within an evolving genre, it is possible to document the codification of tacit knowledge in all its variety.[42] A closer analysis of these volumes reveals not only stylistic preferences of authors and readers but also local differences in practice. Certain topics were given pride of place in the first chapter: for the Spanish, cosmographical definitions, for the English, merchant's arithmetic, and for the Dutch, the calendar and the tides. These should be read as the foundations upon which the authors wished students to build, for, I suggest, they serve as a key indicator of the intellectual tradition to which the author subscribed. A book's first topic was thus important, even if it may not have been difficult. However, when other topics were left out entirely, their absence did not necessarily indicate insignificance. The omission of seemingly relevant subjects, such as the signs of the zodiac, which were routinely included in sixteenth-century texts but omitted from eighteenth-century examples once people became familiar with them, offer tantalizing clues about the base level of knowledge with which these men began their studies.

Textual analysis allows us to identify not only which topics were the most important to early modern navigators—or their instructors—but also which they found most challenging. Authors marshaled a range of techniques to facilitate transmission of these concepts. In some cases, they deployed traditional mnemonic devices or familiar metaphors, but in others they turned to technologies of the book to aid, or substitute for, memorization. Authors and publishers took advantage of the capability of the printing press to reproduce detailed images.[43] Large mathematical and astronomical tables could also be duplicated far more easily, and errors in such printed tables could be corrected with relative ease. (However, if those errors remained undetected, particularly in popular school texts, their deleterious effect could reach much farther.)[44] And yet, even with these more modern options available, authors found themselves returning to paper tools that were less elaborate, such as spinning volvelles anchored to the page that resurfaced from medieval manuscripts.

To understand how early modern mariners learned and worked, the study of material and visual culture also proves important. Drawing and hands-on learning are two underrecognized elements of navigational—and consequently mathematical—education. Classic and recent work on scientific images is instructive for this instance of two-way exchange: here, abstract theoretical concepts needed to be transmitted to a less-educated audience, while the latter group in turn wished to record concrete technical details (e.g., of geographic discovery or shipbuilding) in ways that would be intelligible to men unfamiliar with such topics.[45] These diagrams, maps, and images evolved alongside the practitioner's conception of the world. The manuscript workbooks offer critical evidence here. By studying what worked in terms of technical education—for example, when authors relied upon diagrams, or when they recommended practice problems instead of memorization—we can get a clearer picture of how nonelite early modern individuals learned.[46] This material can help change our understanding of literacy, numeracy, and even the acquisition of more advanced mathematical concepts and can provide examples of how illustrated material could aid cognition. All these lessons about learning, numeracy, and visualization can also inform studies of education beyond the early modern period.[47]

CHARTING THE COURSE

This volume explores the relationship between practice and print across a variety of maritime communities and, in doing so, uncovers surprising nuances. The prologue introduces the Spanish educational model that would be so influential across Europe. From the 1520s onward, the Spanish crown trained its navigators in Seville through a program of classroom lessons, instrument demonstrations, and examinations upon returning home. It published the first manuals. Despite the maritime world's long tradition of apprenticeship-based learning, both the Seville system and the associated pioneering textbooks had their intellectual origins in universities. From this first moment of formalized nautical education, we can see evidence of the struggle between theory and practice.

Chapter one opens in Amsterdam's lively harbor in the first years of the seventeenth century. The Netherlands stood out from its neighbors because of its deep-seated maritime culture. Much of Dutch society was involved in nautical activities, and mariners enjoyed relatively high social positions in a milieu at once urban, commercial, and poised to innovate. Consequently, the market for navigational manuals, instruments, and classes developed there earlier than elsewhere in Europe. The first chapter introduces the traditional tasks that any ship's navigator would be required to master in this early period, for coastal and celestial ("small" and "large") navigation alike. It also tracks the development of the rich Dutch market for maritime books, visiting a thriving bookshop in Amsterdam's central harbor, where wealthy merchants would rub elbows with experienced practitioners as they all purchased these new manuals.

The second chapter moves to the small French port of Dieppe, a hub not only for trade but also for the international exchange of information. This chapter traces the introduction of a new, dramatically more mathematical, navigational textbook and begins to explore the effects of this noteworthy shift. France was in many ways the antithesis of the Netherlands: sailors were of low status, and authors rarely had nautical experience. However, Dieppe was an intriguing exception to these norms. In the 1660s a navigation teacher named Guillaume Denys published two textbooks that revolutionized French maritime education. Unlike the earlier generation of manuals, Denys's little-studied volumes promoted a decidedly mathematical approach. These two texts, which relied on logarithms, trigonometry, and formulas rather than on memory or instruments, brought France into the new era of navigational practice. Denys also introduced a popular solution for the quotidian task of correcting a vessel's uncertain course. This method of "Corrections" was widely used for more than a century in France but was scarcely discussed in other nations, an example of the regional popularity of certain techniques. Denys's influence extended from Normandy across France, for he prepared a generation of instructors for posts at the naval schools that Louis XIV would establish in the 1680s.

Chapter three moves across the Channel to Greenwich, England, in the 1680s, a period of notable reform for both British and French naval institutions. This chapter compares the radically different expectations placed on three groups of English mariners: teenagers at the newly founded Royal Mathematical School (RMS) who were being groomed for the Royal Navy, masters who had gained their experience in the merchant marine, and young gentlemen who hoped to become lieutenants in the Royal Navy. Although trained for similar occupations, these groups were expected to master entirely different skills. The curricular standards were lofty for the RMS students: the logarithms that Denys had just popularized in France had here long been standard practice. At the behest of Royal Navy administrator Samuel Pepys, both Isaac Newton and Edmond Halley weighed in with extensive suggestions about advanced trigonometry, while Astronomer Royal John Flamsteed conducted examinations at the Royal Observatory. The working masters, by contrast, were tested on their knowledge of geography, water conditions, and definitions. Despite the heavily theoretical curricula, a picture emerges of lessons that were surprisingly hands-on.

To probe further the disjunction between idealized and actual practices, Chapter four returns to the Netherlands around the year 1710. Unlike the other countries under discussion, the Netherlands lacked centralized, state-funded institutions. Instead, the market was crowded with private educational options, and Dutch teachers, authors, and entrepreneurs vied energetically for customers. From the latter half of the seventeenth century onward, prominent publishers began including practice exams in textbooks to attract buyers. Initially these sample questions merely reflected contemporary navigational priorities, but as time went on other instructors invariably began "teaching to the test." Although maritime historians have disputed the professional significance of such exams, new evidence from the classroom demonstrates how these printed simulacra shaped curricula, practice, and, ultimately, careers. This chapter draws on a rich set of working notebooks produced by students to re-create how teachers taught and how students learned in the classroom.[48] These manuscripts help to identify key differences between published textbooks and in-class lessons. They also document how teachers handled challenging topics, how they integrated visual materials into lessons, and how instruments could stand in for mathematical competence; such insights can illuminate how technical subjects were taught in other, less well-documented early modern classrooms.

All of these close studies—of classroom instructors, examinations, and textbooks—explore how navigators learned. In many ways, though, the true test of a man's preparation came only when he took to the seas. Thanks to serendipitously preserved archival materials, Chapter five provides a window onto the ship's deck and to the surprisingly high level of mathematics deployed there. This chapter follows the celebrated career of Lieutenant Edward Riou, who in 1789 guided a Royal Navy vessel to safety after it hit an iceberg. Riou's copious logbooks and school workbooks survive and put in high relief the mathematical underpinnings of his technical skill. And yet he did not rely solely upon mathematics. Riou's notes demonstrate that to save his stricken vessel he returned to the tacit skills that he had developed through fifteen years at sea. Expert navigators needed to master not only the latest methods of determining longitude but also time-honored techniques and mnemonic verses.

The book concludes on the deck of another ship, this one with a classroom on board. In this final stage, the theoretical and practical aspects of an early modern mariner's training are unified—with mixed success. By this point, at the dawn of the nineteenth century, the nautical manual has evolved once again. Although now packed with countless rules, practice questions, and tables, it has lost its wider audience of those curious about the sea. In its new, far narrower role, the nautical manual is merely a student textbook.

In our tour of these varied educational settings, we will see that no consensus will emerge; the pendulum of opinion will continue to swing between shipboard practice and classroom study. And yet these diverse programs are united by an ambition to enhance technical education. This long history of improvement — in the face of disparate levels of schooling, social class, and geopolitical circumstance — offers lessons about pedagogy, cognition, and the social history of science that play out far beyond the early modern period. By following the navigator as he negotiated local waters and distant oceans, this study shows the broad reach of the Scientific Revolution, but it also warns against oversimplified narratives of progress. There was no straightforward trajectory from traditional to scientific sailing or from memory to mathematics. Like Van Breen's navigators, these maritime readers would find it both easy and difficult to master these skills — "licht en zwaer," as the topmost banner on his frontispiece proclaims. As mariners flocked to the classroom and expended considerable effort to master complex formulas, they still could not abandon all traditional techniques. A tour through their textbooks, student notes, and logbooks indicates a messy coexistence of traditional and modern skills, memory and mathematics, manuscript and print. By parsing these texts to understand hands-on practice, this study recovers aspects of early modern science that once seemed evanescent. *Sailing School* captures, so far as is possible, the process of transmitting technical ideas — the moment when theory becomes applied.

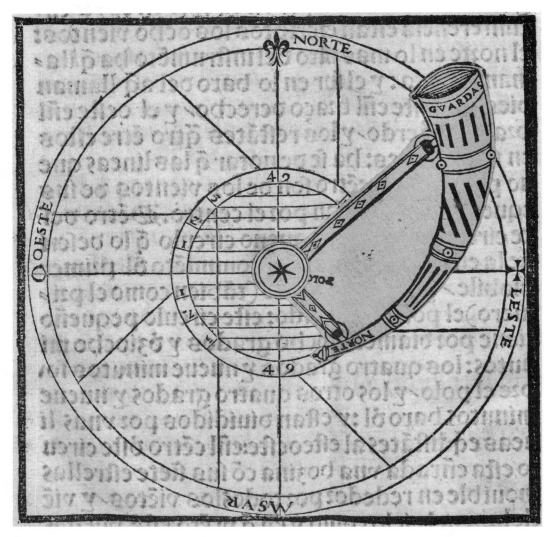

P.1.
Pole Star volvelle. Cortés included two Pole Star volvelles, one of which demonstrates its annual rotation around the pole. The one depicted here is a functional tool, indicating the number of degrees to adjust one's daily observation for any particular time. As one adjusts the "Guard Stars" of the Big Dipper to match their current position in the heavens, the other end of the horn will point to the number of degrees that needs to be subtracted from the observed altitude to obtain the correct latitude. Note that Cortés's decision to ignore the conventional figure of 3° used by working pilots in favor of a maximum of 4.9°—the number suggested by the German astronomer Johann Werner—in fact embedded a serious error of at least 1.5° into his tool. M. Cortés, *Breve compendio de la sphera y de la arte de navegar* (Seville, 1551), f. 83v. Biblioteca Nacional de España, Madrid

Prologue
A MODEL EDUCATION
Seville, ca. 1552

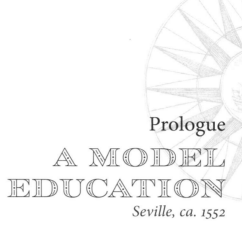

The first Iberian ventures to the New World profoundly changed the balance of power in Europe and beyond. It quickly became apparent to European polities that great wealth and opportunity would be unlocked by those who could safely cross the Atlantic. The Spanish crown pushed to increase the number of skilled *pilotos* who could facilitate this valuable transoceanic enterprise, devising a system to help them master the complexities of the new mathematized, celestial navigation.[1] Generations of navigators would undertake formal lessons at the Casa de la Contratación, the "House of Trade," in Seville. From legislative records and sparse personal accounts, we have a sense of what these sailors were taught in the classroom and how they attempted to demonstrate their competence to the examiners who would evaluate them for promotion.

On August 6, 1508, Ferdinand II appointed Amerigo Vespucci the nation's inaugural *piloto mayor* (master navigator), charging him with examining the pilots who sailed to the West Indies and South America and with approving their maps, instruments, and *derroteros* (their rutters, or sailing directions). Initially sailors had to make independent arrangements to prepare for these exams.[2] (For several years, Vespucci offered private classes but was forced to stop when it came to light that in the examination room his paying students had a suspiciously high rate of success.) Then, in 1552 the Casa de la Contratación established a formal school with a *cátedra de cosmografía* (chair of cosmography). In the centuries ahead, it would serve as a model for maritime communities across Europe.³

Any mariner who wished to earn his credentials as a *piloto* would first apply for the privilege of studying with the instructors at the Casa. The aspiring candidate would then appear before the *piloto mayor* with at least three witnesses. The candidate was expected to be at least twenty-five years of age; have at least six years of experience sailing to the Indies; and be a Christian, a native (i.e., not a foreigner), and a man of "good character" — a term open to infinite interpretations and debate. If the candidate met the biographical requirements, the witnesses, all licensed pilots who had sailed with him, would testify to both the man's experience and his competency.⁴ Having satisfied the master navigator of his bona fides, the candidate was then allowed to attend classes on the art of navigation taught by a learned professor, the *cátedratico*.

Our student then found himself in a classroom with a group of fifteen or so others, listening to twice-daily lectures straight out of a university textbook. The instructor would read for the first half of each class, and during the second hour the students were expected to "ask one another many particulars concerning the art of navigation." The teacher would weigh in on topics that proved especially difficult.⁵ This textual, classroom-based approach was a perplexing strategy for teaching men who were in all likelihood not literate, who had until then effectively acquired their professional skills aboard a ship under the tutelage of more senior mariners, and who were accustomed to learning with eye and hand as well as ear.⁶ However, the university-trained instructors appointed by the king were inclined to replicate the book-bound system with which they were familiar, expecting students to gather in classrooms to learn from written authorities. This mode of teaching stripped away many of the dialectical and embodied aspects of oral communication that had aided these mariners in generations past.

To grasp the complexities of celestial navigation, in the view of the Casa instructors, sailors first needed its theoretical underpinnings. The *cátedraticos* thus turned to the subject that straddled astronomy and geography, the science of cosmography.⁷ The study of "the sphere" had been incorporated into university curricula from the fifteenth century on. Astronomers such as Johannes de Sacrobosco, Petrus Apianus, and Gemma Frisius wrote popular textbooks that extended the cosmographical definitions set out by Ptolemy more than a millennium earlier. These were extensively illustrated with diagrams and movable paper instruments known as volvelles. In the early years of the Casa's school, the instructors were required to teach the first two chapters of Sacrobosco's *De Sphaera*, the standard introduction to the Aristotelian structure of the heavens that defined and explained the relationship between the celestial bodies and their relationship to a moving vessel far below.⁸

The crown-appointed cosmographers soon produced their own textbooks that adapted these celestial concepts for a maritime audience.⁹ Martín Cortés, a humanist with little navigational experience, wrote the *Breve compendio de la sphera y de la arte de navegar* (Seville, 1551). Pedro de Medina, who served the crown as a consultant on maritime matters and taught at Cádiz from 1530 on, authored the *Arte de navegar* (Valladolid, 1545). Medina, in his dedication to Philip II, conjured up a concerning vision of mariners on "long dangerous voyages" where few of them knew the necessities of navigation because "there were no teachers to teach them, nor books to read."[10] He and his fellow cosmographers were happy to step into the breach. Medina's and Cortés's works, which rocketed to popularity, would set the standard for navigational textbooks for years to come.

To lay the groundwork for the subject, Medina and Cortés defined key cosmographical terms. They borrowed this approach from the university "sphere" texts as well as from Euclid's geometry primers. What was a pole? the equator? an azimuth? A reader who understood these concepts, they thought, would develop a mental picture of the globe and, indeed, the universe. Their works each open with a diagram of the celestial spheres and a chapter on the heavens, then move on to discuss the seas and winds, the height of the Sun, the poles, and the magnetic compass. The final two sections treat the Moon and the days of the year and consider their relationship to the tides. This approach — defining the components of the universe in descending order from the grandeur of the spheres down to intervals of days and minutes — is a key trait of the "cosmographical" nautical manual. These works also include astronomical tables — the "regiments" of the Sun and stars — that enabled mariners to determine latitude by measuring the altitude of these celestial bodies. If the Sun were observed at its apex, noon, its altitude could be converted into latitude once the Earth's tilt and the quadrennial leap-year cycle were accounted for. Determining latitude from the stars was somewhat more straightforward, for their positions shifted more slowly than the solar cycle. The rotation of the Pole Star was regular enough to help determine latitude, as long as the appropriate number of degrees was subtracted from the observation to account for its slight offset from true north; for this, the Iberians devised tables and mnemonic devices.

Cortés presented "many newe, delectable, & wytty thynges" to his audience of "Maryners" and "Pilotes" and stressed the "many profitable and certen rules both to reade and understande." In keeping with his humanist background, he embellished his *Breve compendio* with classical references. He discussed the zodiac and provided four pages of tables of the Sun's "true place" and declination. His final section was the most practical: technical instructions for constructing one's own instruments.[11] Here he included volvelles, a feature he adapted from Apianus. Among them was a visual representation of the "regiment of the North Star," a tool that converted the star's observed altitude to the correct latitude (figure P.1).[12]

Medina's *Arte de navegar* includes the same thematic chapters on the heavens and also provides an early definition of rhumb lines or loxodromes. (These spiral lines cross all meridians at the same angle and indicate the path upon the sea that a ship would follow on a constant compass bearing.)[13] Medina — or his publisher — seems to have had in mind discerning, wealthy readers. The *Arte* is handsomely laid out in a folio format and more extensively illustrated than any comparable Spanish tech-

P.2.
Aristotelian cosmography diagram. P. de Medina, *Arte de navegar* (Valladolid, 1545), fol. 1. Biblioteca Nacional de España, Madrid

nical text. With no thought of economizing on paper stock, a full-page emblematic woodcut introduces each of the eight books.[14] The first of these, for the chapter "On the World and Its Composition," is a diagram of the Aristotelian model of the heavens, a dozen concentric circles as if directly from Sacrobosco or Apianus (figure P.2). More than twenty pages are allotted to tables of the regiments, for one of Medina's main goals was to provide these crucial astronomical data in a portable format. However, he also wished his readers — armchair elite and working navigator alike — to comprehend the abstract ideas that enabled navigators to make sense of the heavens. To this end, he employed a variety of pedagogical strategies, both conceptual and graphical.

Medina explained the winds, which serve as the primary directional guide, through a series of incremental figures.[15] To describe the motion of Little Dipper around the Pole Star, he turned to one of the most easily comprehended metaphors, that of the human body. "Imagine a figure of a man at the North Pole," he wrote in book 5. "His head is turned to midday, his left arm to the East, his right arm to the West." Building

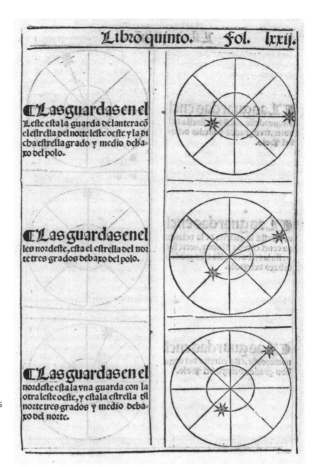

P.3.
Series of Pole Star diagrams, showing the rotating Guard Stars (*guardas*). P. de Medina, *Arte de navegar* (Valladolid, 1545), f. 72. Biblioteca Nacional de España, Madrid

on this clear analogy, readers would find it straightforward to describe the positions and trajectories of the stars: on top of one's head, at its furthest place from the feet. Here Medina used only a plain square to illustrate what his readers could already visualize. However, he also included a series of sixteen diagrams showing the rotating positions of the Guard Stars, the same information Cortés had summarized with his volvelle[16] (figure P.3). Throughout the rest of the volume, the cosmographer included dozens more diagrams to help his readers understand the vessel's course along any given rhumb, and the position of the Sun relative to the equator.[17] Medina's considerable efforts to depict astronomical concepts — and the mariner's relation to the heavens — attest to their importance. By using bodily metaphors as well as extensive visual arguments, he hoped to reach a wide audience.

Both the motives and the impact of these theoretically minded Spanish cosmographers continue to be debated. Were their books intended for working navigators, or would they have been accessible only to educated gentlemen and merchants? (Medina acknowledged the divergent interests and capacities of his readership, preparing two abridgments of the *Arte de navegar* with varying quantities of theory.) In part, they aimed to improve the status of their profession, at a time when such "mixed mathematics" were generally disdained.[18] Few cosmographers had ever completed the sort of lengthy overseas voyage about which they were theorizing, whereas the men in their classrooms had sailed for at least half a dozen years before learning the intricacies of the ecliptic or azimuths. Were these astronomical definitions then truly necessary? Surely, sometimes, economic or political grounds rather than technical ones animated the lessons taught in class — as when Pedro de Medina stipulated that the pupils at the Casa use a particular type of chart to track magnetic variation simply because the *cátedratico* hoped to undermine a monopolistic instrument maker.[19]

Spanish pilots protested the Casa's push away from traditional practices and tools. By the mid-sixteenth century, though, it was already harder to become a credentialed navigator without some familiarity with theoretical constructs. Diego García de Palacio, who in 1587 published *Instrución Náutica* in Mexico City, provided the fullest contemporary descriptions of the roles and desired characteristics of each member of a crew. García de Palacio advised that a good pilot was "one who has practiced and obtained good results at sea." But, he continued, "If it can be verified that he knows astrology, mathematics, and cosmography, he will have many advantages over one who does not know them, and when he does not possess such learning, he must be certain in the latitude by the astrolabe, cross-staff, and quadrant; the moon and tides; knowledge of the land; sea-bottom; and plain dead-reckoning, for pricking a point on his chart."[20] At a minimum, he averred, the ideal navigator needed to know a group of standard tasks (these will be explained in the following chapter), but it was the theoretical learning that would give him "many advantages." In addition to expecting familiarity with instruments *and* science, García de Palacio sought a confident, diligent man who was at once cautious and willing to collaborate. He had far less exacting expectations of general mariners: they "ought to be skilled, modest men who have some resources, have sailed, and [be] good helmsmen." If these nonspecialists should know any rudiments of astronomical navigation — using instruments and charts, telling time and calculating tides — "it will be of much value, and such is reason to give him preference." At a time when a substantial proportion of mariners did not "possess the qualities or abilities that are required for safely conveying such souls, property, and things that are entrusted to them," it was García de Palacio's "intent [with this book] to mitigate this defect."[21]

Already, books and theories were seen as key supplements or even replacements for time at sea. Like the administrators at the Casa who viewed classroom time as beneficial, García de Palacio was confident that "anyone who wishes can learn enough theory that he may become a dexterous pilot with middling experience." Men who hoped to sell more books were often optimists; early in the next century the Dutch almanac maker Robert Cuningham was equally confident: "Nowadays there are many pilots who still cannot read and write, [but] certainly they are decent enough that they can learn."[22] Such views were far from universal, however. The Spanish *pilotos* ranged against the cosmographers were not the only practitioners to mistrust the new book-based learning. Across the Channel, Englishman William Bourne reported that experienced captains were resistant to newfangled tools such as maps, while even Captain John Smith, who penned several manuals for "Young Sea-Men" upon his return from Virginia, was not convinced of the utility of print. As he told his readers, "Get some of those bookes, but practise is the best."[23] Clearly, it would take some time yet before the tide turned in favor of these new technologies.

After two or three months of classes, our studious Spaniard would return to the Casa's examination hall.[24] The room would be filled with, by some reports, up to twenty-five men, including the *piloto mayor*, a clutch of cosmographers licensed by the Casa de la Contratación, and some local licensed pilots. The candidate would state the particular route for which he hoped to be certified — to a specific Caribbean island or, if he were more experienced, to all the major ports in New Spain. From this point on, the exam was very hands-on: the candidate would spread a chart upon a table and, standing before the attending pilots, plot his course — from off the coast just northwest of Cádiz to the Canary Islands to the Indies and home again, for instance. Having done this, he was required to answer questions about tides and the regiments of the Sun and the North Star and to demonstrate knowledge of the landmarks and hazards particular to his chosen route.[25] The navigators in the room peppered him with "the most difficult questions" about how he would handle emergency situations, from a broken mast, leaks, or a failed rudder, to an unfortunate run-in with a "Pirate [who might] leave him destitute of his Chart, Astrolabe, and his [altitude] instruments." Having concluded their examination, the jury then voted, choosing beans of different colors to indicate their approval or disapproval of the candidate. If he failed, the pilot was expected to make at least one more voyage to improve his grasp on the necessary information. If he passed, he needed only to have the authorities inspect his equipment. At last he would receive papers documenting his new status as a licensed pilot. He then would tip his examiners a modest two or three ducats.[26]

In many ways this professionalizing process pitted cosmographers and pilots against each other — setting innovative methods against traditional ones.[27] And yet the examination itself was not antagonistic. Instead, it appears to have been an interactive group process, where experienced colleagues questioned the candidate and then jointly assessed his competence. In many ways, these Spanish exams played to the existing strengths of professional pilots. The expectation that the candidate physically trace the ideal route on his chart, rather than present any calculations, signals that these sixteenth-century authorities continued to value demonstration over description and did not yet require numerical facility. Although certain candidates may well have found charts and compasses novel and daunting, most pilots were likely more comfortable displaying their geographic knowledge using these tools rather than in written fashion.[28] Moreover, although the classes may have introduced overarching theory, the exam questions focused on a single route. Aspiring navigators were not yet being trained as generalists but were responsible only for the primary features of a particular geographic location. At the same time, the Casa exam stressed flexibility, disaster preparedness, and the ability to solve problems in multiple ways — with instruments, tables, or careful approximations. These were the tacit skills that candidates would have developed over years, long before they set foot in a classroom.

The men in the examination chamber at the Casa were thus a diverse group: pilots and examiners, practitioners and teachers, men just embarking on their careers and others whose sailing days were drawing to a close. They represented a wide range of perspectives, varying in social status, levels of literacy, and interest in innovation, to say nothing of experience on the water. Each would have had an opinion about how best to train the next generation, but there was little agreement. How necessary was it to comprehend the theory behind the day's observations? Need a navigator be able to craft his own instruments? Was rote memorization a deft means of coping with information overload or a token of a limited intellect? How would these new techniques change navigation? And how would word of them spread beyond Seville?

At a time when ships often had multiethnic, polyglot crews, and dynastic politics linked distant corners of Europe, it did not take long for this Spanish system to reach the ears of administrators across the Continent.[29] Its main components would soon be replicated in virtually every other maritime community. In Portugal, where the early wave of transoceanic exploration kept pace with Spain, the mathematician Pedro Nuñes was named the first master navigator (*cosmógrafo-mor*) in 1547. Nuñes, whose brief was similar to those of Vespucci and his successors, was expected to teach sailors as well as examine them.[30] Elsewhere, well-connected men worked in conjunction with the printing press to share details of the model.

One early conduit between Spain and England is well documented: in 1558 Stephen Borough, a sea captain from Dover, visited Seville and was admitted to the Casa de la Contratación as an honored guest. On this visit, he viewed the instruments and manuals used to teach Spanish navigators and learned about the examination process. Upon his return to England, Borough presented Queen Elizabeth I with a detailed proposal for establishing a similarly rigorous centralized institution in London. While the queen chose not to support Borough's school, she was interested in his petition. In January 1564 a contract was drawn up appointing Borough England's "Cheyffe Pylott," a post like Vespucci's, responsible for both teaching and examining English navigators.[31]

Borough also brought back a copy of Martín Cortés's *Breve compendio*, which was quickly translated into English in 1561. In the sixteenth century alone, there were at least six English editions.[32] Readers in other countries also sought out the seminal Spanish cosmographical texts. The French read Medina eagerly: by 1633 Nicolas de Nicolai's 1554 translation *L'Art de naviguer* saw fifteen editions. In the same period three Italian editions appeared.[33] The transmission from Spain to the Low Countries went by way of Antwerp, where Michiel Coignet published a Flemish translation of Medina in 1580.[34] These close copies of the cosmographers' texts (which preserved much of the original illustrative material) disseminated the cosmographical approach to navigation to generations of mariners across Europe.

Although these early works did not include details about the final stage of the Spanish training process, the Casa's examinations were of great interest to other nautical experts. The practice of an oral exam would have been familiar to most early modern educators, given the ubiquity of catechisms, Socratic dialogues, and the oral "defensiones" that crowned most university courses.[35] But the hands-on process before an audience of experienced practitioners differed from these traditional scholarly tests. A generation after Stephen Borough visited Seville, the English reading public got additional insight into the Casa's exam room. In 1585 Pedro Dias, a Spanish captain, was captured by the colorful English explorer Sir Richard Grenville. Richard Hakluyt, the compiler of a popular series of travel narratives, solicited from the Spaniard a full account of the Casa's general practices, in both the classroom and the examination hall. While Dias escaped from his English captors in 1588, Hakluyt preserved his account of "the straight and severe examination of Pilots and Masters before they be admitted to take charge of ships," publishing it in 1600. This provides the outlines of the lessons and exams described previously. Hakluyt believed that Dias's account might inspire British administrators: "If they finde [the Casa procedures] good and beneficial for our seamen, I hope they wil gladly imbrace and imitate [them]."[36]

The Spanish approach to "government in sea-matters" would indeed find swift and eager emulators. On the Continent the French appointed their first naval examiner at Dieppe in 1615, while the Amsterdam Chamber of the Dutch East India Company (VOC) did the same in 1619. By 1621 the medieval pilots' guild in London, Trinity House, began examining ship's masters for the navy.[37] As the seventeenth century progressed, the scenario where a humble sailor was required to face a panel of authorities and declaim answers would be replicated not just in Iberia and northwestern Europe but in maritime centers from Marseille to Copenhagen.

The tendency for naval administrators to look across borders for inspiration — a double-edged desire to copy and compete with neighboring states — produced many similarities within the nautical classrooms, textbooks, and examination processes across Europe. The centralized system of the Casa de la Contratación provided a template for training and certification that would hold sway for the next two centuries. But how well did this standard model suit other maritime communities? What happened when inventive sailors and entrepreneurs in different ports blended their own local traditions with evolving theoretical and technical standards? Could books indeed serve as the entrée into this complex set of practices? European sailors at the turn of the seventeenth century shared a common system — but as they strove to master new techniques they developed unique responses to the challenges of the open ocean.

1.1.
D. Vinckboons, "[Nautical company]," (ca. 1610).
The Albertina Museum, Vienna, Inv. 13372

Chapter One

FROM THE WATER TO THE WRITING BOOK

Amsterdam, ca. 1600

At the dawn of the seventeenth century, a group of nautical men assembled in an elegant home on one of Amsterdam's grand canals. They gathered around a large table: nine well-to-do men, wearing the flat-brimmed beaver hats and fitted waistcoats fashionable among merchants and sea captains alike. Muted sunlight glimmered through tall leaded windows, illuminating a wood-paneled room, with a pair of sweeping seascapes hanging on the rear wall. It was summertime, so there was no fire in the ornate fireplace, and one tall casement stood open to let in the damp air. A dog slept underfoot. Two of the men pored over a large globe, while others were engrossed in studying maps, compasses in hand. Two read books[1] (figure 1.1).

The genre painter David Vinckboons completed this ink-and-wash scene in the early years of the seventeenth century. Although he had little personal experience with the sea, he was acquainted with prominent figures in the Dutch maritime community: both the astronomer and cartographer Peter Plancius and the atlas maker Willem Janszoon Blaeu commissioned artwork from him for their publications.[2] A viewer, peering into the gracious room, gains a sense of the material side of a nautical life. The quadrants and cross-staffs hanging from the wall and littering the floor are substantial; many are longer than a man's arm. In the foreground a second large globe sits on the floor, alongside an armillary sphere, several oversized sectors, a mariner's astrolabe, and a small, oblong atlas. While the identity of these men is not known, their tools hint at their occupation: they may be the investors and directors of a merchant company, such as the Dutch East India Company (VOC), or the actual men who would lead the expeditions, the captains and navigators. No identifying details shed light on whether Vinckboons was imagining them preparing for a specific voyage or debriefing after one, but they were clearly sharing information, trading notes on new instruments and routes. To be an expert navigator, a mariner needed to master the equipment that filled this room. As Vinckboons's scene makes clear, such mastery was necessarily hands-on and collaborative.

Each of the boats moored in Amsterdam's harbor would have had at least one such expert on board: a man who could "safely steer a ship from one harbor to another."[3] Herring busses besieged by hungry gulls, low *fluyts* unloading grain and timber from the Baltic, whaling vessels from the North Atlantic, small trading ships bringing wine and wool from France and Spain and carrying cheese, books, and other exports to England and beyond — each of these needed someone in charge of wayfinding.[4] Throughout the sixteenth century, most Dutch vessels plied these shorter routes, staying within sight of the coast, but even such "small navigation" required extensive technical expertise. Vinckboons's scene, however, likely involved men who set their sights further afield — perhaps they were orchestrating a commercial expedition of the kind that had recently returned from the Spice Islands of Southeast Asia or those that would soon venture to Australia, the island of Manhattes in Nieuw Nederland, or Brazil.[5]

How did such expertise travel beyond Vinckboons's chamber? How did people learn to read the charts, wield such instruments, and train their juniors? The Netherlands at the turn of the seventeenth century was a country on the rise, a newly independent nation where commerce and colonial ventures bolstered a booming economy[6] (figure 1.2). Given the geographic circumstances of the "Low Countries," it is unsurprising that so many people's livelihoods were linked to the water. Provisioners, shipwrights, and dockworkers lived (with their wives and children) within a stone's throw of the port. So too did the men who supported the maritime community primarily with intellectual rather than physical efforts: the instrument makers, navigation teachers, and examiners for the local admiralty. (The Netherlands had no central naval fleet; rather, five separate admiralties oversaw military actions.) Then there were the investors who committed personal funds to voyages that offered high rewards at considerable risk.[7] Thus, from an early date the audience for nautical publications and innovative instruments was substantial. For all of these reasons, Amsterdam is an ideal place not only to examine the transmission of technical expertise but also to trace the broader cultural ramifications of navigational knowledge. Members of the maritime community shaped politics and finance, appeared in art and literature, and pushed for new solutions to long-standing, complex scientific problems.[8] If boats and the goods they carried were the lifeblood of the Netherlands, then expert navigators were the brains that kept them circulating safely.

1.2.
Merchant shipping in a bustling harbor. C. J. Visscher II (Northern Netherlands, 1608). Rijksmuseum, Amsterdam

Who were these sailors, both the experts and the trainees? What exactly did they do on board ship? At a time when the navigators' responsibilities were expanding in scope and complexity, what paths did they pursue to learn these skills? Maritime communities are famously diverse, and the men who frequented the quays along the Amstel were no exception. Representing as much as a tenth of the city's male labor force, they came from diverse social backgrounds and education levels, and their socioeconomic status varied equally.[9] Some plied local waters around their home, earning just a few florins each month. Others might return home every second year from distant voyages and were compensated accordingly, generally with carefully governed opportunities to trade on their own behalf.[10] Socially, these men were also diverse. Many were local boys whose fathers and grandfathers had earned their livings in these same boats, in a nearby port town if not at Amsterdam itself; it was common for members of one family to work together in a ship they had invested in. If they did not share a boat, they were at least likely to put their common training to work in the same sector of maritime industry, passing on wisdom about whaling, fishing, or a particular trade route. Other laborers came from inland — immigrants in search of work. These men, new to the sea, were more likely to serve as common sailors and stevedores, providing brute strength for hauling on ropes and loading cargo.[11]

For every ten or twenty green young men just beginning their careers, there was a handful of experts with decades of experience. These men would already have had a deep knowledge of how to use the welter of instruments strewn around Vinckboons's room. They would likely have committed their lives to this profession, working their way up from ship's boy to a position that was respected and well compensated.[12] Smaller ships were often helmed by a master, who combined the responsibilities of both captain and navigator. Such men not only made wayfinding decisions but also managed the crew, supplies, and commercial aspects of the voyage. On larger vessels, these responsibilities were often divided between the captain (*schipper*) and the navigator proper (*stuurman,* mate).[13] The captain concerned himself with business and discipline but deferred to the navigator on daily decisions about the vessel's course. Because the overarching goal was to protect the boat and its valuable cargo, navigators were often the most important figures on board.[14] For this reason, navigational knowledge needed to be disseminated beyond the small group of professional wayfinders. On every vessel there would typically be at least one navigator's mate to assist the head navigator. By the 1610s, chief and assistant navigators (*opper-* and *onderstuurlieden*) were already undergoing examinations to certify their abilities. (In France, sixteenth-century naval regulations stipulated that each vessel have no fewer than three *pilotes*: two navigators who focused on coastal sailing and one for open water.)[15] As commercial and naval fleets expanded, so did the pool of people who were exposed to this knowledge. Many of the young boys and general sailors would have picked up on the rhythm of daily observations and wayfinding tasks. If the midocean hazards and all-too-frequent accidents taught mariners anything, it was that training others was of paramount importance: a ship that lost its navigator to illness or accident would otherwise be in grave danger.

Yet the sixteenth century was a period of transition for maritime practice. Over the course of the century the average length of voyages more than tripled. This change radically altered the tools and skills needed to navigate safely, as well as the expectations for literacy and numeracy.[16] No amount of practical experience could develop all the competencies necessary for ocean sailing. If earlier trips had typically hewn close to shore, they now ventured well beyond the sight of land. Contemporaries distinguished these "small" (or "common") and "large" (or "great") modes of sailing.[17] Consequently, for cues about time and direction, the navigator was forced to shift his focus from land to sky. The Sun and stars increasingly served as both clock and compass. Navigating had always required an excellent memory—for local coastlines, shoals and reefs, shifting coastal currents and fluctuations of high tides, and the types of tacit skills that had been passed down person-to-person for centuries—but the process now increasingly demanded a head for numbers. Geographic knowledge of local harbors and coasts was extended to far more regions, and observations of the heavens now required accompanying calculations. These changes not only pushed the limits of comfortable memorization but also had a profound effect on how maritime men would go about mastering the navigator's complex set of daily skills.

NAVIGATION SMALL AND LARGE: THE EXPERT'S DAILY TASKS

The men who took the helm of the countless boats moored on the Amstel—like those examined at the Casa de la Contratación in Seville—were used to reading the world around them to track their progress. Two basic questions guided the navigator: where was the ship at the moment, and what was the easiest path to its destination? To answer these questions at any given time, even far from the sight of land, an expert navigator drew upon a cluster of clues. The sight of seabirds or heaped clouds signaled approaching land, while wisps of cloud scudding across the sky could help him foretell a storm. Closer to shore, he needed only to glance at the coastline; once he sighted a steeple on a particular headland, he could call to mind the number of leagues to his next port. There were nautical landmarks as well, buoys and lighthouses in strategic locations that local maritime guilds had maintained since the late Middle Ages. A mariner could make his way by memorizing these pertinent landmarks and their relationship to each other, but as time went on, it became necessary to fit these landmarks into a framework—the directions of the compass.

To orient himself, a navigator needed to understand not only the four cardinal compass points but also either twelve or twenty-eight intermediary directions. The compass was perhaps the most fundamental of all navigational devices. One early seventeenth-century English maritime writer likened the ability to recite the thirty-two points to mastering the rudiments of alphabet, the first thing taught in primary school: "First you teach your children to know the letters by name, and so in the like manner we teach our youths, and boyes (which wee intend to make Navigators) the poynts of the Compasse." Because "all Navigators should have the points of the Compasse so exactly ingrafted in theyr mindes," he provided a diagram to facilitate that[18] (figure 1.3). Mariners used this framework not only to orient themselves geographically—with the help of a simple magnetic compass and the stars—but also as a type of notional clock, to keep track of tides in various ports. (A port whose tide would be high at noon on the day of a full Moon was categorized as a "north" port, and one whose tide would be full slightly later would be "north-northeast." Each point on the compass card represented forty-five minutes, the approximate time that high tide would shift in one day).[19]

There were times, even in the period when most shipping stayed near the coasts, when a master, owing to night or fog or storm or a preference for a more direct route, might sail beyond sight of reassuring landmarks. After sufficient experience at the helm of a particular boat, he would become adept at dead reckoning: estimating his current position not with compass or charts but by his eye alone. He did this by tracking a vessel's speed and direction relative to its last known position—whether that was the point of departure or a reprovisioning port. He must constantly monitor the vessel's heading (the direction it was sailing), taking into account the effects of currents and wind. He could then compare that to the rhumb or angle of the intended course and track how long the boat had sailed at a certain speed. An expe-

1.3.
Compass rose: each quadrant has been divided into eight equal points (also called rhumbs or winds). G. de Veer, *Warhafftige Relation der dreyen newen unerhörten seltzamen Schiffart*, ed. L. Hulsius (Nuremberg: Hulsius, 1598), facing p. 3. Rijksmuseum, Amsterdam

rienced mariner could make a very credible estimate. The goal was not necessarily to pinpoint an exact position but rather to keep track of when the boat might be in danger of straying too close to coastal hazards. Overestimating one's progress was not uncommon; it was better for a navigator to anticipate land prematurely than to run aground.[20]

To determine the boat's speed, the navigator would gaze intently at the water as it flowed past his ship's bow. Either by counting steadily as he observed flecks of foam on passing waves or by throwing a small piece of wood overboard and reckoning how quickly the boat passed it, he could assess whether he was covering more or fewer miles than in an average day.[21] The weather conditions, the rhythm of the tides, and the strength of the current could all affect the boat's progress. To compute the distance over a set period of time, navigators used counting rhymes, a sandglass, or their internal metronomes, a sense of time carefully calibrated to recognize the length of half a minute.[22] By the close of the sixteenth century, mariners began using the log and line, tracking how quickly their wooden board slipped out astern by counting knots on the line, an innovation imported from the English[23]

1.4. (opposite)
Traverse board (Amsterdam, 1840–60) A navigator would use pegs to record how far he sailed along specific rhumbs in each watch; he would then tabulate these figures in the bottom section, which would make it easier to convert the total distance (the "traverse") into a numerical position to be marked on his chart. (This example was used as a teaching tool in the Kweekschool voor de Zeevaart.) Het Scheepvaartmuseum, Amsterdam

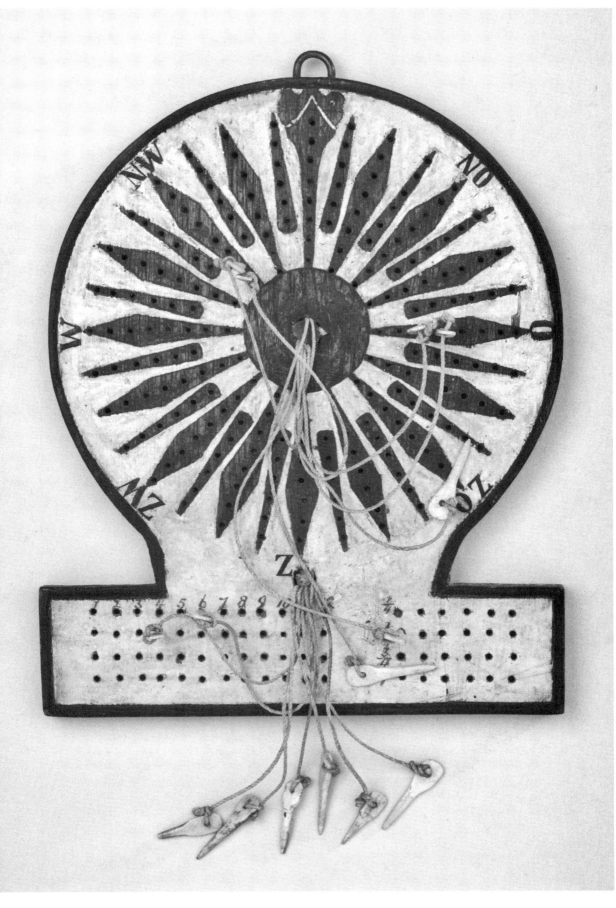

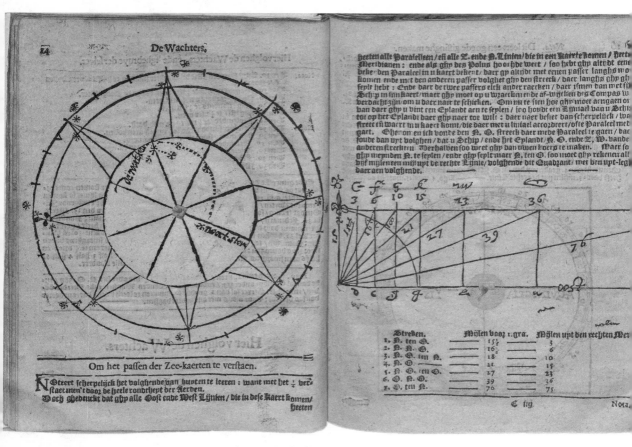

1.5.
Left: Pole Star volvelle. *Right:* Traverse Table (*streek tafel*), with its corresponding diagram to help "understand measurement on a seachart." Sailing with traverse tables did not require astronomical observations or complex mathematics. Navigators would set their course along one of the eight rhumbs in a quadrant (the rectangular diagram could be rotated as needed for the other headings). By reading off the table (*at lower right*) and completing some basic multiplication (with the rule of three), they could determine how far they had progressed in terms of latitude. For example, a boat sailing thirty-nine miles east-northeast would make only thirty-six miles of progress eastward. J. van den Broucke, *Instructie der Zee-Vaert* (Rotterdam, 1609), 14. Leiden University Library, 2003 G7

(see figures I.1, 2.9). To track their progress over the course of a four-hour watch, each helmsman would insert pins into a wooden traverse board, noting how long they sailed on each rhumb (one of thirty-two winds or compass directions)[24] (figure 1.4). Then, with the total distance in each direction known, a mariner would consult a traverse table (figure 1.5). These concise tables noted how far a boat would have to travel on any rhumb to "raise a degree"—that is, sail one degree of latitude.[25] The navigator would use this table in conjunction with the rule of three (a basic formula used by medieval merchants to solve for one unknown when three other quantities were known) to figure out his total distance. Finally, he would note this new position on a chart or in his logbook, two tools that would become more common over the course of the sixteenth century.[26] As the ship approached the shore, the navigator would throw a weighted sounding line off the vessel's side (figure 2.9). When this lead and line hit bottom, he would know whether he had yet reached the treacherous shoal that lay off a particular shore, or if a sandbank had shifted slightly since his last time in these waters. By examining the grains of sand that clung to the sticky tallow packed inside the weight, he would know that he still had at least a few leagues to go—until the tallow brought up darker pebbles, crushed shells, or another telltale marker of an approaching shoreline.

Eventually, all signs would indicate that it was time for the vessel to make the final approach to port. This required its own careful timing, based on the daily cycle of the tides. The navigator would consult with a crabbed table for details about this harbor and then mutter a rhyme while counting on the bumps of his knuckles. But how did that help him navigate? Mariners had a good understanding of the connection between the cycles of the Moon and the tides, which shifted about three-quarters of an hour later every day, but the calculations were nonetheless complex.[27] What day of the month was it, and how full was the Moon? (It was not sufficient to estimate the Moon's age by eye, as being off by just a day could cause one to miss the tide by almost an hour.) To compute this correctly, it was necessary to take into account the lunar as well as the solar calendar. The latter is roughly eleven days longer than the former, and they align every nineteen years (the "Metonic" cycle). Medieval theologians calculating the dates of Easter had devoted considerable energy to these computations. Beginning in the sixteenth century, general almanacs included this calendrical information to help with scheduling medical procedures, markets and fairs, and horoscopes. The current Golden Number may well have been posted in churches, the docks, or other places frequented by mariners.[28] There were three pertinent numbers of interest to navigators: the Golden Number or prime, the epact, and the Sunday (or Dominical) letter.

The Golden Number identifies the position of any given year in the nineteen-year cycle. Once the Golden Number is known, one can calculate the epact—the difference between the solar and lunar years, or the age of the Moon on January 1. For both numbers, the navigator could turn to the "rule of thumb," quite literally counting on his knuckles. Practitioners developed a system of counting on their fingers to arrive at the year's Golden Number, from one to nineteen (similar to figure 1.6). Even more

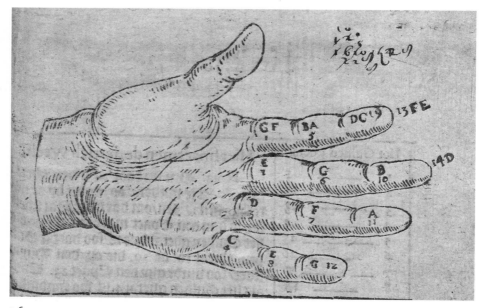

1.6.
"Sunday Letter" hand. If one has forgotten the Sunday Letter for the current year, one must first add 1, divide by 28, and then count off the remainder on one's fingers (beginning at the base of the pointer finger and descending to the pinky). In the case of 1609—for which a contemporary reader jotted his calculation— a remainder of 14 takes one to the tip of the middle finger. According to the diagram, this represents "D," or 4. Counting backward from Sunday, this correctly indicates that the first day of the year was a Thursday. J. van den Broucke, *Instructie der Zee-Vaert* (Rotterdam, 1609), f. 7. Leiden University Library, 2003 G7

simply, they used the joints on their thumb to figure out the epact (figures 1.7, 1.8, 1.9; cf. 1.16). The next step was to determine the age of the Moon in the current month. A mnemonic phrase for Sunday Letter helped with this. Every year, the first day of each month followed a set pattern: "*A[D]am Deed Geen Baet / En Godes Cracht Freest Al Dat Folck*," with the same sequence of capital letters appearing in English manuals: "At Dover dwell George Brown, Esquire; Good Christopher Finch; and David Fryer."[29] The initial letters A–G represent the numbers 1 through 7, a familiar form of acronymic substitution. Therefore, if January 1 were a Sunday (day 1, or A), then both February and March would begin on a Thursday (day 4, or D, counting backward from Sunday). By pairing the epact with these phrases, it was possible to compute the exact age of the Moon for the current day. Returning to the tide table, the navigator used a final series of mental calculations to compute the time when the tide would be full in the port he was approaching. If the Moon was four days old, he would have to subtract three hours (forty-eight minutes rounded down, and multiplied times four). This in turn would alert him that six hours later, low tide might produce powerful local currents among the shoals, or the water might be too shallow to cross a sandbar safely.

36 ❋ SAILING SCHOOL

1.7.
R. Cuningham, *Tractaet des Tijdts* (Franeker, 1605), 38. Collection Maritiem Museum Rotterdam

Calculating the epact with one's thumb. In the Gietermaker image, let the tip of the thumb stand for 0, the middle joint 10, and the base 20. (Confusingly, each of these diagrams offers different values, or inverts their positions.) The navigator should count off the Golden Number from the top to the bottom (e.g., if the year's Golden Number is 8, then you would count down your thumb two and a half times, ending on the middle joint. At that point, you know to add 10 to the original 8, giving you an epact of 18.

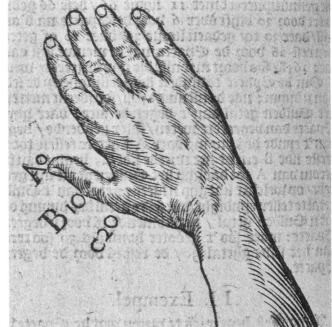

1.8.
C. H. Gietermaker, *Vermaeck der Stuerlieden* (Amsterdam, 1659), 4. Collection Maritiem Museum Rotterdam

1.9.
A. C. Hellingwerf, *Vermaack der Zee-Luy* (Hoorn, 1703), 3. Leiden University Library, 2314 H.20

This ability to track the tide's schedule, along with a deep knowledge of local geography — particularly the underwater sandbanks and deeps — comprised the essential skills of "small" or "common" navigation. Michiel Coignet, a Flemish mathematician, engineer, wine gauger, and author of one the first original works on navigation published outside Iberia, summarized the traditional aspects of sailing thus: "The whole science of common navigation lies in nothing more than simply and certainly knowing all headlands, harbours and rivers; how these appear and are seen in the sea; how far and in which direction these [places] lie from one another; which phase of the Moon brings full and ebb tide; the running and flowing of the currents, and the appearance of these in depths and shallows."[30] While certain commentators felt that small navigation was *more* difficult than sailing on the open water, precisely because of the long process of learning a region's particularities, Coignet was dismissive. He and his contemporaries did agree: long-distance navigators still needed to concern themselves with the "small" tasks. A master required a thorough knowledge of the changeable waters along any route that he was commissioned to sail — including the timing of the tides, seasonal currents, and any distinctive patterns on the ocean floor. Of course, Coignet airily claimed that those tasks required "no other art [*const*] or instruments than experience, and a compass with a sinkline."[31]

To Coignet's brief list, a coastal navigator would likely have added several other tools: a "rutter," a printed or manuscript volume of written directions (from the term *roteiro*); perhaps a portolan chart; and, depending on the navigator's comfort with the written word, a small personal notebook. Rutters and portolans both had medieval origins, and both first saw printed editions in the early sixteenth century. Rutters provided cryptically brief directions between standard trading ports in northern European waters, noting the direction or heading between ports, landmarks visible on the shore, the distance between them, and perhaps a warning about salient hazards. Portolans were charts that depicted coastlines instead of landmasses, inscribed with compass roses to help demonstrate which bearing to take between two ports.[32] A navigator would in most cases have supplemented these general tools with his own notes about his commissions, making quick coastal profile sketches or jotting notes about weather patterns and any of the other details that were revealed as one repeatedly traveled the same route.[33] Still, as Coignet's nonchalance suggests, these were old technologies, descriptive rather than quantitative. His interest was piqued by a different set of navigational tasks: the abstract, mathematical challenges of "large navigation."

With the pace of exploration and commerce picking up steam, ever more helmsmen took their vessels out into the open ocean. This "large," blue-water, or celestial navigation forced mariners to look not to the coast to orient the vessel but skyward. The questions remained the same: What was the ship's location, and how could it best reach port safely? However, further out to sea it was no longer possible to identify one's location by observing familiar landmarks or by testing the ocean floor. Nor could positions be reliably estimated by dead reckoning, as the distances between confirmed locations — and the potential for cumulative errors — increased expo-

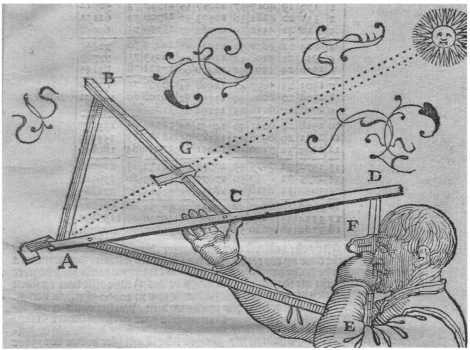

1.10.
Taking an altitude reading with a backstaff. An observer using a traditional cross-staff must look directly into the Sun to measure its altitude. However, to obtain the same reading from a backstaff (here, a Dutch *hoekboog* model), he holds the instrument in front of him, sighting the horizon through the slit at the far end, and the Sun casts its shadow from behind his shoulder. H. Reyersz., *Stuurmans-Praetjen Tusschen Jaep en Veer* (Amsterdam, 1637), 9. Leiden University Library, 2313 F 14

nentially. Instead, mariners began to develop methods of determining their location within an orienting grid of latitude and longitude.[34] Coignet, whose work was published as an appendix to the Dutch translation of Pedro de Medina's *Arte de Navegar* (1580), explained that "large navigation" required not only "the above-mentioned common practices" but also "many other diverse and ingenious rules and instruments taken from the arts of Astronomy and Cosmography."[35] To help with the more complex observations and computations, by the later years of the seventeenth century navigators were adding astronomical tables and some type of altitude instrument to their materiel.

By the sixteenth century, mariners were readily able to ascertain their latitude. Medieval astronomers had recognized that the height of the Sun at noon was equal to the observer's latitude. On days when the Sun was visible, our navigator would take up one of his new tools — either a simple cross-staff, the newer backstaff, or a mariner's ring (a type of astrolabe made sturdier for the rougher conditions at sea). As the Sun climbed to its apex each day at noon, the navigator would "shoot the Sun," measuring the vertical angle between the horizon and Sun[36] (figure 1.10). Once the Earth's current location in its annual orbit was taken into account, this number was equivalent to the latitude. As with his tide calculations, the navigator would turn to his almanac, looking up the current date in the ephemerides (astronomical tables) to determine his latitude.[37]

These observations could be compromised if the sky was overcast or if the sea was choppy, making it hard to observe the horizon line or to hold instruments steady. However, if the weather was uncooperative at noon, the navigator had another chance to assess his progress: once the sky darkened in the evening, he could follow a similar process to "shoot the stars," perhaps peering through a nocturnal, an instrument that could be adjusted to derive the current time from the brightest stars.[38] Over the course of the night, the stars of the Little Dipper turned steadily counterclockwise around the North Star. Mariners soon learned to tell the time on the basis of the position of the stars furthest from Polaris, the Guards (*Wachters*). Every 15° of rotation indicated that one hour had elapsed.[39] Additionally, by measuring the altitude of the Pole Star, and making small adjustments to the observation to take into account a 3°30′ offset from true north, the mariner could obtain his latitude. The fact that the latitude coordinate could be determined with relative confidence led the Portuguese to adopt the strategy of "latitude sailing": a navigator would set his course along the particular heading (recommended in his *roteiro*), and then "run down" the latitude — sail in that direction until he reached the latitude of his destination. He then needed only to sail directly east or west until he reached his goal.[40]

In contrast to latitude, determining longitude was a considerable problem. Due to the Earth's constant diurnal rotation, there was no consistent correspondence between terrestrial position and celestial bodies. Across Europe, states offered handsome rewards for a workable method, and over nearly three centuries mathematicians, astronomers, and seafarers submitted myriad proposals. Most of the schemes were either too technically or too mathematically complex to carry out on board ship. In theory, it would be possible to use celestial bodies like the Moon, the satellites of Jupiter, or even an eclipse as a type of clock, thus tracking how far a boat sailed in a certain period. However, neither existing planetary tables nor timekeepers were yet accurate enough (seventeenth-century pendulum clocks often lost ten or more minutes in the course of a day, and their precision was not improved on the swaying deck of a ship). A time-based longitude solution would have to wait until John Harrison perfected his marine timekeeper in the 1760s.[41] From the mid-sixteenth century onward, the leading theory for pinpointing longitude hinged instead on navigators' tracking the gradually shifting magnetic field. Mariners had long noticed small discrepancies between true north and magnetic north. Michiel Coignet was one of the first to suggest that, if the entire world could be mapped, it might be possible to identify a location from its unique amount of variation. This led to intense research into magnets and compasses, but it resulted in scant practical benefit on the high seas.[42]

If longitude remained perplexing, one familiar tool came into wider use as voyages grew longer: the sea chart. A navigator would use a pair of drawing compasses to get an approximate idea of the distance between ports and to determine the rhumb or great circle course that connected them. He also began keeping a continuous record of his vessel's progress on the open water, using the same sharp compasses to mark his daily position with a "prick" on the chart. As charts became indispensable, experts weighed in on how best to depict the globe's three-dimensional contours on a flat surface. Plain (or plane) charts—which maintained a constant scale on both axes—were simple to create and use, but they introduced dangerous distortions over larger distances. "Round" charts were a more accurate option: Gerard Mercator's graduated projection, devised in 1569, accounted for the decreasing distance between meridians at higher latitudes. However, such charts were confusing to create without elaborate tables and were thus not widely adopted by mariners for decades, if not centuries.[43] In this, as in so much of the navigator's routine, there was a constant negotiation between theoretical rigor and quotidian pragmatism.

Navigation thus comprised a panoply of challenging tasks. Sailors were often derided as marginal figures: illiterate, inebriated, lazy, or untrustworthy.[44] However, in the aquatic society of the Netherlands—which had recently allied with the English to rout the Spanish Armada, and where the majority of trade occurred over water—the mariner was more often viewed as a "stout clever" man rather than an "ignorant" one.[45] And yet how could his expert knowledge be passed on to younger sailors, to say nothing of other sectors of society? The intangible wisdom of the master navigator had never been easy to impart—developing as it did only gradually with years of hands-on practice—but might there be methods to replicate it more quickly? Over the course of the sixteenth century, the Dutch merchant marine saw a sharply increasing need for men who could negotiate the newly discovered regions and crew the ever-larger commercial expeditions. To meet the needs of this burgeoning "Sea Business," men like the Rotterdam surveyor and mathematics teacher Ezechiel de Decker joined the ranks of people offering methods "to improve the Art of Navigation." In 1631 Decker published a modest manual on "The Practice of Large Navigation"—but in it he included his version of the groundbreaking logarithmic and trigonometric tables that would revolutionize how navigators carried out their day-to-day work.[46] Other authors penned new textbooks, instrument makers crafted innovative devices, and teachers offered a variety of lessons. With this flurry of new technologies and educational options, both on the deck and on shore, the Dutch maritime world could pursue a host of new avenues for creating skilled navigators.

TRANSMITTING EXPERTISE: ON THE DECK, IN THE CLASSROOM, AND BY THE BOOK

In previous centuries, young men intent on a life at sea had embarked on some form of apprenticeship. After all, as Coignet accurately declared, "small navigation" was "learned mostly through personal experience and through lessons from old, experienced Pilots." Such lessons regularly occurred at the bittacle—the gimballed ship's compass in the center of the deck. When the chief navigator took his position there shortly before noon to make his daily observations, a select number of junior men would join him. They would also be expected to shoot the Sun and to calculate their best guess at the vessel's position, which would then be compared to that of the others in the group. The navigator and the other men could offer pointers on how to adjust the vanes on their backstaffs or help with computations.[47] Whether such training was guided by a formal contract or overseen by family members or friends, this hands-on model worked for centuries. However, with the evolution of sailing routes and the concurrent changes in navigational techniques, traditional apprenticeships ceased to teach all the essential skills. If simple observation, repetition, and memory helped cement much of the requisite knowledge for small navigation, these practices fell short when it came to teaching the more abstract concepts behind longitude and latitude observations or fostering the degree of numeracy needed for using tables and carrying out long computations.

In the late sixteenth century, pioneering cartographer Lucas Janszoon Waghenaer offered several possible options for becoming competent at large navigation. In 1584 Waghenaer published his *Spieghel der Zeevaerdt*, a landmark volume of maritime charts that contains a wealth of detailed tips for the maritime community. Unsurprisingly in a book that promoted maps as a new, portable reference, Waghenaer's first instructions for "Apprentises of the Art of Navigation" are cartographic; he tells his reader to compare the coastline he was observing with the map in his hand and then to note discrepancies, draw landmarks, take soundings, and enter all the foregoing in his "Compt-book," the personal "account book" that would soon become the standardized logbook. Such first-person observation was often the best way to understand and memorize things. However, even better than using these maps would be for "young beginners in this Art" to consult with the learned men on board ship. "The Master of the Shippe, and other men exercised in this study" would instruct them in local geography, advise on the best routes, and teach them how to use the standard instruments ("the Crosse staffe, and Astrolabe" alongside the "Compasse") for "safe and skilfull seafaring."[48]

Clearly, for Waghenaer, it was better to learn from an experienced helmsman than from the chapters he himself had authored. As a former sailor, Waghenaer respected "experienced captains, pilots, navigators and others," relying upon their geographic observations to improve his maps. (In this already competitive market, he and his publisher offered to "pay more than double" for any tips they passed to him or to his Amsterdam publisher, Cornelis Claesz.) Waghenaer admired these men as society's leading experts on geography and nautical science. Not only did their daily activities make them ideal contributors to his cartographic project, but their participation also served to certify the authority of his printed work.[49]

Unfortunately, Waghenaer was forced to recognize the shortcomings of such a collegial training plan. Not every skilled sailor was willing to be a good and generous teacher. When he revised his atlas eight years later, Waghenaer included a new prefatory letter — this time, for the "old capable Navigators." In this "Exhortation," he implored these more experienced men not to "keep [their knowledge] all to themselves" but to share it. It was their responsibility, he rebuked them, to "take all the young youth and teach them everything about the science of the sea." As Waghenaer and other commentators pointed out, "beastly" strategies — such as willfully hiding the records of their calculations "to make themselves indispensable" — risked a vessel if something should befall the lone navigator.[50]

Even if such proprietary instincts were not ubiquitous among older navigators, learning from one's colleagues could be challenging. Navigational practice was becoming inexorably more difficult. One English captain called it "an art beyond most of others, not to be snatched at, at idle times and on the bye, but rather requiring so full a taking up of a man in the learning of it, as for the time nothing else is to be looked after."[51] As ships sailed to higher latitudes, plain charts dangerously distorted any record of the voyage; as they descended below the equator, they sailed beyond view of the Pole Star and its familiar company. Cosmographers and pilots set out to modify existing techniques and invent new ones. They devised innovative solutions, but these required new levels of precision, patience, and mathematics, skills less easily conveyed to new trainees on board ship. These changes explain the popularity of guidebooks such as Waghenaer's and Coignet's. In the absence of senior mariners familiar with the latest techniques and willing to pass along their expertise to others, the "young beginners" turned to alternative venues to learn these new, more mathematical tasks: schools, public lectures, and especially books. Each of these options presented the standard set of skills in a slightly different way.

David Vinckboons once again provides insight into what such new lessons might look like, drawing a scene similar to the one with which this chapter opened, depicting nautical men poring over the tools of their trade. In this case however, he crafted a stylized frontispiece for one of the most popular nautical works of the period, Willem Blaeu's *Licht der Zee-vaert* (1608) (figure 1.11). This atlas, which includes a similar, brief introduction to the "Art of Seafaring," displaced Waghenaer's works and saw twenty-five editions in as many years.[52] Vinckboons created a striking allegorical image that differed in two key aspects from his other sketch. Here, fourteen figures perch on benches beneath another grand seascape, flanked by imposing statues of Aeolus and Poseidon, rulers of the winds and the waves. A lantern illuminates the lesson metaphorically as well as physically.[53] These men are clearly less elite than those in the other image; they wear the soft caps of mariners rather than the broad brims of their social betters. Two young boys appear in the corner of the scene, the one in more rustic garb clutching a book or writing board. The presence of these youth, not unexpected in a didactic text, signals an important change in the type of participants in formal maritime education. Vinckboons also adds a significant central individual, a keen-eyed, bearded teacher holding forth on the technicalities of the jumble of instruments. No longer is maritime knowledge communicated solely by means of informal shipboard lessons or even within a sequestered community of

1.11.
A hands-on, collaborative navigation lesson. W. J. Blaeu, *Licht der Zee-vaert* (Amsterdam, 1608), frontispiece. Universitätsbibliothek Marburg

equals. Rather, this motley group of men is poised, mouths open in enthralled conversation, with their hands on the tools of their trade, ready to learn from the wise instructor—a schoolroom master rather than a shipboard one.

Vinckboons's—or, more properly, Blaeu's—mariners also learn from each other. As the teacher points to the polar regions on the terrestrial globe, one man takes a measurement with his compasses. Two others do the same in an atlas, while another pair studies an unfurled chart behind the instructor's shoulder. Blaeu's students are not just holding instruments but teaching each other how to use them through glance and gesture. One man on the left side of the room assiduously points out how to adjust a mariner's astrolabe, while across the way another demonstrates how to use a cross-staff. Here, even in a book devoted to teaching theoretical matters, there is great respect for the lessons that could be gleaned by watching an expert manipulate his instruments. Vinckboons's two images of nautical expertise, similar in detail but dramatically different in tone, demonstrate the emergent Dutch approach to technical education, a hybrid of paper and memory, instruments and mathematics.

By the turn of the sixteenth century, numerous educational options were available to the curious mariner. Each training regimen depended somewhat on personal resources: literate men who had money could engage a tutor or enroll in a series of lessons. The model established in Iberia earlier in the century was of a state-run school in which the instructors produced textbooks and prepared students for examinations. In the Netherlands, however, the most common venue in this early period was the small private class. (Formal public lectures would not be introduced until 1711.) Independent instructors, "mathematical practitioners" who may or may not have had seagoing experience, offered in their own homes classes on the basics of arithmetic and geometry, the use of the instruments, and navigational practices in general.[54] Such classes probably resembled the setting of Vinckboons's first image more closely than Blaeu's stylized tableau, as students would likely have been seated at a table to take notes or work through exercises (see figure 4.4). Unlike the professors at the Casa de la Contratación in Seville, these teachers were not officially sanctioned and were rarely of high social status. However, cities in the Low Countries had a strong tradition of mathematical tutoring, particularly for men who needed to master the basics of commercial arithmetic.[55] As a consequence, educators occupied a well-respected entrepreneurial niche. After developing their reputations as private instructors, these teachers often obtained official positions within the maritime establishment, as professors or examiners for one of the merchant companies or branches of the navy. These institutions multiplied as the seventeenth century wore on. Similar independent teachers in France, England, and elsewhere would soon find rivals in the state-funded schools that were established in the latter part of the century. The Dutch, however, did not turn to such centralized education, largely because — as we will find in chapter four — the small schools were proving so successful in training skilled mariners.

Books were another salient means of learning the new techniques. Part of the general sixteenth-century explosion of print culture in the Netherlands, they also signaled a shift in how sailing was practiced and transmitted.[56] Introductory manuals began appearing on the Dutch market at the same time as venues for formal instruction did. This synchronicity was not coincidental: navigational handbooks were published only when people were willing to pay for them, after professional obligations pushed them beyond their technical comfort zone. Cornelis Anthonisz. produced the earliest recorded such work, a volume that contained extensive sailing instructions for the Baltic Sea alongside an introduction to the art of navigation. In that oblong quarto, Anthonisz. explained how to use a trio of fundamental tools for determining place and direction: the compass, the cross-staff, and the quadrant. He described four methods "to know the latitude," how to construct a chart (*paskaart*), and how to estimate the speed and the direction of a current.[57]

New nautical works appeared only slowly until the final years of the sixteenth century. Although independent instructors were beginning to offer lessons in Amsterdam and elsewhere, they published few titles during the chaotic years of the "Dutch Revolt."[58] The maritime world was also far from convinced that books were a reasonable way to disseminate this subject matter. Skeptics still fretted about the reliability of print. Willem Blaeu went to pains to reassure his readers that his printed maps were at least as accurate as hand-drawn manuscript charts. At the same time, Blaeu mistrusted printed tide tables—despite including one in his atlas—and recommended that mariners memorize their courses and the times of high tides rather than rely on almanacs.[59] Moreover, some observers felt that books were inappropriate for sailors. Stereotypes continued to circulate about illiterate, "very ignorant" seamen, though more often outside the Netherlands. Even if it remained true that "these days many *Pilots* can neither read nor write," more and more Dutch authors were optimistic that they "still were smart enough to learn."[60]

Writers, publishers, and booksellers began to recognize that large segments of Dutch society were interested in maritime matters. The book industry not only contributed to the already popular genre of travel narratives. It also produced stirring accounts of naval victories and pirates and added nautical content to entertaining dialogues (*tafel-spelen* or *praatjes*) and political pamphlets. Bumptious sailors often served as comic relief in political or social commentary.[61] In addition to this wave of nautical material that was inundating the general public, the Dutch maritime and mercantile world would also demand more and more technical publications: blank books ruled for journal keeping, updated mathematical and astronomical tables, printed copies of company regulations, and international signaling codes.[62] But no nautical publication was more ubiquitous as the sixteenth century drew to a close than the manual explaining the basics of navigation.

Over the next two centuries, thousands of nautical manuals would be published across Europe, some in dozens of editions. But early on, not yet assured of a commercial market, Dutch printers for the most part produced modest books in small print runs. As far as budgets allowed, they included illustrations, dialogues, helpful mnemonics, and other features to make them accessible and engaging. Rather than targeting a specific, narrow readership, authors cast their net wide, addressing aspiring youth, working sailors, enthusiastic gentlemen, shipowners, and even merchants who had little prospect of sailing. Lucas Waghenaer, for instance, intended his handsome folio-sized *Spieghel der Zeevaerdt* for "Mariners, Masters, and Marchants."[63] More than simply offering advice to apprentices and soliciting feedback from "old, capable Navigators," he was holding up a "mirror" to illuminate a fascinating and vital subject for merchants and other curious readers.

The Netherlands had a sizable audience of these curious enthusiasts, or *liefhebbers*. Early published Dutch works automatically included sailors in this category of "lovers" of the art of navigation (the sobriquet had the same connotation as *amateur*).[64] A popular early seventeenth-century dialogue between Jaep and Veer—a wise captain and an inept one—addressed the reader who was "eager to learn." The author, Heyndrick Reyersz., clearly felt that practitioners belonged in this group, claiming that his text would be "useful for all enthusiasts of navigation, and primarily for those who wish to attend to steersmanship with good judgment and comprehension." These professions of love and curiosity were noticeably more elitist in other countries, and even in the Netherlands they would grow more perfunctory by the eighteenth century, but in the early years sailors were a significant and well-respected segment of the audience.[65]

The printing press ushered in notable changes in mariners' practice as well as their training. Just when sailors began traveling regularly beyond familiar waters, adventures that required them to learn exponentially more geographic and astronomical information, it was becoming far easier to replicate dense tables and engraved charts.[66] These printed materials were promoted as new and convenient memory aids, stepping stones toward traditional mastery. Volvelles, which had been included in early cosmographical books, saw a resurgence not only for budgetary reasons but also for the conceptual benefits that came from manipulating the tools. In the late sixteenth century, English captain John Davis described a volvelle's pointer as "a necessary Instrument for the better understanding of such things as are required to the finding of the Pole[']s Elevation."[67] The rotating dials helped readers make sense of the swirling constellations and other related concepts and also provided concrete answers to questions about dates and vertical angles (figure 1.12).

Already, in this preliminary phase when nautical knowledge was initially migrating into print, sailors risked getting swamped by so much data that it was difficult to discern what was most important. Certain techniques might be necessary for a typical voyage, while others might be useful only in an emergency. Experts of all stripes — authors, teachers, and administrators — stepped in with advice, each taking a different approach to this question. Some, like Coignet, viewed their volumes as repositories of basic information. These authors stressed the rudiments of navigation, especially definitions and the form and function of instruments. They approached the subject systematically, progressing logically from simple ideas to more complex ones. Other authors focused more narrowly on challenging topics, or what they deemed essential. Such books could venture into sophisticated theoretical grounds but more often remained bluntly practical. These authors admitted that their books were not comprehensive but suggested that their brevity made their works more effective and appropriate. In his *Licht der Zee-vaert,* for example, Willem Blaeu declared longitude too complex for sailors and omitted the topic entirely. Writing a decade later, the entrepreneurial customs clerk Jan Hendrick Jarichs van der Ley chose to focus *solely* on longitude, contending that so many other authors had ably covered the basics.[68]

Thus, at this moment in the late sixteenth-century Netherlands, we see a genre that was not yet codified. While experts agreed to a remarkable extent about the set of topics "necessary for a navigator," there was considerable variation not only in the principles and tools that authors deemed most significant but also in how they expected the reader to learn.[69] But there was one common feature: in almost every case, authors offered their readers multiple strategies for solving basic problems, tacitly preparing them for emergency situations large and small. Navigators could be tossed by tempests, unable to see the horizon due to fog, beset by pirates, or faced with a broken vane on a cross-staff — and they needed to keep track of their vessels despite all these

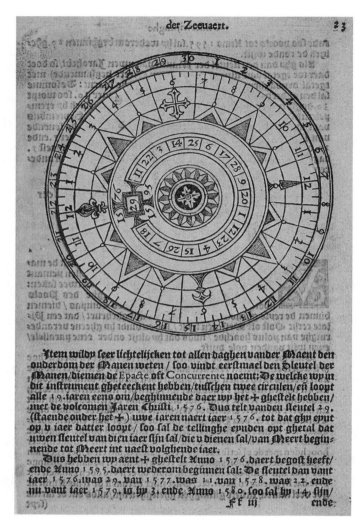

1.12.
Age of the Moon volvelle (valid between 1576 and 1595) M. Coignet, *Nieuwe Onderwijsinghe, op de principaelste Puncten der Zee-vaert* (Antwerp, 1580), 89. Courtesy of the John Carter Brown Library at Brown University

hazards. In certain situations, it might be possible to use their favorite instrument, but in another they might need to carry out computations or return to traditional methods that eschewed tools. At times, they would need to combine several techniques. By including a panoply of solutions as insurance against the vicissitudes of the sea, nautical manuals effectively preserved traditional techniques alongside newer options, ultimately shaping the evolution of navigational practice in unexpected ways.

IN CORNELIS CLAESZ.'S BOOKSHOP

Aspiring navigators and curious *liefhebbers* could easily access these pivotal volumes by visiting bookshops like that of the maritime publisher and bookseller Cornelis Claesz. His store was "on the water by the old bridge" at the sign of "the Writing book" — on the Damrak, the mouth of the Amstel River in the heart of Amsterdam. Few details are preserved about this particular bookstore, but it was likely Cornelis Claesz.'s family home, with the living quarters upstairs, several printing presses in the back room, and a substantial retail operation in the front, facing the oldest part of Amsterdam's harbor. Just as at other publishers, the press room was a whirl of activity. Production took place amid cases of type, under freshly printed pages hanging from racks near the rafters. Young apprentices dampened sheets of paper and prepared ink, typesetters clicked pieces of type into place between the woodblocks for diagrams, and one or more bespectacled proofreaders pored over proofs looking for errors. As the master printer, Cornelis Claesz. would have supervised the room himself, but he also kept busy meeting with authors, translators, and his clientele. In the retail space of 't Schrijfboek, a curious reader could have perused a large selection of titles, sold either in plain vellum bindings or, more likely, as loose quires.[70] He could also probably buy pens, ink, reams of paper, and blank notebooks — the "writing books" for which Cornelis Claesz. named his business. Men who came to purchase maritime volumes would also exchange tidbits of news, learn about new instruments, and forge connections with the wider community (figure 1.13).

Over the course of nearly three decades (from 1582 to 1609), Cornelis Claesz. became the most important bookseller in the Netherlands. He began as a general-interest publisher, printing Bibles and translations of Calvin's *Commentaries* (1582), political tracts on the war against Spain, medical and legal textbooks, maps, and popular literature. In the 1590s, however, he narrowed his focus to nautical imprints. Unlike rarefied master printers who edited Latin humanist works (such as Christoffel Plantin, who had published Waghenaer's *Spieghel* in Leiden), Cornelis Claesz. simply had an eye for what would sell.[71] Visitors to his establishment in the late 1590s could find something to suit any budget, career trajectory, and appetite for detail, from a pocket-size set of astronomical tables or a Dutch-language edition of some intriguing foreign text to an elegant atlas. Cornelis Claesz. specialized in less-expensive reprints of important, relatively technical titles. For instance, he purchased the old plates of Waghenaer's first famous volume of nautical charts, the *Spieghel der Zeevaert*, issuing a cut-price German version just when Plantin was printing the updated *Thresoor der zee-vaert* in Leiden.[72] He reprinted Martin Everaert's Flemish translation of Pedro de Medina's 1545 *Arte de navegar*, including Michiel Coignet's important appendix, three times in a decade (1589, 1592, and 1598). Recognizing its utility to maritime readers, he also reprinted an Antwerp edition of Apianus's *Cosmographie* (1598 and 1609). He shepherded other foreign titles into Dutch, bringing the latest Spanish and English works to the Dutch market.[73] Cornelis Claesz.'s strategic decision to offer a robust list of variously priced titles ensured that the already substantial nautically minded audience would expand beyond the jostling harbor.

The volumes spilling from the shelves at 't Schrijfboek offered their readers a range of different approaches to the science of navigation. Let us focus for a moment on two contrasting texts, both reprinted by Cornelis Claesz. in the same year: Waghenaer's *Spieghel* and the Medina-Coignet volume. Suppose a well-off merchant or

1.13 Customers rub shoulders in a well-stocked Dutch bookshop. Note the globes, cross-staff, and rolled charts on the upper shelves. Dirck (or Salomon) de Bray (Haarlem, 1650–55). Rijksmuseum, Amsterdam

a captain with a steady commission wished to buy an atlas of charts. Perhaps he had heard of Lucas Waghenaer's lavish *Spieghel der Zeevaerdt* (1584), but rather than making such an expensive purchase, he might choose to acquire the slightly more affordable German translation, *Spiegel der Seefahrt* (1589), which Cornelis Claesz. sold for six guilders and ten stuivers.[74] From Waghenaer's prefatory notes, we know that he intended his atlases chiefly as working tools — graphical rather than textual instructions for navigating the "eastern" and "western" oceans. However, as he also wished to attract the interest of merchants, he explained several topics that would already have been well known to men with nautical experience.[75] Waghenaer opened the *Spieghel* with details about the calendar, the Moon, and the Sun's declination, preferring to describe these verbally rather than to depict them in diagrams. He paid comparatively little attention to noncartographic instruments, touching only briefly on the cross-staff before reaching the meat of his endeavor, the sea chart. The only thing Waghenaer deemed as interesting as his new *paskaarten* were the stars; he included a dauntingly detailed two-page table, followed by a brief list of the six most useful stars. Also, he boasted, he had committed every known star to memory.[76] For Waghenaer, mastering this astronomical knowledge was not only a crucial aspect of sailing but one of the credentials that made him a reputable teacher.

Waghenaer was one of the few authors in this genre to include any discussion of local geography — perhaps because of his own experience as a navigator and surely because of his desire to demonstrate the utility of his new charts. (Waghenaer, equally concerned about accuracy and self-promotion, compiled an extensive list of errors in a predecessor's rutter. Ironically, twenty years later, his competitor Blaeu would critique Waghenaer on the same grounds.)[77] These maps offer graphical depictions of details previously recorded succinctly in rutters and jotted in captain's notebooks: the course of navigation channels, beacons, and other landmarks such as church spires and houses visible from the sea. A cluster of depth figures appears next to most ports (see plate 1). Surprisingly, Waghenaer moved away from most of these graphical innovations in the 1592 *Thresoor der zee-vaert*, responding to navigators' demands for more of the textual details that had been standard in conventional rutters. He reduced the size of the maps and reintegrated the traditional coastal profiles that the *Spieghel* had omitted—the familiar outlines of headlands were more identifiable than the small-scale bird's-eye views of modern maps. Until they learned how to read the latter, sixteenth-century mariners doubted that Waghenaer's visual tools contained as much useful information as the conventional written ones (see plate 2). These adjustments made the *Thresoor* smaller and consequently somewhat more affordable. (The two Vinckboons drawings both depict a volume in the oblong format so typical of these portable Golden Age atlases.)[78]

Waghenaer clearly valued visual knowledge. He invested time and resources into each of the maps in his atlases, and yet the introductory portions of each volume were sparsely illustrated. In the *Spieghel*, he depicted three instruments, while the *Thresoor* included only one (a diagram showing how to graduate a cross-staff) in addition to a set of coastal outlines at the end. In each of these volumes, he devoted numerous pages to tables. Notably, he also created two large circular diagrams that conveyed much of the same information in a creative graphical form: he combined a rhumb table with a chart of high-tide times—one woodcut page instead of five pages of dense tables. Waghenaer also devised a large engraved volvelle that, like Cortés's horn, allowed the navigator to adjust the observed altitude of the Pole Star[79] (figure 1.14). Here, Waghenaer's personal experience at sea, coupled with his sophisticated understanding of the heavens, inspired him to repackage textual data into a more user-friendly form. These two figures, small components in his larger pedagogical project, are an early example of how inventors and practitioners could alternate seamlessly between numerical and instrumental tools in their quest for efficient, foolproof means of completing daily tasks.

Naturally, not all readers were interested in pricy nautical charts. A reader more curious about the general concepts of navigation might prefer to invest in the new Amsterdam edition of Medina's *Arte*, the *De Zeevaert oft conste van ter Zee te varen* (1589). Cornelis Claesz. published a new edition of Martin Everaert's translation, which had appeared in Antwerp eight years earlier. This substantial quarto volume included the twenty-five-page appendix *De Nieuwe Onderwijsinghe*, the "new lessons on the principal points of navigation," by the Fleming Coignet.[80] Everaert's translation of the *Arte* is a milestone in the circulation of technical knowledge: it represents the moment when the Iberian mode of thinking about navigation was first made accessible to Dutch-speaking navigators. Although it is likely that early Spanish navigation texts would have been available in the Habsburg-administered Low Countries from the mid-sixteenth century on, few sailors working in the region

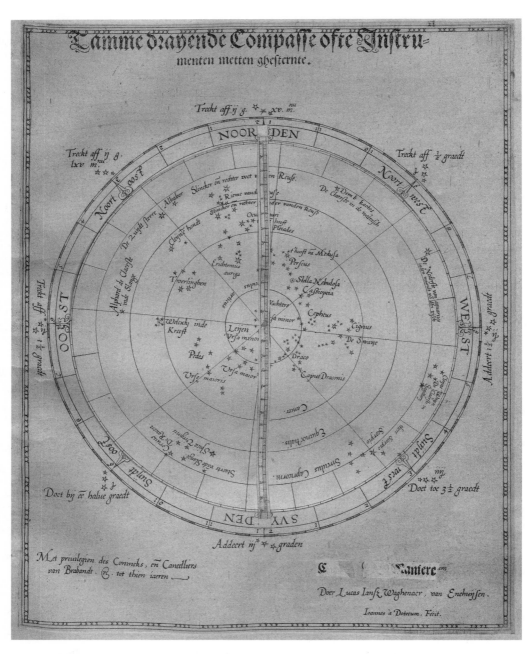

1.14.
Pole Star volvelle. By rotating the volvelle to match the position of the Guards (*Wachters*) and then adjusting the graduated crossbar according to the observed altitude, one can find the observer's latitude. This working instrument was also sold as a separate broadside in Antwerp. L. J. Waghenaer, "Tamme drayende Compasse ofte Instrumenten metten ghesternte," from *Spieghel der Zeevaerdt* (Leiden, 1584). Utrecht University Library, MAG : P fol 111 Lk (Rariora)

would have been literate in Spanish. Cornelis Claesz.'s decision to reprint the *Conste* twice, in 1593 and 1598, indicates that, even after the United Provinces declared independence from Spain in 1581 and Philip II's Armada was embarrassingly decimated off the Scottish coast in 1588, these Iberian techniques continued to be of interest.

Coignet had initially conceived the brief but important *Nieuwe Onderwijsinghe* as a companion to Medina's *Arte*, so the Antwerp polymath saw no need to tackle the big concepts that the Spaniard had already explicated. Instead, after a list of "principal points" and their definitions, Coignet turned to instruments. Each chapter features a different piece of equipment, from the compass and maps to the astrolabe. In 1580 even the most basic nautical instruments were unfamiliar to most readers and not readily available commercially, so Coignet and other writers often provided instructions for constructing one's own. Few of the tools Coignet described were new; instead, he presented modified versions of standard instruments. (For instance, his own "marine Hemisphere," which ostensibly allowed mariners to find "the time and the latitude at any time," was a type of three-dimensional astrolabe.)[81] To further help his readers, Coignet added moving disks and pointers to woodcut images of the astrolabe and the nocturnal. He also included a volvelle that enabled the user to adjust for the year's Golden Number and find the age of the Moon — certainly a simpler solution than counting on one's knuckles.[82]

Of course, Cornelis Claesz. had no monopoly on the sale of maritime texts; as the market developed, swelling numbers of Dutch publishers found it profitable to print nautical works, and more and more men were willing to write them.[83] Many — though not all — in this widening circle of authors had some maritime experience, while others billed themselves as *liefhebbers* and "lovers of mathematics" (*philomaths*).[84] Some of these entrepreneurial men wished to promote instruments of their own invention and hoped to catch the attention of a wealthy patron — or at least to persuade buyers to pay for lessons on how to use their contraptions.[85] Certain men, particularly the examiners, continued to write comprehensive introductions to the art of navigation. However, readers were often eager for smaller, more focused texts: for instance, Adriaen Veen in Amsterdam wrote a pamphlet that explained how to practice "nautical bookkeeping," the first formal instruction on how to keep a logbook.[86] In the university town of Franeker, Robert Cuningham produced an almanac modeled on two important volumes of tables published by Cornelis Claesz. (figure 1.15). Cuningham included clear instructions on how to apply the calendrical information. (He, like Waghenaer, wrote primarily for mariners but felt this technical knowledge could be useful for many others — medical doctors, ministers, and teachers.)[87] By the mid-seventeenth century, Dutch authors began to include these ephemerides in the standard textbooks, making it easier than ever for sailors to calculate their positions at sea. (These presented declination data in four tables, to account for the extra day in every leap year. The solar tables became outdated quickly and were useless if not updated regularly. By contrast, the same values held approximately true for the Pole Star for at least twenty years.) It would take seventy-five years before the French royal publisher would begin annually issuing separate, portable volumes, and then another

1.15.
Tide tables. R. Cuningham, *Tractaet des Tijdts* (Franeker, 1605). Collection Maritiem Museum Rotterdam

eighty-seven before the British did the same. In the meantime, mariners relied on general almanacs, hoping to replace them before they were too out of date.[88]

Not all mariners or enthusiasts would buy their books from a shop like *'t Schrijfboek*; they might instead purchase a textbook from their navigation teacher during lessons at his home. By the first decade of the seventeenth century, instructors had begun publishing their own notes — and, indeed, over the coming century they would be the main contributors to the genre. That a teacher should publish a textbook was relatively new (for many centuries, students had simply made manuscript notes from classroom lectures), but as literacy and schools both grew, students came to expect a published manual, a device that aided study and signaled their teacher's expertise.

Jan van den Broucke was one such teacher. Originally from Antwerp, he taught first in Vlissingen, and then moved to Rotterdam in 1609. That year, Van den Broucke published his *Instructie der zee-vaert* (with a second edition the year after).[89] The *Instructie* was much more pared down than Medina's text or even Coignet's — it treated some of the same instruments, but none of the theoretical background. (Van den

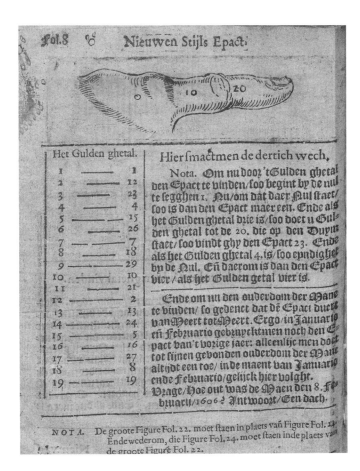

1.16.
Epact thumb (for the Gregorian calendar), and Golden Number table. J. van den Broucke, *Instructie der Zee-Vaert* (Rotterdam, 1609), fol. 6. Leiden University Library, 2003 G7

Broucke was familiar with other authors, local and foreign, and asked the Leiden mathematician Willebrord Snellius to double-check his solutions for more advanced puzzles, such as how to solve cube roots geometrically.) Although hastily printed on inexpensive paper, the *Instructie* contains copious illustrations and eight volvelles. Like other early manuals, Van den Broucke relied on figures to clarify the technical details of instruments, but he also added other images: one of the earliest printed diagrams of the "rule of thumb," to help calculate the epact, as well as a schematic of a hand to determine the year's Golden Number (figure 1.16; see figures 1.6–1.9). These images, the first steps to figuring out the tides, would soon replace cosmographical diagrams at the beginning of every Dutch textbook.

Van den Broucke expected eager "Leerlings" to work through the book and then to commit twelve key points to memory by reviewing the list *every evening for a year*.[90] (figure 1.17). As a prerequisite for all of the tasks, he instructed his pupils to memorize the twelve signs of the zodiac and the seven planets and their angles of conjunction. "Understanding the ephemerides" was first on the list, followed by the standard calendrical concepts: the Golden Number, Sunday letter, and epact. The sixth task

FROM THE WATER TO THE WRITING BOOK ✤ 55

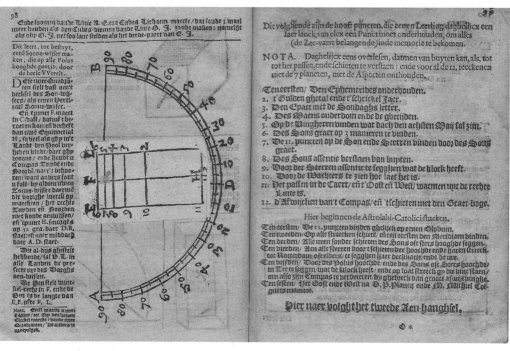

1.17.
Right: Van den Broucke's list of twelve essential points to "read over once a day." (A type of sun dial with a volvelle appears on the facing page.) J. van den Broucke, *Instructie der Zee-Vaert* (Rotterdam, 1609), 98–99. Leiden University Library, 2003 G7

was to find the Sun's altitude, a process so important that Van den Broucke expected the reader to master three different methods. After another question relating to the Sun's altitude, he asked about its ascension — essentially, the Sun's longitude. Points nine and ten focus on using the stars as astronomical clocks. The final points were more hands-on: plotting a course on a chart, understanding compass variation, and using a cross-staff.

Van den Broucke offered something quite different from his predecessors. Medina, for his part, presented a big-picture view of the art of navigation. To explain everything that influences a ship's movement, he investigated the finer points of why water flows the way it does, for instance, and the logic behind dividing a year into days. He hearkened back to Euclid and Apianus, building a foundation of cosmographical definitions to help a mariner understand his place in the world. Waghenaer and Coignet were more selective. They focused on basic tasks but also on issues that were particularly difficult, unresolved, or unintuitive, such as why the Pole Star and plain charts were more complex than they appeared. Jan van den Broucke fell into a third camp: he wished to simplify a sailor's life. Rather than attending to foundations or challenges, he offered shortcuts. He explained that he "necessarily taught the shortest and most minimalist method, because the seafaring people learning this art in a lengthy way gain a hatred for it." He opened the volume with a twelve-verse "zodiac song," an extended mnemonic invented by one of Amsterdam's earliest navigation teachers to simplify the vast astronomical data that mariners needed to memorize.[91] Sailors who read the *Instructie* could "solve spherical 'things' without calculations" and, when taking latitudes, avoid fiddly instruments, be they compasses, cross-staffs, or even the sky itself. Van den Broucke, like Blaeu, felt mariners could manage without a "Declination Book" and included mnemonic phrases that helped not only with the Golden Number but also with the Sun's altitude; in a pinch, the navigator could use the standard number for any given month as a reasonable approximation of the Sun's position.[92] Later in the book, he also depicted a physical instrument that could provide an equally accurate number. After presenting these shortcuts, Van den Broucke offered an alternative for those students who quailed at the thought of spending a year memorizing his key points: he promised to teach his readers all they needed to know in one month or less.[93] Van den Broucke, like many of his contemporaries, seems to have been conflicted about the best means of learning navigation. Were printed rules and paper instruments effective, or should junior navigators seek out face-to-face lessons?

"FROM THE BOOK TO THE READER"

In these early Dutch imprints, we see several different models in play: the academic framework of the Spanish tradition juxtaposed against the more practical, entrepreneurial Dutch usage, the one beginning with cosmographical definitions, the other with pragmatic calendrical data. The authors were of varied backgrounds — Waghenaer, a sailor; Coignet, an engineer and wine gauger; Van den Broucke, a teacher — and thus each brought different interpretive skills to the same material. Each chose different things to illustrate, striving to elucidate complex material in creative ways. Their choices show us what was important to these men. What did they find difficult? What did their readers already know?

When it came to what most mattered to effective navigating, time and distance, we find that the audiences of these books must have been starting from different perspectives. The compass rose, for instance, widely deemed a sailor's foundational tool, was rarely depicted in these early texts. Medina carefully described the rationale behind the thirty-two subdivisions of the compass and furnished four increasingly segmented circles to illustrate this point. Coignet and Waghenaer agreed that it was one of the most important pieces of equipment but chose not to illustrate it. (Waghenaer's large tide calculator volvelle was superimposed on a basic compass dial, but he saw no need to comment on it in the text.) Van den Broucke provided a list of the thirty-two directions (*Compas zijn streecken*) to memorize but no diagram.[94] This absence is perplexing in an otherwise richly illustrated volume and suggests that his readers already knew this fundamental tool. Note that near-contemporary French manuscripts — produced for elite readers rather than working navigators — all included compass diagrams, and over the course of the seventeenth and eighteenth centuries, when mariners produced nautical workbooks in the classroom, many opened their manuscripts with a compass rose[95] (see figures 4.2 and E.3).

Such omissions are perplexing. Does the absence of a topic signal ubiquity or insignificance? When something widely recognized as essential does not appear in introductory guidebooks, we can infer that sailors must have encountered it elsewhere. Lacunae in textbooks can thus point to an alternative, shipboard curriculum. By the same token, these volumes were silent on other foundational tasks — the practicalities of tying knots or predicting the weather. Although masters were expected to know how to "Work a Ship with safety in respect to Wind and Weather," preferably "foreseeing changes in the weather before they occur," the only discussions of meteorological signs in sixteenth- and seventeenth-century nautical texts were recycled from Pliny and Ptolemy.[96] Mariners must have learned these crucial details on board ship.

If the authors (or publishers) of introductory navigational works felt it unnecessary to devote space to illustrations of the compass — because their audience already knew it well — they were far more thorough about the stars. Waghenaer was not alone in deeming them essential. Like Jan van den Broucke, Heyndrick Reyersz. included lengthy zodiac songs in his popular dialogue. As the wise captain Jaep reminded readers, it was not safe to rely entirely on the North Star when it could be so easily obscured by clouds. There is little evidence that working mariners actually sang the "zodiac liedjes" on the tossing deck, although they did feature in classroom lessons. Reprinted in at least half a dozen textbooks over the course of the seventeenth century, they testify to how early modern people wrestled — playfully at times — with large quantities of cosmographical knowledge.[97]

As for the regiment of the North Star, this *was* clearly essential for many Dutch navigators (even those whose routes took them below the equator would use it in waters close to home). Virtually every manual illustrated these stars, focusing on their utility as positional aids rather than as timekeepers. In the Iberian literature, Medina carefully documented their positions in his sixteen-image series. He also used a corporeal metaphor — head, arms, and legs — to orient the observer.[98] Cortés, for his part, included the horn volvelle (see figure P.1). Dutch texts also devoted space to the Guards: Waghenaer included a list of the *Wachters* as one of his many tables, in addition to producing the large volvelle, which he sold separately as well as inserted in the atlas. This device not only taught readers the stars' positions but also served as a functional tool — although in order for the printed stars to line up with those in the sky, it was necessary to hold the book upside down.[99] Van den Broucke created yet another volvelle, showing the star offset from true north (see figure 1.5). All these graphical efforts suggest that readers needed a higher level of assistance to learn how to deduce latitude from the Pole Star. And yet very few authors took up Medina's metaphor of the head and arms, or his sequential woodcuts. Nor did they replicate Waghenaer's detailed upside-down volvelle. Later generations of authors and sailors preferred the clear visual message and calculating power of Van den Broucke's simple dial.[100]

Spanish and Dutch authors seemed to have reached similar conclusions about the primary tools for time and direction: the compass was essential but already well understood, whereas the regiment of the North Star was both important and challenging. However, there was one topic that Iberian and Dutch authorities viewed very differently. While Spanish texts emphasized the fundamentals of astronomy and cosmography, beginning with diagrams and definitions about the structure of the world, Dutch books had an alternative focus: almost every book written by a Dutchman — from those of Waghenaer and Van den Broucke to Claas Gietermaker's preeminent textbook later in the seventeenth century — began with a lesson on how to calculate the tides.[101] The Dutch were less interested in the physics behind the tides than in how to compute their timing. Given the North Sea's shallow coast and strong tides, tracking the daily cycles was crucially important, for miscalculating by an hour could mean that a fleet would be trapped in port or, worse, stranded on a sandbar. By contrast, the Iberians had only trivial tides to deal with, especially in the predictable Mediterranean. Thus, their nautical experts had considerably less reason to concern themselves with the calendar and the Moon. (The English, for their part, focused on the challenging passage through the Channel.)[102] In Dutch texts, particularly in the early years, authors saw it as imperative to teach mariners how to use the tables but also to equip them to compute the tides if the tables were outdated or absent. Thumb diagrams like Van den Broucke's became ubiquitous. A printed talisman of medieval practice — and one of the most successful body metaphors — the "rule of thumb" remained in published Dutch texts and manuscripts until the late eighteenth century. While later authors grew less interested in the Pole Star, the fixation on tides never waned. Numerous variations on the calculating volvelle appeared in pocket almanacs and standard textbooks alike[103] (figure 1.18).

Just as omissions can shed light on authorial priorities, so too can the first pages of a textbook. The topics that these authors chose to begin their books are simple but also indispensable. They gesture back to the tradition in which the texts are rooted and set the trajectory for subsequent lessons. From the pride of place of the thumb diagrams and Sunday letter rhymes, we can infer that each Dutch navigator's workday would have started with these calendrical calculations. Although tides were far from the most significant factor once boats were out in the open Atlantic, Dutch sailors had to negotiate the treacherous shallow waters of the North Sea on each voyage and thus viewed these calculations as the cornerstone of their practice. Spanish mariners, on the other hand, sailed within a framework that they had committed to memory from their first days in the classroom, one built from the ground up with Euclidean definitions and from the heavens downward to their individual ships. These contrasting approaches — computation and memory, the rhythms of the Moon and tides versus those of the stars — emerged from traditional priorities based on local maritime geographies, which were then encoded in print.

We must not be too quick to divide these complex practices into two camps. Just as contemporaries recognized that small and large navigation were two halves of the same coin, so too were memory and mathematics.[104] Similarly there is no clear progression from old to new. Traditional methods continued to coexist with newer techniques — counting rhymes alongside new dialed instruments, medieval volvelles

1.18.
A volvelle for calculating the age of the Moon, one ostensibly valid in perpetuity. *Eeuwig Duerende Almanach. Beginnende A:1681* (Amsterdam, [1681]). Collection Maritiem Museum Rotterdam

repurposed to solve new problems. And yet this early seventeenth-century moment does herald a shift in how Dutch navigators deployed their memories. In the centuries before print, memory was the tool that facilitated local navigation. Now, practitioners were increasingly persuaded of the power of the press. Able to easily access tables that helped with computations and diagrams that helped with comprehension, they were free to deploy their memory in different ways. Some might memorize star songs, but larger numbers worked to commit to memory the regiment of the Pole Star, the steps for computing high tide, or the way to derive one's position from the graduated vane on a backstaff. Rather than memorizing their lessons as apprentices on a fishing boat, sailors had to memorize mathematical formulas in a classroom.

Amsterdam was a maritime hub, as Seville had been earlier in the sixteenth century. Its denizens had a fervent interest in nautical information and unparalleled resources to satisfy their curiosity. This vibrant entrepôt fostered teaching, instrument making, and innovation. Nautical experts — both oceangoing men and theoreticians — enjoyed cachet. The books that poured off Dutch presses may have had one principal goal: to help mariners get a handle on celestial or "large" navigation, especially the abstract concepts and the increasingly complex computations. However, they also had a secondary effect: by taking an ecumenical stance toward their readership — including everyone who wished to "improve the science of navigation" — these authors exposed much of the nation to new mathematical tools and visual modes of abstract thinking. As members of the expansive Dutch maritime community worked their way through these texts, they learned the definitions of theoretical terms; the functions and structures of assorted instruments, both old and new; and sophisticated mathematics along with shortcuts to avoid it. The flood of guidebooks published by Cornelis Claesz. and others, to say nothing of the notebooks produced by land- and ocean-based *liefhebbers*, suggest that Dutch mariners had started to believe that they could, in fact, learn to sail from a book. This transformation had another consequence: as the seventeenth century rolled on, the days were numbered for the mariner who could not read. The same shift was about to occur for the navigator who could do no trigonometry.

2.1.
Trigonometric and logarithm tables. The first poem notes that while the *Sinus* tables may be "propped up with few letters," they are more useful "than a thousand books." For their part, the second poem declaims, logarithms help make hard work easier (*het sware werck verlicht*). C. J. Vooght, *De Taeffelen,* bound after *Zeemans Wegh-Wyser* (Amsterdam, 1695). Leiden University Library, 587 D45

Chapter Two
"BY THE SHORTEST PATH"
Developing Mathematical Rules
Dieppe, 1675

Imagine the act of navigation as a "mysterious chariot," the French abbot Guillaume Denys proposed to the group of students gathered in his home. Each wheel of this chariot stands for a different element, he continued: the ship's route, its direction, its latitude and longitude. Consider what would happen should one or two of these wheels get stuck in the mud; could you still steer the chariot?[1] Denys deployed many such metaphors to catch the attention of the readers of his textbooks and his students at the royal navigation school in Dieppe. Latitude could, he suggested, also be likened to an eagle — one would need to soar up to obtain it — while the route (*chemin*) could be represented by a lion, the direction (*rhumb de vent*) by a man's face — because that element demanded a human's "prudence and circumspection" — and the fourth component, inscrutable longitude, was a bull. (The abbé chose these animals in part because of their associations with the four Evangelists.) He viewed metaphors as "theoretical" preliminaries, reminding readers that logically "theory must precede practice." At heart, he was a mathematician, convinced of the power of numbers and rules and confident that they could improve a sailor's day-to-day practice at sea.

Head of the royal school at Dieppe for nearly half a century, and author of five works on navigation, Guillaume Denys had influence well beyond the modest port city. He trained many hundreds of mariners, as well as at least half a dozen men who would take up teaching posts at new state-run navigational schools that were established in the 1670s and 1680s.[2] He had the ear of Minister of the Navy Jean-Baptiste Colbert, and his books set a pattern for a new outpouring of nautical publications by French authors. He advocated for systematic journal keeping and popularized a new method of correcting one's charts at sea.[3] Most importantly, Denys pushed to introduce French mariners to the most significant navigational advance of the seventeenth century: trigonometry.

Mariners, not just in Dieppe but across Europe and beyond, had long thought about navigation in triangular terms. As we have seen, they tracked their daily progress on radiating traverse boards (see figure 1.4). However, it remained tedious and challenging to calculate a vessel's course without an instrument or a table. Then, at the close of the sixteenth century, mathematical practitioners developed two related techniques that would revolutionize sailing. The first was trigonometry, "the doctrine of triangles," followed closely by logarithms. The term *trigonometria* dates from Bartholomaeus Pitiscus's 1595 work, one of several in the period to reinterpret classical Arab work on triangles. The earliest published explanations of the "three speciall right lines belonging to a circle, called signes, lines tangent, and lines secant" were not yet directed at mariners.[4] In his 1595 *Seamans Secrets,* Captain John Davis wrote admiringly of trigonometry as "another most excellent way for the finding of the sunnes declination at all times." However, he conceded that "because Seamen are not acquainted with such calculations," he would stick with the "plaine way before taught," by which he meant the process of consulting the regiment of the Sun under the current date to find out how far the Sun would have moved by noon. In his 1614 English translation of Pitiscus's seminal *Trigonometria*, Ralph Handson made the innovative decision to add "certaine nauticall questions . . . for the mariners use."[5]

Also in 1614, another mathematical landmark was published in Edinburgh: John Napier's *Mirifici Logarithmorum canonis descriptio.* Logarithms made it possible to convert trigonometry's complex and laborious multiplication into much more manageable addition and subtraction. From the beginning, Napier saw logarithms as directly relevant to sailors, speeding the process of computation and counteracting "weaknesse of memory."[6] In order to put Napier's conceptual invention to use, however, practitioners first required a new physical tool: logarithmic tables. Memorizing the six- to eight-digit numbers that represented each function was impossible, so practitioners consulted tables. Such tables—dozens of pages and up to 100,000 numbers—were labor-intensive to produce, but the best of them were affordable, portable, and relatively error-free[7] (figure 2.1).

Many readers, and not just mariners, found these tables daunting. People were thus inspired to find instrumental work-arounds: the English mathematician Edmund Gunter devised the principle of the widely used Gunter's scale, or rule, in 1623,

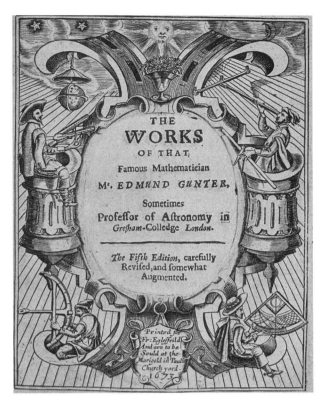

2.2.
Edmund Gunter devised several eponymous instruments to aid navigators with measurement and trigonometric calculations: (*clockwise from top left*) the sector or rule, a variant type of cross-staff, a quadrant, and a bow. The sector was by far the most popular of his tools; using a dividing compass to transfer quantities from one limb of the scale to the other, a mariner could multiply, divide, and even work with the Mercator projection. E. Gunter, *Works* ([London], 1673). Courtesy of the John Carter Brown Library at Brown University

while French inventors developed a variety of graphical quadrants[8] (figures 2.2, 2.3). These precursors of the slide rule — which Guillaume Denys demonstrated in his introductory classes — allowed users simply to measure out the known distances or angles and add them along the various graduated lines along the instrument. These instruments were based upon certain simple equivalences: one degree of latitude equals sixty nautical miles, an hour of the twenty-four-hour day is the equivalent of 15 degrees of a 360-degree circle, and each heading on the compass rose is 22.5 degrees (90 degrees divided by 4).[9] Even mariners who had not memorized these basic relationships could quickly "reduce" a course, using a drawing compass or simple string to convert the day's mileage into geographic coordinates: the position of the compass tip or string's knot on the board revealed one's longitude and latitude.

Over the course of the seventeenth century, many sailors turned to these instruments as a way of avoiding tedious calculations. However, many sailors were won over by the powerful simplicity of logarithms and carried out their days' work using trigonometric calculations. The rapidity with which mariners in certain regions incorporated trigonometric terms and logarithmic computations into nautical practice testifies to their utility. Trigonometry streamlined questions about the vertically oriented triangles of astronomical observations, such as the Sun's declination, as well as the vessel's course, a horizontally oriented triangle comprised of the angular distance

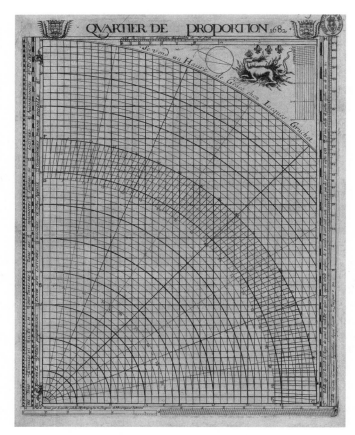

2.3.
This sinical quadrant allowed mariners to resolve "sailing triangle" problems with few calculations. They would track their vessel's course with a thread attached in the lower left-hand corner, and easily compute their new position using the engraved trigonometric scales. (Many quadrants were made of wood or brass, but this plate was sold as a separate broadsheet, perhaps to be glued on a board or used for classroom lessons.)
S. Le Cordier, "Quartier de Proportion" (Havre de Grâce: J. Gruchet, 1682). Bibliothèque nationale de France

between the intended course and the cumulative bearing of the course sailed. Richard Norwood published the first English trigonometry textbook explicitly for sailors, *Trigonometrie. Or, The Doctrine of Triangles,* in 1631. Within a few years, other English-language texts followed, illustrated with increasing numbers of diagrams[10] (figure 2.4).

In the Netherlands, meanwhile, the prominent author and VOC examiner Cornelis Lastman was already computing with logarithms in the 1620s.[11] By 1657, Lastman opened his exceptionally popular *Beschrijvinge van de Kunst der Stuer-luyden* with trigonometry tables and devoted the first chapter to right-angled triangles, which treated the sea as a flat surface. He then moved on to spherical trigonometry. This allowed for the globe's curvature, enabling far more accurate calculations at northern and southern latitudes, but was significantly more complex to solve.[12] This was a radical departure from the standard order of topics in Dutch textbooks, which by and large began with the calendar rather than the boat's course or any mathematical lessons. Lastman's decision to give pride of place to spherical trigonometry would later draw criticism — Adriaan Teunisz. Veur, an eighteenth-century teacher, did not believe that spherics were necessary for successful navigation, and viewed them as

2.4.
In this comprehensive navigational textbook, all the basic trigonometric functions are illustrated on one page. Jonas Moore, *A New Systeme of the Mathematicks* (London, 1681), facing p. 84. John Hay Library, Brown University

entirely too difficult for the average sailor — but most instructors and practitioners seemed quite willing to engage with fairly sophisticated calculations.[13]

By the mid-seventeenth century, almost all Dutch and English authors unambiguously endorsed trigonometry. Lastman's rival author-examiner, Claas Gietermaker, devoted a volume explicitly to "Triangle Calculating," as did Dirck van Nierop, a sophisticated mathematician who offered classes in the small northern Dutch town of Nieuw Niedorp. These manuals typically included full trigonometric and logarithmic tables so that mariners could easily consult them at sea.[14] Navigation teachers in Spain, by contrast, saw little reason to include trigonometry in their books until the mid-eighteenth century.[15] The French took a middle road; after a slow start to nautical publishing, Guillaume Denys's mid-seventeenth-century efforts in Dieppe effectively spurred French mariners to adopt this powerful tool.

MATHEMATICS LESSONS FROM DIEPPE'S "GOOD PRIEST"

For most of its history, France was more agrarian than nautical. Despite France's considerable Mediterranean and Atlantic coastlines, large sectors of French society paid minimal attention to the sea. To be sure, the port towns on the Normandy, Breton, and Basque coasts were as salt-tinged and bustling as the smaller Dutch cities. These regions had a long — if little documented — maritime tradition of crossing the North Atlantic to fish the Grand Banks for cod and whales. (The Mediterranean, certainly a historic site of commercial and naval activity, required different vessels and navigational techniques, so its mariners had relatively few connections to their Atlantic counterparts.)[16] Despite these pockets of marine activity, only a small proportion of the general French population comprised sailors.[17]

The Iberian system — where a *piloto mayor* oversaw lessons and examinations, and the manuals published to serve as teaching tools also informed a broader public — had marked influence on navigational training in the Netherlands and elsewhere but arrived late to France. The crown appeared to be familiar with these components, making legislative gestures toward them: an *ordonnance* from 1584 first mentioned mandatory examinations for pilots and masters, and in 1615 an examiner was appointed in Dieppe.[18] In 1627 Louis XIII created the position of "Grand Maître de la navigation," responsible for the nation's navigation and commerce, and in the same burst of idealism he laid out plans for navigational classes for gentlemen and poor youth alike. However, if such examinations or lessons ever took place, no records survive.[19] In the first decades of the seventeenth century, a handful of instructors began offering lessons in Dieppe, several decades earlier than most other regions of the country.[20]

The French placed increasing emphasis on overseas colonial and commercial aspirations. After Jacques Cartier's expedition to New France (Canada) in 1534 and Nicolas Villegagnon's to France Antarctique (Brazil) in 1555, by the seventeenth century they had established toeholds in the Caribbean, Africa, and Asia. These long-distance voyages, in conjunction with existing coastal trade patterns, led to a need for considerable numbers of navigators (*pilotes*). According to a 1634 Ordonnance, each naval ship required three pilots: two responsible for coastal navigation and one for large, oceanic sailing. (Each would take consecutive watches, reporting his progress to the others and alerting the captain should they need to change the route.)[21] If captains abided by these guidelines, the French fleet—twenty-five hundred vessels in 1664—would have required at least that many long-distance navigators.[22] And yet these thousands of mariners must have developed their skills exclusively through apprenticeships, for France was surprisingly slow to produce accessible maritime publications or develop schools for nautical training. Although French merchants and elites occasionally referred to the Gauls' nautical exploits or invoked Poseidon in royal pageantry, they relied comparatively little on the sea, and consequently had scant motive to investigate nautical lore or techniques.[23]

With its lively intellectual and technical maritime developments, Dieppe was thus exceptional. Tall white cliffs flank the deep, natural harbor that marks where the Arques River flows into the English Channel. From at least the twelfth century, this small city was a fishing and shipbuilding center, and it would develop a thriving trade in spices and ivory. In the early sixteenth century, Pierre Desceliers and half a dozen other chart makers constituted the "Dieppe school" of cartography, creating a series of important manuscript maps and atlases.[24] In the first decades of the seventeenth century, with the advent of maritime teaching and professional examinations, the city also became a hub for translations of nautical works. Well before any French authors wrote original works on navigation, Nicolas Acher, one of the area's first printers, produced several French translations of English and Portuguese texts.[25]

Guillaume Denys was born in this educational and publishing center in 1624 and began teaching mathematics—specifically navigation—in his early twenties. Like the first generation of entrepreneurial Dutch teachers, Denys ran his school out of his home. The bases of his mathematical and maritime expertise are obscure, although he reportedly made at least one voyage to New France.[26] He was ordained as an Oratorian priest in 1647 and the following year published the first of five handbooks on the "art of navigation," a preliminary attempt to explain the rudiments of trigonometry for sailors embarking on short-distance voyages. After a pause, he published three titles in the 1660s: a volume of solar declination tables, and a pair of works on the *Art of Navigation,* the first devoted to magnetic variation, the second to trigonometry. His final work discussed different methods of determining latitude. They were all first published locally by Nicolas du Buc and saw several reprints in Havre de Grâce as well as Dieppe.[27]

Over the course of two decades "le bon prêtre" built a reputation sufficient to attract attention from the court. In 1664, secretary of state Jean-Baptiste Colbert became minister of the navy and founded the Compagnie des Indes. That same year, the soon-to-be superintendent of the navy, the duc de Vendôme, offered Denys a prestigious position as the kingdom's commissioner-examiner of pilots and professor of hydrography. The following year, Colbert placed Denys in charge of the École

Royale d'hydrographie in Dieppe, with instructions to "teach pilotage to young people and also prepare mariners for service to the king and the country." Enrollment would be free, and Denys was to "harness his zeal to instruct [*former*] the largest number of pilots possible." For this he would receive twelve hundred livres a year. By this point, Denys's school was already well established; he was able to open the royal school officially on September 30, 1665, just one day after the General Assembly approved the plan.[28] The school brought a certain cachet to the maritime quarter of the city, and gentlemen — including the son of royal mistress Mme. de Sévigné — were known to drop by Denys's house for occasional lectures.[29]

As was his habit, the bureaucratic mastermind Colbert kept close tabs on the new school, requesting updates about the number of students and precise details about the classroom schedule.[30] Denys reported in with some regularity. His correspondence with Colbert covered a range of topics, including a request for a clock (*un pendule*) to keep track of the length of his lessons, and a quarrel with an intemperate local councilor — after the said M. Berrier insulted Denys and menaced his students, the abbot had no wish to associate further with him. However, he could not avoid him as Berrier was the local source for compass magnets (*pierres d'aimant*). Denys, like most contemporary educators, wished to be better compensated: he had paid five hundred livres to renovate his school, but found the location less than optimal. Situated as it was near a public fountain where lace makers sang over their work, his students had a hard time concentrating on their lessons, or he on his own studies and daily prayers.[31]

Denys fretted over the publishing schedule of his first textbook but always found time for glowing reports about the students — whom he claimed were all "very good scholars" (*de forts bons escholiers*).[32] Over his forty-five-year career, Denys taught many hundreds of sailors. Working almost nonstop, he estimated that in one year he typically "prepare[d] one hundred and twenty to one hundred and fifty young men to become pilots in their subsequent voyages."[33] About twenty men, ranging in age from thirteen to thirty, attended each day's class. Denys distinguished between two types of pupils: those aspiring to a pilot's license and free auditors. Not all completed the course on their first attempt, so some would return after a hiatus. These educated mariners had good career prospects; many would sail with the Compagnie des Indes or other merchant concerns, while at least some — the crown hoped — would join the navy.[34] Colbert, like naval administrators elsewhere, was perpetually concerned that men trained at these state-run schools gravitated toward the more lucrative merchant marine. In an effort to guarantee sufficient numbers of naval recruits, he introduced the widely criticized "system of classes," a type of conscription where every French sailor was required to serve in the navy every third year.[35]

Denys's reports to Colbert provide tantalizing insight into his classroom. Each hour-long morning class, he explained, included "a quarter hour of propositions," when he would present theorems and related practice questions.[36] He often stayed late to correct the work students brought with them and to give additional demonstrations to weaker ones. Classes always concluded with a prayer for the king. Denys made efforts to engage his more advanced pupils, lecturing not simply on the basics but on "more unusual practices." To prevent these stronger students from departing in boredom, he started teaching them trigonometry. He found this no simple task but was gratified that his students found it so useful that they began teaching other men in turn.[37] This early success would inspire him to prepare his important 1668 textbook devoted to trigonometry, *L'Art de naviger par les nombres*.

The young French corsair Jean-François Doublet was one of the capable students who eagerly embraced Denys's trigonometry lessons. In the lively memoir he published late in life, Doublet recorded tantalizing snippets about his classroom experience in Dieppe, a rare record of Denys's teaching. Doublet was born in 1655 in Honfleur, one hundred kilometers west of Dieppe and just across the Seine estuary from Le Havre (de Grâce), the other key Norman port. He first went to sea at the age of seven, stowing away to New France on a vessel commanded by his father, and made several other eventful transatlantic voyages before his tenth birthday. After briefly studying Latin — at his mother's insistence — Doublet went back to sea. A relative taught him the rudiments of navigation on a fishing journey to the Grand Banks in 1669. The return journey was challenging; after a month of contrary winds prevented his boat from entering the Channel, Doublet reported that the captain of his fishing vessel and twenty others sought the guidance of an expert pilot, René Bougard. Despite his knowledge of the Channel, Bougard misjudged the tides, and Doublet's boat ran aground on a sandbank near Portsmouth. Doublet's subsequent reverence for navigational skills may have dated from that incident.[38]

In Doublet's teen years, one Captain de Latre took him under his wing. In 1673 he appointed Doublet second lieutenant on the navy frigate, the *Droite*. The precocious young man was already confident about his abilities as a navigator, convincing de Latre over a glass of wine that, he, Doublet, could teach his mentor the "principles of navigation" in just six weeks. At this early stage in the Franco-Dutch War, de Latre was charged with protecting French ships returning from the Indies from Dutch privateers.[39] Doublet exuberantly described taking "every care possible to fulfill my duty," providing navigational assistance to the noncommissioned officers of the *Droite*. He "took soundings four and five times a quarter, punctually wrote depth readings and noted the bottom to track the flow of the tides." His superiors warned him that such assiduous behavior would tire him out; though they found Doublet's knowledge of the ocean floor and timely note-taking impressive, they advised him to "leave that work to our pilots who are paid to do it." Only three weeks into that particular tour under squadron chief François Panetié, Doublet encountered in the North Sea a convoy from Hamburg "with whom we were also at war." One hundred forty-eight men were lost in the ensuing skirmish, and Doublet's right arm was "broken in two by a shot that hit my side and tossed me from the forecastle."[40] He and other survivors abandoned their ship.

After this misadventure, the young corsair needed time to recover from his wounds. Rather than taking on a new privateering commission, Doublet decided to pursue lessons at Denys's famous school. Although he was already comfortable with the basics of the art of navigation, he wanted "to perfect himself with [the help of] such a skilled master." As Doublet later reported in his vivid *Journal*, he considered this deeper theoretical grounding an insurance policy should he be injured in a future privateering battle — for then he could run a navigation school of his own. In the winter of 1675, the corsair paid Denys fifty livres a month for room, board, and the necessary books — presumably Denys's own manuals.[41] Initially Denys offered the young man the standard introductory lessons: "He started me with the principles of the sphere, the tides, heights [*hauteurs*, altitudes], the sinical quadrant [*quartier de réduction*] and Gunter's scale [*l'echelle angloise*], etc." However, Doublet was already up to speed on those concepts — as well as "sines, tangents and logarithms." Pleasantly surprised to have a new pupil equally comfortable with theory and practice, Denys moved on to spherical trigonometry and Euclid's *Elements*. Doublet seems to have completed his lessons within three months, at which point, Panetié recalled him to his squadron.

Denys, however, intervened, persuading Doublet to stay on in Dieppe. Denys expressed concern that Doublet had "devoured" the concepts in his lessons too rapidly to retain them, and the abbot felt that additional time in Dieppe would provide a more solid grounding. He offered the young corsair the position of classroom assistant: "You would oblige me infinitely by staying, for you would ease the puzzle of this [large] number of students, the majority of whom have heads as hard as stone." Doublet agreed, crowing over his windfall: "six months of room and board for which I paid only three."[42] According to Doublet, Denys also paid for him to take the pilot's exam at the conclusion of his stay, arranging a panel of "four experienced captains and four pilots, who questioned me from all sides, and at their approbation I was registered in front of [the Messieurs] of the admiralty." After a celebratory feast with Denys and his sister, Doublet took his leave with great affection (*des amittiez et tendresses réciproques*). He never needed to try his hand as a sailing teacher, returning instead to the adrenaline-filled corsair's life. His navigator's education stood him in good stead his entire life: upon returning from a final voyage to South America in 1711, he proudly noted how accurately he had estimated the route. He finally retired that same year and, at the urging of friends, composed his autobiography. He recreated his life events as best he could — in spite of having lost many of the records of his voyages. He died in 1728 at the age of seventy-three.[43]

After Doublet's departure in the fall of 1675, Denys remained at the helm of the *école royale* for nearly fifteen more years, until his death in 1689. He was respected by his students, serving as godfather to at least four of their children. His teaching was much lauded. During an inspection tour in 1681, the naval superintendent Arnoul deemed Denys's method of teaching "extremely good, even extraordinary."[44] But even more extraordinary than his teaching were his textbooks. These pioneering mathematical works carried his ideas beyond Dieppe, in the process effectively throwing open the sluice gates of French nautical publication.

LESSONS IN PRINT

Within the tidal wave of maritime works that were flowing off presses in Spain, Portugal, the Netherlands, and England in increasing numbers over the course of the sixteenth century, French titles were curiously absent. (Voyage accounts were an exception, with printed collections finding a wide readership from the 1560s on.)[45] Despite the early embrace of printing throughout France, maritime works continued to circulate in manuscript form. Working sailors and elite audiences alike read these manuscripts: some surviving copies are salt-stained, like the pocket tide tables produced in the 1640s by Guillaume Brouscon, a cartographer of the Dieppe school, while other exquisitely decorated cosmographical works surely never went to sea.[46]

The reach of these manuscripts would have been limited, but printed works — portable and affordable — would soon find a far larger readership. And yet the French market developed slowly. In the Netherlands and elsewhere, introductory nautical publications emerged in tandem with educational venues, so in light of the delayed development of French educational infrastructure this paucity of publications is unsurprising.[47] In the first half of the seventeenth century, French mariners and enthusiasts seeking to supplement their in-person lessons must have learned primarily from foreign texts. Pedro de Medina's *L'Art de naviguer*, which saw fifteen editions in Nicolas de Nicolai's translation, was the lone Iberian manual to circulate widely in France. Those members of the maritime community who adhered to the prevailing cosmographical approach to navigation could choose from dozens of editions of works on the sphere.[48] In addition to the foreign titles translated in Dieppe, a handful of Dutch works came out in French later in the seventeenth century — but these translations, produced by Protestant refugees for enterprising Amsterdam publishers, often stumbled over technical vocabulary.[49] French authors began publishing a few specialized technical treatises in the latter half of the sixteenth century.[50] The earliest texts that can be considered French navigational manuals are the *Compost manuel* (Rouen, 1586), a volume of calendrical tables and concepts, with a brief overview of the rudiments of navigation, written by the Norman professor of mathematics Jean de Seville, and Samuel de Champlain's *Traitté de la marine et du devoir d'un bon marinier*, a concise treatise on "the duty of a good mariner" published as an appendix to the final edition of his *Voyages* (Paris, 1632).[51] These modest texts signal a shift in the nature of French maritime publications. If readers had previously sought out accounts from distant lands, they were now increasingly interested in how to reach those places.

When a native French author finally set out to write a comprehensive introduction to the art of navigation, it was nothing like the traditional or updated cosmographies published by Medina or Coignet. In 1643 Georges Fournier, S.J., published the encyclopedic *Hydrographie* in Paris. The learned Fournier, who had been a *maître de mathematiques* at La Flèche before being named royal hydrographer, took a broad view of the "theory and practice of all the parts of navigation." His patriotic seven-hundred-page opus included excursions on magnetic compasses, the history and natural properties of the lodestone, religious homilies, and accounts of France's nautical superiority. He did not even provide a definition of navigation until the volume's eighth book.[52] Fournier had served as a navy chaplain, sailing to Asia, and thus had a modicum of maritime experience. This allowed him to explain certain tasks in technical detail, but his descriptions of sailors themselves dripped with condescension. The feeling was mutual: mariners found Fournier's book too expensive, heavy, and off topic — "full of many useless things."[53] Nonetheless, it did present navigation as a subject of great political significance and, consequently, one with which educated Frenchmen should be familiar.

There is a striking contrast between Fournier's learned tome and the accessible, applied manuals on various aspects of the "art of navigating" that Guillaume Denys produced less than a generation later. In his modest, affordable textbooks he tackled a trio of topics essential for every modern sailor to understand: magnetic variation, trigonometry, and the determination of latitude. The content of Denys's works was not original — he owed a good deal to foreign texts — but they revolutionized the French maritime world. Denys promoted his own publications astutely. Not only did he use them with his students, but he also sent them to the most influential bookshop in Paris, secured a review in the *Journal des sçavans,* and gave half a dozen copies to Colbert himself.[54] As an educated cleric, Denys wove together copious learned citations and elaborate metaphors. Such flourishes notwithstanding, he provided clear, practical advice in a tone his students would have recognized from the classroom.[55] The well-read teacher crafted a program designed to improve French sailors, reducing their reliance on foreign navigators and charts by sharing with them new techniques from abroad. The most important of these was that "mysterious chariot," trigonometry.

Denys opened his final book, *L'Art de naviger dans sa plus haute perfection, ou Traité des latitudes* (1673), with his extended metaphor equating the elements of navigation with a chariot's wheels. "If, unfortunately, two of these four wheels are bogged down in the mud," Denys explained that "as long as two are rolling properly it is possible to unblock and free the two others and make them work together." For Denys, this chariot was a clear metaphor for the process of solving an equation for two unknowns. He, like many other writers and mariners, subscribed to the view that "navigation is nothing more than a right triangle."[56] The boat's point of origin, its point of arrival, the direction of travel, and the distance covered could readily be configured into a triangle, often with a right angle to simplify computation. If a mariner knew any two components of this "sailing triangle," trigonometry would help him figure out the other two in terms of latitude and longitude. Depending on how far the boat was traveling and how accurate the calculations needed to be, the sides of the sailing triangle could be straight or curved.

How did Denys go about introducing the "mysterious chariot" of trigonometry to his students? Only two decades previously, the supercilious Fournier had dismissed it as too challenging for sailors: "Because the worker is hardly curious, and not very diligent, [using a table based on the doctrine of triangles] remains very difficult."[57] Denys,

L'ART DE NAVIGER
PAR LES NOMBRES

DANS LEQVEL TOVTES LES REGLES DE
la Nauigation sont resoluës par vn Triangle rectiligne
rectangle, comme dans les Cartes hydrographiques.

AVEC

La Table tant des Sinus Tangentes & Secantes, que
Sinus, & Tangentes Logarithmiques, & les Logarithmes depuis l'Vnité iusques à 10000.

ENSEMBLE

Vne Table des croissantes Largeurs, pour Nauiger confor-
mement à la Carte reduite, laquelle bien que construite
selon les voyes Geometriques, & censée iusques a
present pour la plus accomplie est neantmoins
demonstrée defectueuse comme ne reuenant
point au Globe pour la diminution des
degrez de la Longitude.

Par M. G. DENYS Prestre, Enseignant pour le Roy la
Nauigation en la Ville de Dieppe.

* * *
* *
*

A DIEPPE,
Chez NICOLAS DVBVC Imprimeur-Libraire & Graueur
deuant l'Hostel de Ville.

M. DC. LXVIII.
AVEC PRIVILEGE DV ROY.

2.5.
Guillaume Denys's influential 1668 textbook, which introduced trigonometry into French navigational education. Bibliothèque Sainte-Geneviève, Paris, photo N. Boutros

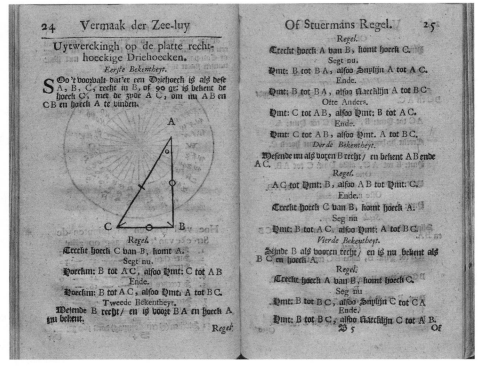

2.6.
The six types of triangles schematized. B. Martin, *Young Trigonometer's Compleat Guide* (London, 1736), 66. John Hay Library, Brown University

2.7.
Solving plane, right-angled triangles was the most straightforward aspect of trigonometry—but it was necessary to master the rules, four of which appear here. On this diagram, Side AC, angle C, and angle B (the right angle) are marked with tick marks; these are the given values. The values to be determined—angle A, side BC and side AB—are each marked with a small circle. A. C. Hellingwerf, *Vermaack der Zee-Luy* (Hoorn, 1703), 24–25. Leiden University Library, 2314 H.20

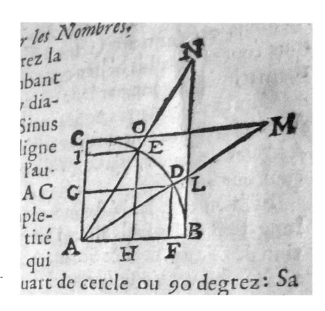

2.8.
This small diagram (reproduced from the woodcut that appears four times in Denys's textbook) shows many of the key relationships of the "doctrine of triangles" in a single figure. "To demonstrate to your eyes all that I've just presented, since it is the foundation of all that we have to say, look at the accompanying figure." G. Denys, *L'Art de naviger par les nombres* (Dieppe, 1668), 33. Bibliothèque Sainte-Geneviève, Paris

however, was convinced that at least his stronger students were ready for it. To engage his readers in the relatively dense subject matter of his *L'Art de naviger par les nombres* (1668), Denys offered another one of his poetic metaphors: "If our Sine Tables are the nursery [*pépinière*] of all the most beautiful mathematical secrets, [then] Trigonometry is the gardener, where the master . . . [helps these secrets] produce the hoped-for fruit." Denys felt it was his duty to teach a wider audience how to make effective use of these powerful tables. "I constantly guarantee that he who possesses Trigonometry can easily solve all the thorny problems that one encounters in Mathematics and possesses the key to them." Trigonometry was not only faster but also more accurate. In his discussion of how best to determine longitude, Denys asserted that arithmetical calculation was "much more exact" than using an instrument "because Numbers are always more precise than Geometrical ways"[58] (figure 2.5).

Denys modeled his introduction to trigonometry upon earlier mathematical books, emulating their structure and pedagogical approach. Denys began — as Euclid did in his geometric work, and Apianus in his cosmographical one — with definitions. Here he was emphatic: it is just plain stupid, even beastly, to discuss trigonometry without defining the terms. In the mode of inspirational teachers everywhere, he also insisted that his reader not just memorize the definitions but actually comprehend them.[59] Denys, again following the Euclidean model, then presented a series of rules or cases, one for each possible combination of sides and angles known about a given triangle. Studious readers were expected to commit to memory all possible cases — four or six for plane trigonometry, sixteen for spherical — so they would know which functions to employ in a particular situation. Early modern authors presented these cases in various formats, designed to help the learner memorize all the cases: tables, ratios, or brief statements modeled on formal logic[60] (figures 2.6, 2.7).

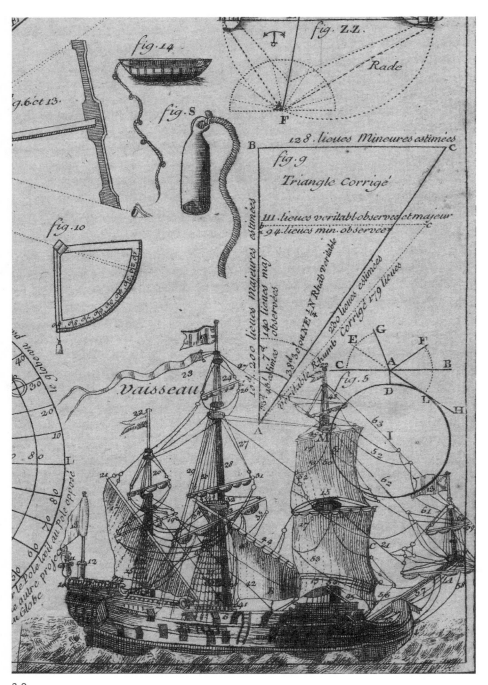

2.9.
The "corrected" sailing triangle, visualizing a vessel's course in trigonometric terms (*fig.* 9). Also note the knotted log and line (*fig.* 14) and the weighted lead and line (*fig.* S). Nicholas Bion, *Traité de la construction . . . des instruments* (Paris: Brunet, Ganeau, Robustel & Osmont, 1725), pl. 24. Utrecht University Library, U-MUS : C 20 BIO 1#OUD

Denys chose to present the "analogies" in brief, italicized stanzas, explaining the relationship between the mathematical entities and the navigational ones. He made the significant assumption that his audience would have no trouble comprehending these complex new terms from his verbal descriptions.

In general, Denys's handbook followed the conventions of contemporary works on trigonometry, which generally included only a modest number of diagrams. There was some recognition that diagrams could help make unfamiliar concepts accessible: one of Denys's later rivals, the Jesuit C.-F. Milliet de Chales, noted that "figures greatly relieve the imagination, and they make a work more intelligible, and more pleasant."[61] However, they were not yet the primary means of conveying information, especially in early French-language nautical books. (This abstemiousness compared to the plethora in Spanish or Dutch texts of similar date might stem from their origins in smaller publishing houses in port towns.) Denys made two unusual decisions in the visual program of *L'Art de naviger par les nombres*. Rather than illustrating the primary definitions with a simple triangle, he chose a diagram that demonstrated all the important concepts of trigonometry at once (figure 2.8). Then, somewhat perplexingly, he repeated the same diagram three more times; it is the sole image in the 260-page volume. (Denys may have used this image in his teaching, or perhaps his publisher had scant budget for cutting new woodblocks.)[62] On the third iteration of the image, Denys made an important link between the labeled diagram and the components of the sailing triangle: "Take the triangle ACM. . . . A represents the place of departure; . . . the leg AC will represent the latitude (by which the Boat advanced North), the other leg CM will be longitude (by which it advanced East), and the hypotenuse AM represents the Boat's route to the NE¼E: angle CAM will be the rhumb."[63]

Here, as in his list of the six cases or "analogies," he makes a clear effort to describe the overlapping triangles in terms his maritime readers would understand. Denys's strategy is striking compared to that of other navigational textbooks, where the relationship between the boat's course and trigonometric functions remains abstract. Most authors gave only terse descriptions, and virtually never illustrated the place of the boat on the various triangles[64] (figure 2.9). If Denys's restrained use of diagrams was typical for the time, this dearth of "intelligible and pleasant figures" also suggests that most French sailors could work through such a book without the aid of images. They were evidently so accustomed at thinking of their course in triangular terms that they already knew the boat belonged on the hypotenuse.

The remainder of Denys's pioneering if unprepossessing volume included sixty pages of tables; he clearly intended to equip his readers with the tools they would need to carry out trigonometric calculations. He included in an appendix an extended discussion of the four main propositions (relating each to the respective questions in Euclid) and then demonstrated how to solve a series of problems. He wanted pilots to work through each type of question using at least two distinct methods, choosing from trigonometry, logarithms, traverse tables, or graphical instruments.

Case I
Apply the First Correction:
Adjust the distance sailed

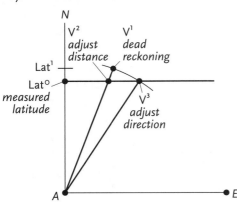

Case II
Apply the Second Correction:
Adjust the direction (course angle)

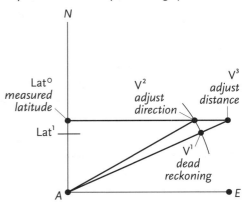

2.10.
How to adjust a dead reckoning estimate using the Corrections. After a day's sail (from point A along a certain course), a navigator will estimate his vessel's position by dead reckoning (V^1 at Lat^1) and take at least one latitude observation (Lat^o). How should he reconcile any differences that arise between Lat^o and Lat^1, and where should he mark the vessel's revised position (V^2) on the chart? In case I, for a course that runs approximately north-northeast, the rules call for the navigator to apply the first Correction: adjust the distance sailed (*chemin*, AV) and maintain the direction (*rhumb*). In case II, for a vessel sailing east-northeast, the navigator will apply the second Correction: he will maintain the distance of the course but modify its direction. In each case, the rule aims to minimize the overall distortion. If the navigator were to make the opposite adjustment, notice how much more dramatically the position of the vessel would shift east (V^3). Choosing the inappropriate Correction would consequently have a dangerously large effect on the already tenuous longitude estimate. Drawing by Margaret E. Schotte with Joel Silverberg

He offered each set of options without comment, leaving his students to choose the ones they found more straightforward.[65]

In addition to providing his students with multiple problem-solving strategies to equip them to face unexpected challenges at sea, Denys also professed a lofty intellectual goal. Like many of his contemporary authors, Denys felt that navigation should be considered a "true science." His books are peppered with promises of "infallibility" and expressions of admiration for exactness. But it was proving vexingly difficult to systematize the practice of sailing, no matter how many sophisticated theorists were devoted to the task. Even as observations and calculations grew more accurate, navigation remained dependent on the navigator's skill at estimating speed and position. "Seeing that the majority, or I'd even say almost all the rules of Navigation are based on and applied to estimates," Denys explained, "one should not be astonished if the conclusion . . . one draws is an estimate at best."[66] Ultimately, a course charted using trigonometric calculations could only be as good as navigator's underlying skill at that old art of dead reckoning.

At this point in the mid-seventeenth century, despite the refined instrumental and mathematical methods of calculating the Sun's altitude and the ship's course, navigators continued to estimate their daily position without looking at the sky. They would then compare that position to their ostensibly more exact instrumental observations. Problems arose when these two calculations differed — as was almost inevitable, given the nonstandardized instruments, inhospitable observation conditions, and the navigator's own quirks. When the course tracked on paper did not match the observed latitude, how was the navigator supposed to adjust the route? Which leg of the sailing triangle should he extend? His decision would differ if the boat had been carried off course by a storm or an unexpectedly strong current, or if some other factor had led the navigator to mistrust his estimate.

In his 1668 volume, Denys gave an example of two Dutch navigators on board the same ship who, upon finding that their estimated position varied significantly from their latitude observation, used divergent strategies to correct their charts. The first retained the course he had estimated for the vessel but extended what he calculated to be the distance run to place her at the observed latitude. The other pilot had more faith in his estimated distance and so decided to shift to a more northerly compass bearing. Consequently, his new position on the chart, although at the same observed latitude, was considerably further westward. For Denys, this demonstrated not only the obdurateness of sailors — both men "maintained that they had corrected reasonably because they were now at the observed height [latitude], and no one could convince them [otherwise]" — but also the need to ground long-distance sailing on something more rational than the whim of the individual navigator.[67]

Denys felt that there should be a more systematic plan. The answer ultimately lay in trusting latitude, that most reliable wheel of the vaunted chariot. In the second half of *Naviger par les nombres,* Denys introduced his French audience to a concept he had come across in foreign mathematical texts, the "three Corrections."[68] One of the many authors who would promote them to subsequent generations defined the Corrections as a set of "prudent Rules founded on the fundamental observations of Navigation, with which one can adjust [*réforme*] the route's direction, as well as the distance and the longitude."[69] In the event of a discrepancy between an observation and a ship's estimated position on the imaginary sailing triangle, there were three options: to extend the distance (*chemin*) and maintain the direction (*rhumb*) for the first Correction, or to do the reverse for the second (figure 2.10). The choice depended on the proximity of the course to the north-south axis. If the vessel sailed exactly halfway between any two cardinal directions, the navigator applied the third Correction — a combination of both first and second — and used the average (figure 2.11).

These "infallible rules" had been devised in Holland half a century earlier for a completely different purpose. In 1612 the Dutch entrepreneur Jan Hendrik Jarichs van der Ley filed a patent for a new "general ground rule" (*Generale Grond regel*), which he proposed as a solution to the longitude problem[70] (figure 2.12). (He tested his system, which relied on transparent protractors and specially prepared paper for charting northern latitudes, on a state-sanctioned voyage to Iceland and Canada in

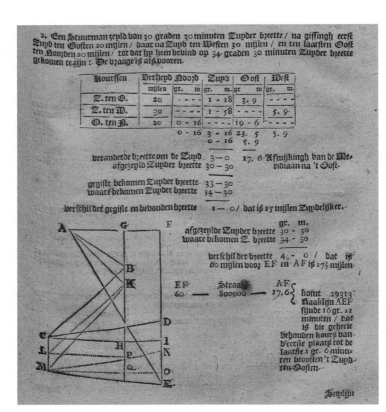

2.11.
Vooght walks readers through the process of "correcting" a course. As happened all too often, the course tracked on paper did not match the observed latitude; the most common advice in Dutch textbooks was to split the difference. Vooght, *De Zeemans Weghwyzer* (Amsterdam, 1695), 183. Leiden University Library, 587 D45

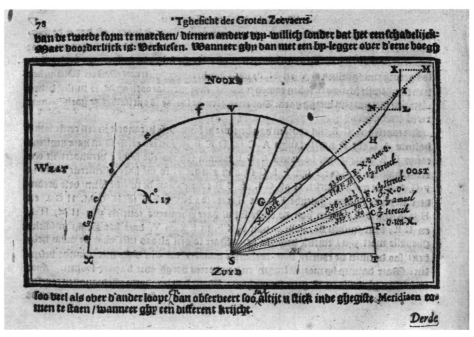

2.12.
The Corrections were originally conceived of as a solution to the longitude problem.
J. H. Jarichs van der Ley, *'T Ghesicht des grooten Zeevaerts* (Franeker, 1619), 78. Bayerische Staatsbibliothek München, Res/4 A.hydr. 46 m, p. 78, urn:nbn:de:bvb:12-bsb10871873-0

1618.) A great opponent of "inconsistency," Jarichs van der Ley wished to replace intuitive estimates with rules; at a time when so much of sailing was approximate, he railed against guessing.[71] Although the contest judges ultimately deemed Jarichs van der Ley's longitude proposal ineffectual, contemporaries quickly recognized the potential of his method to resolve any discrepancies that accrued over a day's sail. Two Dutch mathematicians, Adrian Metius and Willebrord Snellius, discussed Jarichs van der Ley's concept at length a few years later, and it was through their Latin treatises that Denys learned of the Corrections.[72]

Denys devoted a substantial portion of his book to explaining how and when to apply each of these rules to improve one's charted course and offered sophisticated analysis of the subtly different opinions of Metius and Snellius. He made no attempt to explicate the mathematical logic behind the three rules. Cornelis Lastman, the early adopter of logarithms, was one of the few authors who provided any explanation for why the first Correction applied to north-south courses and the second to east-west ones: Lastman explained that if one sailed approximately south or north, the "mis-estimate" of distance sailed would cause a large difference in latitude, while an error in the angle of the course would cause a much less drastic difference. The converse would occur if one were sailing on an almost eastward or westward course. Lastman did not discuss longitude, for there was not yet a reasonable way to assess it. Following his recommendations, however, would minimize the east-west distortions as well — and it was this that made the Corrections so powerful and popular. In the absence of any environmental clues about his location, a navigator should select the adjustment that would have a less distorting effect or "burden" (*belast*) on the triangle, and thus on his boat's course.[73] This cautious choice, similar to the strategy of slightly overestimating one's position to avoid running into the shore unexpectedly, would ultimately result in the best possible estimate for the otherwise unobservable longitude.

Few mathematicians or navigational experts initially accepted all the details of Jarichs van der Ley's system. Metius was particularly doubtful that any navigator could apply such rigid rules amid the tossing waves and shifting heavens of long-distance sailing.[74] And yet Metius was in the minority with his absolute rejection of Jarichs's rules; most recommended adopting them in modified form. Lastman, in his early configuration of Jarichs's system, advocated applying his "common rule" only if one could not identify any other reasonable cause for the distortion, such as tides or current.[75] In England, John Collins spent ten pages discussing "How to Rectifie the Account when the dead Latitude differs from the Observed Latitude," which he learned of from "Maetius, a Hollander." Like others, he accorded great authority to latitude observations ("when an error ariseth, the sole remedy to be trusted to, is the observation of the Latitude"). But as Collins made clear, even the relatively straightforward latitude observations could be distorted. The ultimate arbiter must be the navigator's own "sound and experienced judgement."[76]

By the end of *Naviger par les nombres*, Denys himself had moderated his tone. He admitted that — despite what Jarichs van der Ley had hoped — the Corrections were not completely fail-proof. "Even after all the Corrections that one can make, there remains even still always room for doubt." He noted drily that "at least they can produce more certainty than if we had no knowledge."[77] In the book's concluding passage, he apologized to his readers: You are astonished, I'm sure, that after having given you a practice established on certain and infallible foundations, I have saved the Corrections for last, to crown our Art of Navigation, [even though these] clearly show and uncover the imperfections and faults of Navigation."[78]

This admission of uncertainty did not concede defeat. Instead, Denys's main message for his readers was one of balanced realism: "Sailing is not as exact as many imagine."[79] He was proud to teach his students the latest mathematical methods, demonstrating how one could achieve impressive precision. But in his view the traditional techniques also remained important—for there could well be occasions when a navigator would need to choose efficiency over accuracy. For instance, in his treatise on latitude, he presented five different methods for obtaining altitude readings. A savvy navigator knew when to choose each one based on weather conditions and the instruments at hand. In *Naviger par les nombres*, Denys dismissed the method the Dutch used to calculate longitude as too time intensive. Even his pet project, logarithms, were not *always* the preferred solution. He cautioned that, for certain problems requiring multiplication (as opposed to division), it could take much longer to look up the logarithms in their tables than simply to multiply. In each case, he declared, "Nature, like a good mother, teaches us to [reach our goal] by the shortest path."[80] Once again, Denys wished to teach navigators a variety of solutions so that they could be prepared for any situation on the unpredictable seas.

This notion, that a navigator would have to choose from a range of techniques—quickly, when faced with risky or suboptimal conditions—was exactly the type of intangible expertise that could develop only with time. It was here that Denys ran into difficulties with Colbert. On October 21, 1666, only a year after he opened the *école royale,* Denys wrote a strongly worded response to one of the minister's regular missives:

> It gives me great displeasure to learn . . . that you are complaining that the royal school is not supplying pilots to sail the Channel. Not only do you know that all pilots are not good in all places, and that in these [types of] small navigations you need a long time to acquire the knowledge of the coasts and any number of other details[;] it seems to me that I already told you in earlier letters that after I have given the precepts of this art to my students, they need to go to sea to become good at estimating (*se former dans l'estime*), which they'll need later to direct vessels.[81]

Denys's umbrage is palpable. How could Colbert expect these recent graduates, after only a few months of lessons, to be able to safely navigate the Channel—which stymied countless skilled navigators. (Even the navigator Bougard, who was familiar enough with the Channel to publish a chart depicting it, had run aground on a sandbank.) Although Colbert's desire for skilled Channel navigators was understandable, particularly midway through the Second Anglo-Dutch War, this was in many ways beyond Denys's remit.[82] He had been hired to introduce the applied tasks of large navigation — instruments, cosmography, and mathematics — rather than to impart the local knowledge of small navigation. Colbert seemed to expect these new trainees to have the familiarity with a certain geographic area that could come only with years of experience. (The defensive Denys could not resist a curt comment about the caliber of the students: "Those who were at our place [the school] were not of such moral fibre" — evidently they were not such strong scholars after all.)[83]

Denys's impassioned reply to Colbert underscores his view that sailors needed time to absorb the mathematical formulas and to become comfortable at discerning the disparate cases they were dealing with in order to efficiently choose which observational technique or trigonometric function to apply. They also needed time to get their sea legs.[84] These processes not only could be lengthy but, as Denys reminded Colbert, also required peaceful waters in which to practice. In 1667, at the tail end of the latest war, Denys noted that he had "many [*pilotes*] ready to put into practice what I've taught them; we just need peace with the English to reestablish commerce" — after which point his students could find safe positions on merchant ships.[85] Thus, no matter how successful Denys may have been at conveying the latest techniques, he recognized that his classes alone did not suffice to create expert navigators, nor could they prevent disaster on the high seas — a lesson the French navy soon learned the hard way.

CODIFYING DENYS'S LESSONS

On the evening of May 11, 1678, nine French vessels under the command of the vice-admiral Comte Jean d'Estrées were wrecked on the "Île des Aves" reef in the Antilles.[86] At the peak of the Franco-Dutch War, d'Estrées sought to capture the Dutch island of Curaçao. Contemporary accounts vary: some reported that three Dutch ships lured the fleet onto the reefs of Aves de Sotovento. Others claimed that an advance party of small ships had ventured ahead, only belatedly noticing the reef. When these ships tried to alert the remaining vessels to the shallow waters, the rest of the fleet believed it was a signal to advance — leading seven men-of-war and two other ships to be stranded. Fortunately, most of the crew survived the wreck (twenty-four men drowned in an intoxicated state), but it was an inopportune moment for d'Estrées to lose half of his fleet.

The unpopular comte was exonerated upon his return to France, and the navigators were blamed for the incident. Evidently, they had been unable to agree on their position on the morning of May 11 and could not fix their latitude, likely due to weather conditions obscuring the sky. The navigators' disagreement led the entire fleet to strike the reef. And yet, in a statement soon after the event, the captain of the flagship *Terrible,* Nicholas Lefèvre de Méricourt, insinuated that the vice-admiral's pride had caused the wreck. Captain Méricourt claimed to have asked for a knowledgeable pilot to help guide the *Terrible* through the unfamiliar waters but that d'Estrées angrily contended that he had enough such men. Local pirates, warning of frequent losses in those waters, volunteered their expertise. They agreed with Méricourt: experience was crucial, and furthermore, celestial navigation was ineffective in the shallow waters. But d'Estrées declined any assistance.[87] The navigators accused of this debacle were little different from Denys's inept trainees floundering in the English Channel. Both faced a situation that required deep local knowledge, but in each case they had insufficient time to learn it. This points to a more systemic problem as well: in the clear hierarchy of the navy, there was a sharp disjunction between the responsibilities placed on the shoulders of expert navigators and the respect accorded them.

This embarrassing episode, like other newsworthy shipwrecks, launched a program of reform. In the last five years of his life, Colbert redoubled his efforts to improve the French navy. He and his son, Jean-Baptiste-Antoine Colbert, Marquis de Seignelay, who succeeded him in 1683, introduced a pivotal pair of marine ordinances that would guide the French maritime community for the ensuing half century. The first, the "grande ordonnance" or the "code de la marine," signed in August 1681, laid out a new mandate for the merchant service, while the second, introduced in 1689, addressed the navy.[88] This legislation, which would hold sway for nearly a century, steadily eliminated idiosyncrasies among local programs and imposed a model of instruction that was mathematical and test-driven. In a period when few experts agreed on how best to develop and transmit the science of navigation, this was a surprisingly strong vote for theory over practice. However, it was not yet clear what this would mean for the mariners themselves.

In his heated response to Colbert a dozen years before the Île des Aves incident, Denys had underscored the need for more experience at sea and reminded the minister that practice was essential — not only to gain local knowledge or hone specific technical skills but also to develop a navigator's ability to cope with uncertain situations. Even in his textbooks, Denys declared that it was better for his students to apply their knowledge than to read or listen to his lectures: "In one moment, one learns more from practice and use than in all the discussion one can have about it."[89] While Denys seems to have persuaded Colbert of the utility of trigonometry, he could not convince him that practice was equally important. The minister continued to view theory as paramount. He delivered a statement a few months after the d'Estrées disaster: "Regarding navigators [*pilotes*], it does not suffice that they have experience and practice, they also need to have theory and even, if possible, good sense and judgment, because it is principally in that officer [the navigator] that resides all the safety of the voyage and the preservation of the boats."[90] Ironically, good judgment appears to be an afterthought. This academic focus led Colbert to implement a national curriculum, found more schools, and hire more teachers. (While some of the next generation of instructors had trained under Denys or were recruited by Colbert from the Netherlands, in 1686 Louis XIV expressed a preference for Jesuits; soon most chairs of *hydrographie* were filled by clerics).[91] Such investments in infrastructure, in Colbert's view, would not only produce a more imposing fleet but also foster better navigators.

The curriculum that Colbert enshrined in the ordonnances placed significant weight on formal scientific fields. In 1681 the "Science of Navigation" was defined as "consisting of the knowledge and practice of several noble Sciences, notably Cosmography and Mathematics." The instructors—abruptly elevated to the status of professor of navigation—were expected to be competent in an extensive list of subjects: astronomy, geometry, trigonometry, meteorology, arithmetic, mechanics (physics), a knowledge of "natural things" including lodestones and magnetic variation, and drawing. They needed above all to have "good and solid judgment." The prospective navigators, for their part, needed to be at least twenty-five years old, with at least five years of experience on merchant vessels or two naval campaigns, and must be deemed "capable and experienced" during a formal examination. Rather than assessing their good judgment, they were expected to be "men of probity and good habits."[92]

For all its emphasis on selecting model educators, the 1681 Ordonnance offered little guidance on how they were to teach the ambitious curriculum. Over the next few years, Colbert solicited reports from the regional professors and other naval authorities about their teaching methods and their students' performance. Many of these recommendations were then incorporated into the 1689 document.[93] Colbert's "Instruction and Exercises for the Officers and Marine Guards" stipulated the details of lesson times, examinations, and even one of Denys's primary emphases, journal keeping. The crown devised a set of penalties and rewards to promote assiduous studying—or at least regular school attendance.[94] In addition to expanding access to the basics of navigation by means of free public lectures, there was a new suggestion from one of Colbert's correspondents: that prospective teachers publish textbooks, which the state would then market at reasonable prices.[95]

The experts who submitted their opinions to Colbert had divided views on how much theory was suitable. Père Milliet de Chales, S.J., who dedicated his 1677 volume to Colbert, felt that the inaccuracies inherent in traditional sailing could be overcome with the aid of science and education: "We must borrow from Astronomical principles several practices that experience alone cannot teach."[96] Another Jesuit, Claude-Joachim Thoubeau, was very positive about classroom lessons. Thoubeau, a professor of hydrography at Brest who taught alongside one of Denys's students, M. Coubert, in the 1690s, issued a loud call not to underestimate the abilities of untrained youth. Far from being incapable of intellectual achievements, he estimated that out of two hundred gentlemen only five or six "souls could not learn all these sciences"—and because one could not identify those five or six in advance, the teacher should compel everyone "to work with ardour and constancy." When faced with the challenge of teaching a difficult subject to desultory pupils, the answer for Thoubeau was more school, not less.[97]

The Sieur de Viviers weighed in on the opposite side. Viviers, a naval captain who had been charged with reviewing and improving the state of the naval galleys in the Mediterranean, scoffed at the thought of focusing on theory: "Everyone knows that most of these men are so rough that it would be a waste of time to want to teach them anything but practical elements of a trade whose theory demands many years of study, and a genius for mathematics that few possess."[98] Instead, Viviers felt that education should be more practical, along the lines of Dutch shipboard schools. He proposed that "one young sailor destined to be a Navigator" should embark on each galley to spend the winter learning as part of the crew and then continue his studies upon returning to shore. He also suggested that the prospective *pilote* "might even have to go to sea himself . . . in order to make the young men practice in the summertime what he had taught them in the winter, since in this profession Theory is completely useless without practice."[99]

This idea of having formal lessons on board ship became increasingly popular. The year after receiving Viviers's proposal, Louis XIV laid out a clear curriculum for lessons during naval campaigns. (Colbert presumably aided in this regulation, which closely presaged the 1689 ordonnance). "During the course of the campaign," the *gardes de la marine* should spend four hours each day split among "hydrography, musket and cannon exercises, military movements and [nautical] maneuvers." This would help them "maintain the knowledge they gained in port." One of each ship's navigators was also expected to teach "pilotage and hydrography" to the *gardes* so that "they [could] learn something of his profession."[100] A decade later Thoubeau prescribed the same four lessons for his students to pursue at sea. He suggested that each summer (during times of peace) one of the smallest royal frigates should be dispatched fully armed to Brest so that guards at the college where he taught mathematics and hydrography could practice exercises on fine afternoons. Samson Le Cordier, Denys's successor at the Dieppe school and author of a series of popular, acerbic textbooks, shared Denys's view about the value of applied learning. He contended that "a single campaign of five or six months" could "teach a young navigator" better than "all the books he could see on this subject."[101]

The skills that a green navigator could learn on these vessels extended far beyond the niceties of sailing triangles. And yet having a degree of familiarity with both books and the real world could save his life. During the Nine Years' War,[102] Marshal Anne-Hilarion de Costentin de Tourville wrote a dialogue intended to quickly familiarize young officers with the opaque technical vocabulary of working sailors. At a time when naval officers were criticized for knowing nothing about sailing, this pocket volume included verbatim transcripts of routine shipboard phrases, a cheat sheet that might prevent its readers from getting knocked overboard in the midst of a naval maneuver.[103]

French naval administrators, after some debate, seem to have absorbed Guillaume Denys's insistent message that practice was an essential step toward learning the science of navigation. There was no agreement about how much time would suffice — and, indeed, the efficacy of any of these policies is difficult to determine.[104]

Denys was equally convincing about the benefits of mathematical rules. Whether due to Denys's decision to publish his teaching texts, Viviers's call for practical textbooks to replace Fournier's inaccessible tome, or Colbert's receptiveness to both of these ideas, the 1680s ushered in the heyday of French maritime manuals. A welter of new textbooks poured forth from presses in Dieppe, Le Havre, and other seaside towns. Minimally illustrated, pocket-sized, and affordable, they were aimed at practitioners. Many of the authors were former students of Denys, jockeying for teaching posts in Colbert's new schools.[105] Their textbooks were more basic than those written by Denys, but they do show how his priorities disseminated throughout France's maritime communities.

Not every author in this second generation of French nautical manuals adopted the trigonometric approach — but the tide was shifting quickly. The 1689 Ordonnance included trigonometry in the curriculum, although only for the most advanced students.[106] As the French gentleman innovator Guillaume Blondel, sieur de Saint-Aubin, declared in no uncertain terms in his 1673 *Trésor de la Navigation*, "At the present, one is not a good Pilot if one does not work with Sines, and with Logarithms." He viewed this as "the current Method, and the most correct [*juste*]." (Despite Blondel's commitment to numbers, his first book, about "The True Art of Navigating with the Sinical Quadrant," enjoyed far more success. That text — which made it possible to avoid trigonometry completely — was republished at least fourteen times over the following century.)[107] In some quarters, trigonometry continued to be viewed as an elevated method, the province of officers rather than common seamen, for the latter could get by using graphical instruments.[108] By contrast, Denys's other innovation, the Corrections, were taken up far more widely.

French navigators and their instructors, in their quest for safer, more accurate sailing, eagerly welcomed this set of seemingly objective rules. Although the Corrections also appear in manuals from other countries, they were particularly popular in France — and would remain a prominent aspect of the daily routine for at least a century and a half.[109] By the eighteenth century, they were so well known that another author did not even bother to define them; "those who practice navigation were familiar enough with these corrections . . . that I need not repeat them here."[110] The new books streaming from port presses explained the three rules, in terms of either compass headings or degrees.[111] Following Denys's model rather than Lastman's, these authors attempted neither to explain the mathematical underpinnings of these rules nor to illustrate them. Readers were expected to learn how to solve each type of problem, with either a trigonometric formula or a graphical instrument such as the *quartier de réduction*.[112] If their lesson started by reading the book or listening to their teacher, they were then expected to visualize the distinct cases, even in the absence of clarifying diagrams. The final step was to work through the practice questions that textbooks featured in ever greater numbers.

Le Cordier offered one of the more thorough — and opinionated — discussions of these rules, devoting forty pages to them in his 1683 *Journal de navigation*. Le Cordier, whose career in the navy was ended by a cannonball, deemed the first Correction useless. He included it in his books only because it was taught in the state hydrography schools — a sign that Denys's preferred methods had indeed been incorporated quickly. Le Cordier felt that the minimal longitude distortions on north-south courses made it scarcely worth even formulating a rule and preferred instead to teach the "prudence and good sense that govern a Pilot."[113] Le Cordier's emphasis on the discretion of professional navigators was unusual at a time when other teachers simply expected sailors to memorize the three different situations and apply each rule when it obtained.

French navigators seem to have embraced the Corrections largely because they offered a strategy for extracting certainty from approximation. In its original formulation, Jarichs van der Ley intended his 1612 "general rule" to systematically replace estimates with "more trusted" observations. He felt that latitude always trumped the rhumb and the distance run — but he nonetheless retained an inherent respect for estimation. No matter which Correction a navigator chose to apply each day, erasing either his compass heading or his guess at the vessel's speed, *one* of his two estimates remained central to the boat's position on the chart and in the ocean. The French, by contrast, would have preferred to do away with uncertain estimates entirely. This view, that personal opinions were less trustworthy than instrumental readings, related more generally to their attitude toward experts.

In the Netherlands and England, experienced mariners retained some authority. While Jarichs van der Ley wished to minimize the effect of stubborn captains, in 1629 Lastman assigned them a significant degree of autonomy: they should rely on a blanket rule only when they had considered all other factors that could have affected the ship's course. For Englishman John Collins, writing in 1659, a pilot's judgment might cause mistakes — "the judgement erres in supposing the distance run to be too much" — but "a sound and experienced judgement" could still extricate him from a problem.[114] Collins articulated a hierarchy of certainty regarding not only the components of navigation but also the men who practiced it. These early descriptions of the Corrections emphasized that they were to supplement the pilot's personal judgment, but by the time they disseminated into later French texts, such nuance had disappeared. In his search for "infallibility," Denys ardently promoted the rules set by Jarichs van der Ley, convincing generations of French sailors to follow them. However, his students became so focused on the power of mathematical rules that they inadvertently divorced the Corrections from the mesh of opinions and skills in which they had been conceived.

By the close of the seventeenth century, Denys's successful educational program, disseminated by his strongest students and preserved in printed form, had effectively transformed navigational practice in France. And yet the pendulum continued to swing between the poles of theory and practice. Men like Doublet, who first went to sea at the age of seven, had ample time to absorb traditional methods and the latest technologies, honing their understanding both in the classroom and on the ship. But for a man whose experience was limited to only a few months of lessons, or one or two brief naval tours in an interlude between wars, many questions remained: Was he equipped to face a crisis? How would he demonstrate his knowledge? At what point did he make the transition from trainee to expert?

3.1.
A blue-coat boy: "One of the Children Educated in [Christ's] Hospital."
P. Perkins, *The Seaman's Tutor: Explaining Geometry, Cosmography, and Trigonometry* (London, 1682), facing p. 1. Courtesy of the John Carter Brown Library at Brown University

Chapter Three

HANDS-ON THEORY ALONG THE THAMES
London, 1683

In England, the navigator's examination that was such a pivotal moment in the life of a sixteenth-century Spanish mariner gradually took on even greater weight. By the final decades of the seventeenth century, at least three groups of British sailors were required to face examiners before moving on to the next phase of their career: young pupils at the Royal Mathematical School who hoped to become certified as masters in the Royal Navy, experienced masters who wished to improve their wages by qualifying to guide larger naval vessels further afield, and young gentlemen who sought their first naval commission as lieutenants. All of these men were preparing for the same task: safe navigation on the open ocean. Despite the similarities of their goals, the content and emphasis of these three formal examinations could not have been more distinct. The questions posed to candidates ran the gamut from very theoretical to extremely hands-on. Although the English model of nautical training held within it not one but two theoretical traditions — one very ancient and Euclidian and the other related to the Royal Society and the empirical new sciences — in every case a practical element remained: a mariner could not do his job without both head and hand.

In the fall of 1670, one Charles Hadsell, recently returned

from the Caribbean, decided to seek his certification as a ship's master.[1] He presented himself at Trinity House, the seaman's guildhall on the bank of the Thames, near the royal dockyard and just upriver from the new royal observatory in Greenwich. Trinity House had been responsible for examining masters since 1621, charged with "examin[ing] and report[ing] to the Navy Board, if desired, the Fitness of Masters for the King's Ships; and certify[ing] what Rate the Ships are they take the Charge of. And giv[ing] Certificates and Testimonials to the said Masters."[2] These men vied for permission to sail to a specific location — perhaps to the Mediterranean, the Caribbean, or the North Sea — and had to show their familiarity with the conventions and hazards germane to that route.[3] Hadsell prepared to demonstrate his competence to a panel of half a dozen "senior brethren." To be considered for the position of master, each candidate needed at least three years of experience on a naval vessel; Hadsell was well qualified here, as he had spent more than ten years at sea, serving as a master on the London-based ship *Prosperous* until 1660. However, in that year he was captured and spent the ensuing decade sailing with the notorious Captain Henry Morgan, seeking retribution against Spanish ships.[4] Given this recent company, it is uncertain whether he would have been able to present the requisite letter vouchsafing his character.

During the next phase of Hadsell's exam, the examiners peppered him with a series of "moderate questions in Navigacon." These oral questions seem to have gone beyond geographic knowledge to assess more abstract concepts, such as latitude, magnetic variation, charts, and tides. Hadsell "could not give any Considerable answeres" to these questions, and so the committee declined to pass him — not once, but twice. Hadsell appealed this decision with the Admiralty Board, but the senior brethren stood by their decision. The examiners wryly conceded that "it is possible by his many yeares useing those parts [the West Indies], hee may have [gained] some knowledge of those Coasts" — but in their view, Hadsell's familiarity with the coves and reefs of the West Indies did not make up for his lack of broader theoretical knowledge required of a master. His examiners suspected his expertise would not translate beyond the Caribbean Sea: "Wee doe not find that hee hath any competent skill in the Coasting of the Channell of England and doubt of his Capacity in nav[i]gating a shipp on the greate Ocean."[5]

Hadsell's memory of home waters had apparently faded over the decade, and he failed to display adequate facility with standard instruments. The reformed pirate may never have had any formal instruction and thus been unfamiliar with the conventional vocabulary — those codified definitions that formed such a key part of every exam. The board would not have asked Hadsell about his seamanship or manual skills, as in a man of his background they would have taken such things for granted. There is no way of knowing whether Hadsell's poor results were due to embarrassing errors in the exam room, or whether he had simply earned the mistrust of his examiners because of the company he had kept in the Caribbean.[6] However, his repeated failure makes clear that a master needed to demonstrate competence in both small and large navigation — local English waters and the open ocean — to say nothing of the realm of personal character.

At Trinity House on March 29, 1683, two young students found themselves facing a similar process: Richard Crockett and Nathaniel Long stood up straight and tried to anticipate the questions their examiner might have for them. These two aspiring masters were "blue-coat boys," part of a class of forty at the Royal Mathematical School (RMS) who were preparing for service in the Royal Navy[7] (figure 3.1). Charles II had founded the school in 1673 in connection with Christ's Hospital, a charitable foundation at Newgate that educated poor but promising children.[8] Earlier that winter, their mathematics teacher Edward Paget had glibly deemed Long and Crockett ready for the formal examination that marked the end of their eighteen months of classroom study. Providing that they could demonstrate "competent proficiency" in the "Art of Navigacon and the whole Science of Arithmatique," they would embark upon a seven-year apprenticeship, with an eye to careers as masters in the Royal Navy.[9]

The blue-coat boys' examination was radically different from Hadsell's. Rather than answering for specific geography or even navigational terms, the young men were drilled on a long list of mathematical questions. John Colson, who himself taught "the Mathematical Sciences" in the riverside parish of Wapping,[10] set out to determine if Long and Crockett were familiar with the "principles of geometry" — the standard Euclidean definitions. Could they determine the age of the Moon (its current phase) or use Gunter's scale? Colson probed Long and Crockett's knowledge of more than a dozen topics, reporting that at least half of these "they underst[oo]d competently well." He took care to note that the youths "have been taught ye use of the Davis Quadrant, ye Nocturnall, and ye Com[m]on Land Quadrant to find the hour of the day or night," the most common tools for taking altitude readings and time telling. Regarding the remainder of the decidedly mathematical curriculum, however, they fared less well. Colson found that they were "not well versed in Decimall Arithmetick," were "somewhat deficient in the Application of Oblique Triangles in plaine sayling," and "understand but are not very ready in findeing the Moones age." If the first two of these were advanced concepts, the "Moon's age" was one of the most basic requirements of coastal navigation, the first step toward "shifting one's tides."

After reading the critical report, diarist and naval administrator Samuel Pepys concluded that the "sayd Children were in Severall particulars deficient in what . . . they ought to have been instructed in."[11] On the basis of Colson's examination, the brethren of Trinity House failed them. This reaction likely had more to do with the school board's trepidation about the latest in a series of underwhelming instructors: Paget, a Cambridge graduate fluent in Latin and Greek, had come highly recommended by Isaac Newton and his peers but had never been to sea.[12] Members of the board, with Pepys at the forefront, worried that in his first six months on the job Paget had overlooked certain topics on the admittedly ambitious curriculum. Long and Crockett fell victim to these behind-the-scenes concerns. The two young men were sent back to the classroom for more lessons, and the merchant captains to whom they would have been apprenticed sailed to India without them.[13]

In 1692, yet another young man faced a panel of naval examiners—this time at the Navy Board Office on the grounds of the Royal Dockyards at Deptford. George Gale, the son of a wealthy north country merchant, had ambitions of becoming a lieutenant in the Royal Navy. Pepys, as Secretary to the Admiralty, had purportedly persuaded Charles II to introduce these exams in 1677.[14] Called into the exam room, Gale presented his journal from a recent voyage (one of the newer mandatory technical requirements) and "two certificates from Capt. Arthurs and Capt. Boteler of his ability & good behaviour." Having sailed extensively from the age of thirteen—from the Irish Sea north to Norway and across the Atlantic—now at twenty-two Gale easily satisfied the requirement of three years at sea. He had served on both merchant and naval vessels, moving up the ladder from midshipman to master's mate, and his journals and testimonies from his superiors were in order.[15]

For young gentlemen such as Gale, the Navy Board took "good behaviour" largely for granted but were keen to confirm his ability to "work a ship in sailing." His certificate thus itemized the practical skills he was required to demonstrate: "knot a shroud, splice a Rope, Reef a saile, bring a ship to saile, work her in sailing[,] take an observacōn, keep a reckoning of a ships way, . . . take an amplitude & find ye variation of ye compass."[16] The first five of these tasks—showing competence with handling ropes and sails and getting a ship to move where one wished—were plain seamanship. While such menial shipboard tasks were often deemed below the dignity of officers, critics warned of disrespect or even insurrection if officers did not understand the basics of maritime vocabulary and the myriad components of ships and their maneuvers. To remedy any shortcomings, junior appointees could consult the spate of manuals and glossaries that appeared in the late seventeenth century, some of which were aimed directly at officers.[17] But it was harder to learn rope work and sail handling from a book; textbooks ignored these topics until the late eighteenth century.[18] Thus, by requiring a test-taker to demonstrate his bowlines and sheepshanks, examiners could verify that he had applied himself during his three years of service rather than simply cramming from a book. The navy's insistence that its lieutenants be able to tie humble knots should be seen as an attempt to simulate in novices the tacit expertise of more practiced men.

The exam's remaining tasks fell under the heading of "large," open-ocean navigation. Gale and other hopeful officers were expected to be able to "take an observa[ti]on, keep a reckoning of a ships way, . . . take an amplitude & find ye variation of ye compass." Other lieutenants' passing certificates extended the list: "Find the Prime, Epact, Moon's age, Shift [the] tydes," and sail by both the "Mercator and plain charts."[19] Here were echoes of the tasks on the syllabus at the RMS. These navigational tasks had evolved since the time of Pedro Dias's experience at the Casa de la Contratación. By the second half of the seventeenth century, navigators were more likely to consult tables than to have memorized the regiment of the Sun and the North Star. They were also expected to know several methods for taking their daily observations: with an astrolabe, cross-staff, or Davis quadrant, at noon, or — with more difficulty — at the moment the clouds cleared, or using a nocturnal at night (figure 3.2). Rather than simply noting the ship's progress on a traverse board, they now calculated courses trigonometrically, tracked them in a logbook, and plotted them on a chart.[20]

Seventeenth-century lieutenants' exams included a component too new to have featured in the early Spanish exams: determining magnetic variation by taking amplitudes, the position of a celestial body when rising or setting relative to the east or west point of the horizon.[21] At the same time, all lieutenants needed to know how to "shift," or compute, the tides. It was this critical calendrical task that took pride of place in the first chapter of every Dutch manual. If the examiner told him the time of high tide at a particular port on the first day of the year, could he compute when to expect it for the current date? Although it was ever easier to find the time of high tide thanks to the proliferation of almanacs, these were rapidly outdated, and English navigators were still expected to turn to the "rule of thumb" to calculate the date in the absence of current tables.[22]

It remains unclear whether the lieutenants' exams were entirely oral or if they included a written component. If the mathematical questions entailed simple right angles and round figures, men may not have needed to work through computations on paper. A physical demonstration with one's chart and dividers or other instruments may also have been required, as in the Spanish archetype. Because the exam took place at the Royal Dockyards, it seems possible that men may have taken actual observations of the heavens or reefed real sails, or they might simply have made use of model ships to point out the parts of naval vessels.[23] Records from unsuccessful exams corroborate that candidates were put through their paces on manual or computational tasks — and might fail if they could not compute high tide on the Thames or "knot a shroud."[24] (Men who failed on a technicality — whose previous service was on merchant ships rather than naval vessels, or who had no logbook to submit — often appealed to the Navy Board and might sometimes win their certificates.)[25]

Ultimately, George Gale must have proved satisfactory in all domains, for he was deemed "thorowly qualified to do ye duty of an able seaman & Midshipman." He seems never to have pursued a career in the navy, though. It was not unusual for promising young lieutenants to end up following a civilian career path, particularly during periods between military conflicts when there were few commissions to be had. Over the ensuing decades, Gale continued to crisscross the Atlantic with his family's tobacco ships, making lifelong use of the applied elements of navigation.

These three groups of practitioners all aimed to achieve competence at the same tasks but took very different routes. This differentiation sheds light on how in late seventeenth-century Britain authorities conceived of the "art of navigation," and how much theoretical, mathematical knowledge was required to qualify as a navigator. Did unlettered practitioners still have a place within the field?[26] Masters can be seen as representatives of the traditional method of learning — they honed their skills through hands-on practice aboard ship. They typically would return to Trinity House

3.2.
Boy with Davis quadrant. Jonas Moore, *A New Systeme of the Mathematicks* (London, 1681), facing p. 248. John Hay Library, Brown University

every few years to get certified for progressively larger ships, a certification process that reflects considerable regard for personal experience.[27] However, this traditional route could lead to blind spots: Charles Hadsell, despite years of experience with shipboard and wayfinding procedures, had not memorized the standard definitions that would have signaled that he had been groomed for the profession. As navigation was framed in increasingly mathematical ways, practitioners who did not at least pass through a classroom were increasingly left behind.

Aspiring lieutenants, for their part, faced examinations that focused far more on manual tasks. With such men, naval authorities were more assured of the candidate's pedigree than his manual dexterity. Considering the elite social origins of commissioned officers, the Navy Board could assume that they had some exposure to mathematics; by their late teens, most of these young gentlemen would have completed basic arithmetical education — through the rule of three — with private tutors or in grammar school.[28] There was still a steep learning curve ahead for these men to acquire a minimal competence at sea. Ideally a navigator would develop an intuitive sense about how boats sailed, but these men lacked the time for such a gradual process. Instead, their training prioritized leadership skills and military strategy, leaving them to pick up the basics of knot tying and nautical terminology in their first months at sea. If the French navy sought shortcuts for certain tasks, here Britain's Royal Navy was seeking shortcuts for entire groups of officers.

On the face of it, the program for the RMS boys seems the least effective or relevant to the tasks at hand. After perhaps eighteen months of lessons, they could recite geometric definitions like any university student reading Euclid but had yet to master the basic topics most necessary at sea, such as shifting their tides, making an altitude observation, or using a chart. The instructors evidently presumed that these young men would learn the practical tasks in their subsequent apprenticeships. And yet despite the stark divide between the theoretical and applied portions of the blue-coat boys' training, a version of their curriculum would become the standard method for teaching navigation. Instructors focused on a foundation of arithmetic, geometry, and trigonometry, with the expectation that the practical tasks could readily be solved with suitable formulas.

This numbers-heavy approach to teaching British mariners in the 1680s signals the emergence of a new conception of navigation. Instead of regarding navigation as a problem of astronomy, for which you needed to understand your place relative to the stars, it was now distilled into a series of mathematical equations, for which you needed basic arithmetic. This new framing was not limited to England: as we have seen, Guillaume Denys had recently introduced the French to the new trigonometric tools that would radically change navigators' practice, and the curriculum promulgated in the ordinances of the 1680s had a similar focus on mathematics. The numeracy expected of seventeenth-century Dutch mariners was, if anything, higher than anywhere else. However, for several important reasons the turn to mathematics happened in the English maritime world earlier and more decisively than elsewhere.

THEORIZING AT THE ROYAL MATHEMATICAL SCHOOL

In the closing years of the seventeenth century, maritime culture in Britain was in flux. Although an island, England had not always defined itself as a maritime power. After the mixed success of sixteenth-century colonial efforts in Virginia, Massachusetts, India, and the Caribbean, by the Restoration overseas ventures had become central to foreign policy. In 1651, Oliver Cromwell's Rump Parliament introduced the first of the Navigation Acts, which stipulated that three-quarters of the crew on every English vessel must be English-born. This legislation identified key colonial products (including tobacco, sugar, and cotton) that could be exported only to England rather than to foreign ports, a policy that profoundly changed the nature of short- and long-distance trade.[29] Whereas the Dutch had developed a substantial carrying trade in the first half of the seventeenth century, these Navigation Acts tipped the balance toward England. They also accomplished another goal: they greatly increased the number of skilled seamen, a vital step in building up naval strength. The number of Englishmen involved in maritime activities, including coastal trade in home waters, long-distance trade, and the navy, roughly tripled over the eighteenth century.[30] All the while the merchant marine was flourishing, the crown, always aware of the success of its rivals, wished to improve and expand the Royal Navy. Charles II felt he could correct the "want of able pilots" and create more skilled officers by establishing the Royal Mathematical School. As the sea and sailors began to feature more often in patriotic rhetoric, it became clear that the sailors who would help buoy England to greatness needed to be numerate.[31]

England, like the Netherlands, had seen a healthy maritime book market develop in the final decades of the sixteenth century. The earliest titles to appear were decidedly cosmographical in organization: William Bourne's *Almanacke and Prognostication* (1567), which included his "rulles of navigation" and a regiment of the Sun (similar to the Iberian examples from the 1530s), and the English translation of Pedro de Medina's *Arte of Navigation* (1581).[32] A spate of original English works followed in the 1590s: Thomas Blundeville's *Exercises* (1594), which were intended to support the "Furtherance of [the] Arte of Navigation"; Edward Wright's *Errors in Navigation* (1599); and John Tapp's *Seamans Kalendar,* a nautical almanac that appeared regularly from 1602 on. Occasionally an English author would emulate Dutch handbooks, opening his book with calendrical concepts, "for the better finding the time of the Tide in any place," as Thomas Addison wrote in 1625.[33] However, by the end of the sixteenth century, British authors had already begun to shift their emphasis away from the movements of the heavens toward the functional tools of mathematics. In 1594 Thomas Blundeville described the requisite skills for English gentlemen who wished to travel by sea, declaring it "unpossible for any man to be perfect unles he first have his Arithmetick, and also some knowledge in the principles of Cosmographie." By 1636, when Captain Charles Saltonstall wrote his prefatory poem (in Latin and English) "In Praise of Arithmaticke, and the Arithmeticall Navigator," he could point to half a dozen "divers Books already extant that will Instruct him" in "all manner of Arithmeticke [and] Geometry."[34] These new English nautical textbooks — in which mathematics preceded astronomy — differed significantly from examples produced elsewhere during this period. Increasingly they treated topics that were familiar from business arithmetics: tables of accounts, foreign weights and measures, and gauging barrels.[35]

In many ways, it is unsurprising that this change occurred first in English manuals. The status of mathematics was on the rise across Europe. Universities began to include the subject in the undergraduate course from the middle of the sixteenth century.[36] At the same time, mixed mathematics began to enjoy more prestige in commercial centers, particularly Italy and the Low Countries. London, like Amsterdam and Venice, developed a vibrant community of independent mathematical instructors, instrument makers, writers, and publishers that thrived in proximity to the city's bustling maritime scene.[37] The move to include basic mathematics within navigational texts signals the importance that English authors put on maritime trade: to improve one meant placing equal emphasis on the other. This hybrid model — the business arithmetic *cum* navigational text — did not arise elsewhere. In Italy few theoretical techniques were required to safely cross the Mediterranean, so navigation manuals were uncommon, whereas in the Netherlands the audience for mercantile texts was as healthy as for nautical ones; thus, the two genres flourished in parallel. After a lull in English nautical publishing during the turbulent years of the Protectorate, the 1680s and 1690s saw a boom similar to that which had occurred in the Netherlands in the early seventeenth century.

Two men associated with the Royal Mathematical School produced a pair of textbooks specifically for the blue-coat boys that conformed to the new, arithmetical approach. The eminent mathematician Sir Jonas Moore, a member of the school's board of governors, and Peter Perkins, the well-respected mathematics master from 1678 to his death in 1680, both produced nautical manuals that began not with cosmographical definitions but with the rudiments of arithmetic and geometry. Moore embarked upon his mammoth *New Systeme of the Mathematicks* (1681) at the behest of the Christ's Hospital board of governors, but he died in 1679 before completing it. The volume starts with chapters on arithmetic and geometry, only moving on to cosmography after one hundred pages.[38] Perkins wrote the subsequent sections on navigation, Euclid, and algebra and shepherded it into print. His own, more accessible work, *The Seaman's Tutor* (1682), opens with basic Euclidian diagrams, followed by a brief history of mathematics. This volume, "Compiled for the Use of the Mathematical School," introduced readers to logarithms before gesturing toward cosmographical essentials. Perkins had a grand view of his subject: "This Art of Traversing and Caravanning over *Neptune*'s vast Dominions, is not a simple or single Science; but a Complication of the three most noble Sciences, *Arithmetick, Geometry,* and *Astronomy*"[39] (figure 3.3). These three muses were depicted on the silver badges worn by

3.3.
Royal Mathematical School badge (1673): The muses of Arithmetic, Geometry, and Astronomy gather around a blue-coat boy. Perkins explained the iconography in his textbook: The art of navigation "makes use of Arithmetick for Computation; Geometry for Delineation and Demonstration; and Astronomy for Observation.... [This] is clearly expressed in his Majesties Silver Badge worn by the Blew-Coat-Children of the Royal Foundation in Christ's Hospital, whereon are three Women for these three Sciences." [Perkins], *The Seaman's Tutor* (1682), 175. Christ's Hospital

the blue-coat boys, but Perkins treated only the first two sciences. His decision not to "meddle with much of Astronomy" was very much in keeping with the new approach to educating navigators. No one associated with the Royal Mathematical School, or indeed the Admiralty, would have disputed the idea that mathematics took precedence over astronomy in this new systematic approach to the art of navigation.[40]

The Royal Mathematical School had a contradictory mission: while it primarily served poor youth only a few steps above "common seamen," the institution was shaped by an elite group of naval administrators and governors who stressed math and science. Many of these men had multiple ties to the Navy Board, the Royal Observatory, and the Royal Society (Samuel Pepys, for instance, was involved not only in the Admiralty Board, the Royal Mathematical School, and Trinity House but also in the Royal Society).[41] Subscribing to Francis Bacon's view that scientific knowledge should be useful, the men strove to develop a technical body of knowledge that would benefit society more broadly. What was initially simply a venue for quickly training skilled navigators was soon imbued with loftier goals.

Pepys, as a very active member of the RMS board, felt the school had a mission to "carr[y] the science of Navigation beyond what our best mere tarpaulins, as they are now qualified, are ever likely to do."[42] He admitted that the students were of uneven intellect and behavior (some, he noted, sold their schoolbooks, refused to wear their silver badges, and ran away). But he felt that much of the problem lay in the design of the curriculum. Critical of the fact that the school's curricula were originally the province of individual instructors, he called for a new "Method of Mathematicall Learning." An "orderly introduction into the knowledge of so much of each part of the Mathematicks" would help the students memorize the material and pass their exams. Pepys recommended that this program be designed with the "joynt advice and Approbation not only of some of our ablest M[aste]rs, but of some principal Commanders and Experienced M[aste]rs of ships."[43]

While few naval officers weighed in on the school's curriculum, the RMS board sought input from an illustrious group, including the Astronomer Royal John Flamsteed, his successor Edmond Halley, Robert Hooke, and Isaac Newton, natural philosophers with unabashedly theoretical ambitions. Flamsteed and Newton felt that "speculative" — or theoretical — elements of navigation should precede the "practical." William Oughtred, who invented several mathematical instruments, including an early slide rule, was surprisingly dismissive of hands-on instrument training; only "vulgar Teachers" chose the "preposterous course" of beginning "with Instruments and not with the Sciences."[44] Newton, with whom the board consulted a number of times about the curriculum, had a particularly idealistic vision of what school could do for these mariners-in-training. Like Oughtred, he disdained nontheoretical lessons. He believed that the "Mathematical Boys" should be trained in "inventing new things and practises, and correcting old ones, [and] in judging of what comes before him," for it was that type of innovation and critical thinking that would "furnish the Nation with a more skilfull sort of Sailors, Builders of Ships, Architects, Engineers and Mathematicall Artists of all sorts, both by Sea and Land, then France can at present boast of."[45]

Of course, not everyone felt that these rough youth could handle so much theory — or that mathematics was the best entree into this traditional subject. Flamsteed, for instance, softened his endorsement of theory when it came to practical seamanship, the "best way of working a ship." Flamsteed averred that "this is commonly found out by experience and tryall, and a discreet saylor is a better judge of it then ye best read man of ye Age."[46] One of the naval experts that Pepys consulted contended that "art" — in the sense of man-made artifice, or theory — was dangerous: Thomas Phillips, a chief engineer, felt there were "more men of art lost than of plain illiterate seamen, the latter minding their lead and marks and minding their reckoning." Furthermore, he added that the Dutch, "who are more exact in their learning and charts than any other people, do make more faults than any other, and therefore that many masters who know not how to write or read, have fewer losses than the more skilful."[47] Such a statement may, however, have had more to do with rivalry than actual fact.

Several experts suggested that RMS offer a tiered education, as many other schools did. This could be stratified by aptitude rather than social status. Robert Hooke proposed that a trio of new textbooks be written for "Learners," "Proficients," and "Artists in Naviga[ti]on" — although the school seems to have deemed Moore's text suitable for all levels. Newton also suggested two different paths for learning algebra and speculative geometry, depending on the abilities of the "Schollars." (He would assign the more advanced group several Euclidean texts, as well as Mercator's *Astronomy* for an introduction to "the Sphere.")[48] Although certain late seventeenth-century authors continued to feel that trigonometry was unnecessarily complex, it was increasingly considered a basic tool, which in turn required greater comfort with arithmetic.[49]

The institution's curricula got ever more technical as candidates competed for the post of mathematical master there. (In addition to letters of support from various members of the Royal Society, many of these men submitted printed courses of study as part of their job applications.)[50] The committee placed little emphasis on actual nautical practice, preferring to debate how much Latin the boys should study. Over the years, they consistently voted for prospective teachers who had university educations rather than maritime experience. The school's board suggested that Paget's total lack of sailing experience could be rectified by sending him to sea for twenty days, a term Pepys, in his role on that board, found ludicrously inadequate. It must be noted that at the end of Paget's troubled tenure, they finally chose a nonuniversity man, Samuel Newton, whom they soon found equally unsatisfactory.[51] All of these masters adhered to conventional teaching strategies, very much for the same reasons that the Spanish cosmographers chose lectures and textbook dictation to pass on knowledge to their illiterate pilots: for them it was the path that worked.

Royal Navy administrators did have some knowledge of foreign navigational training: in addition to the early reports on Spain by Stephen Borough and Pedro Dias, they had access to English translations of landmark French legislation, and English translations of Dutch atlases and manuals were widely available in English bookshops.[52] Pepys was one of the most assiduous researchers into conditions at home and abroad. As part of his preparatory work for a long-planned history of British nautical science, he investigated the origins of specific innovations — from the magnetic compass to maps and atlases. He was dismayed to find that, compared to those elsewhere, Britain's maritime experts paled. Far from being the incontrovertible ruler of the seas that it would become in the eighteenth century, in Pepys's view England was barely keeping afloat.[53]

By contrast, Pepys found ample evidence of the stellar accomplishments of England's chief rivals — the Netherlands, Spain, and France. He was quick to acknowledge England's debts to the Dutch, from a manner of sheathing hulls to nautical terms England had borrowed. He envied not only their techniques and cartography but also their personal habits (it was "notorious how much neater they generally live on shipboard than we," and, he reported, they "are ever better clad" and "can bear drink better").[54] Pepys had an even greater respect for England's closest rival, crediting the French with numerous naval innovations from ship designs and exemplary regulations (the *Ordonnances*). Without a hint of acrimony, he conceded that "it appears that we have learned many useful things from the French." For Pepys, the most significant imports from each of these neighbors were their inspiration for more rigorous naval training. "We have borrowed also from the French, and they I think from the Spaniards, the King's late institution of the Mathematical School."[55] He made notes to investigate the methods of the "sea-seminaries" in "Holland and its fellow-provinces," as well as their "schools to teach the mechanic [physical] part of shipbuilding, where strangers may come and learn as well as natives." He was also fascinated by the Spanish "Sea-Nurseries," seeking out publications about the Casa de la Contratación during his 1683 expedition to Tangiers.[56]

Pepys was a status-conscious London gentleman and woefully prone to seasickness; despite or perhaps because of this, he had a deep admiration for maritime practitioners. In the contemporary debate over which type of men would make the best naval officers, "gentlemen" or "tarpaulins," the meritocratic Pepys sided solidly with the tarpaulins, those men who rose through the ranks due to experience.[57] He clearly saw the value in hands-on learning, enthusiastically poring over the latest mathematical instruments — from telescopes to pendulum watches — to understand how they worked.[58] Although he knew virtually nothing about the sea when he was appointed Clerk of the Acts, he turned to written accounts and knowledgeable experts to become familiar with shipyards and naval terminology. Similarly, despite his Cambridge education, he had learned very little mathematics. He thus engaged the services of Richard Cooper, the one-eyed ship's mate of the Royal Navy's flagship, to teach him the rudiments; their lessons began with the multiplication tables.[59]

Pepys's research made him both anxious and motivated. Not only did he come across century-old texts documenting "the illiterateness of our sea-masters" and the "imperfectness of our navigators at that time," but he had alarming conversations with one of the early RMS examiners. Pepys reported that "Old Norris, our mathematician," claimed "that the Dutch do exceed us in their books of navigation, and that in a little time the French are like to do so too by the encouragement given by their King, while (he [Norris] says) we daily grow worse." Pepys deployed such dramatic rhetoric and statistics in an effort to convince the king and Parliament to fund the navy and the RMS.[60] At the same time, he developed his own strategies to improve England's odds at sea. He made plans to translate foreign textbooks to aid the general maritime community.[61] He considered free public lectures the first step in staving off encroachment from literate French and Dutchmen, and he even toyed with plans for naval knighthoods and other marks of honor to encourage more investment in nautical matters.[62] He clearly felt that formal lessons would lead to the greatest improvement and so redoubled his efforts to reform the RMS.

The educational infrastructure that developed in late seventeenth-century maritime England braided contrasting strands: a tradition of local pilots who learned from their elders at Trinity House in a commercial center that fostered entrepreneurs, against a theory-filled intellectual model from abroad. Given that the elites who ran the navy favored university-educated instructors, the less-learned proponents never unseated the presumption that classrooms were the best place for transmitting theoretical knowledge. Thus, the familiar pedagogical triad of lecture, textbook, and examination lived on. And yet English mathematical masters put their own spin on these formal lessons, infusing a practical sensibility into what, on the surface, seemed excessively theoretical.

APPLIED THEORY: RMS CLASSES

The mathematical masters at RMS dutifully wrote down their curricula, but about how they would transmit those lessons they were largely silent. To have a sense of what Nathaniel Long and Richard Crockett would have experienced in their classes, we must glean hints from the school's financial records and administrative reports and survey the practices at similar institutions on the other side of the Channel. This composite picture reveals how the blue-coat boys likely encountered new concepts — and then progressed toward understanding them.

The RMS class, with forty pupils, was fairly sizable — although French state-funded schools were larger. Instead of more conventional two-hour-long classes, RMS regulations stipulated that students were to study for four hours every weekday morning and the same again after lunch. (They made occasional excursions to lectures at Gresham College.) The master was responsible for ensuring that the pupils repeated bible verses along with their nightly homework.[63] Teachers were prohibited from taking on additional pupils, which would have been their preferred means to augment their salary, but they might delegate some classroom responsibilities to an assistant — one of their strongest recent graduates. (When Denys employed Doublet, this star pupil could expect to continue his studies — with Euclid rather than anything nautical — and would help keep order in the classroom. This strategy ironically reduced the likelihood of the brightest candidates going to sea.)[64] In most schools, students were required to bring their own writing supplies to class, but the RMS provided all the basic materials for the boys: paper, notebooks, "ink, quills, rulers, and many other necessaries." (They had pen knives for sharpening their quills, as well as slates and drawing compasses. When they were ready to be apprenticed, the graduates received their own Davis quadrants.)[65]

Many of the day-to-day practices and pedagogical techniques would have been familiar from standard grammar schools. Dialogues, memorization, and dictation were chief among the strategies for transmitting fundamental concepts. Dialogues were perceived as helpful tools for young pupils and more advanced students alike — and thus featured in numerous textbooks throughout the period.[66] They were useful preparation for oral examinations but also more generally for memorizing the material. Students were drilled regularly on the content of recent lessons "in order to better recognize and retain what follows."[67] If early seventeenth-century instructors like Jan van den Broucke wanted their pupils to review calendrical and astronomical details daily to help "understand measurement and observation" (*passen en schieten*), later in the century English experts recommended a different first step: "You must perfectly get the following Table by Heart" — the multiplication tables. (Pepys did well to heed Richard Cooper's lessons.)[68] Memorizing the "rudiments of arithmetic" would aid "young learners" and more experienced men alike. By the eighteenth century, authors condensed longer texts into pocket-size "epitomes." With little in the way of diagrams or explanatory text, readers were supposed to commit entire series of rules to memory.[69] Written works now functioned as external memories — instead of memorizing key features of a coastline, one could simply consult an atlas. The same held for tide tables, star charts, and other aspects of seamanship but was particularly important for the new, more elaborate data associated with celestial navigation. And yet in many ways a strong working memory remained vital for these newly mathematical sailors; to make use of those dense tables, they had to memorize when and how to apply myriad mathematical procedures.

The RMS instructor usually spent much of the lesson dictating directly from a published textbook, a practice long deemed to have intellectual as well as economic benefits.[70] The school used Perkins's text for many years, until Samuel Newton replaced it with his own.[71] In addition to instructional texts, trainees also transcribed documents pertinent to their specific situations to carry with them to sea: descriptions of trade goods and glossaries of foreign terms for men shipping to Asia, or handbooks on maneuvers and signaling.[72] (By the eighteenth century, when pocket-sized textbooks were affordable and abundant, instructors and administrators alike objected more vociferously to the time-consuming practice of dictation, but it never entirely disappeared.)[73] (See plate 3.)

The standard lessons at RMS were lengthier than elsewhere: four hours in the morning and in the afternoon. The teacher would leave time after the dictation for "answers to questions and objections that the [students] think to pose."[74] In the early years of the seventeenth century, Sir William Monson, an admiral whose career included stints as a privateer and member of Parliament, bluntly contended that "ordinary mariners are oftentimes ignorant of what they shall read, as not understanding either word or sense." He felt that they would have an easier time grasping "what they shall hear delivered by mouth"; then, "when they shall have it demonstrated to them, and the hard words and meaning made plain, they will both conceive what they heard and be able to put in practice when they understand it."[75] Thus, the teacher's role went beyond simply defining "hard words" for his students; he needed to "demonstrate to them" a lesson's core ideas.

These nautical classroom "demonstrations" seem to have consisted not only of verbal exegesis or a step-by-step walk-through of a mathematical calculation or proof but also of physical gestures and hands-on procedures using instruments.[76] Although there are few hints about the hands-on practices in the RMS schoolroom, contemporary French instructors were quite explicit about how they used these techniques in class. One strategy was to cement knowledge by working through practice questions; teachers expected students to read "pen in hand" in order to do several examples of each type of question, which again was considered a valuable step toward comprehension.[77] It was best for students to take charge of their own learning—particularly when printed textbooks might still be riddled with typographic errors. One of Guillaume Denys's more outspoken trainees, Samson Le Cordier, agreed with Monson: young people were not used to following lectures when they were keyed to diagrams, because not only were there too many labels but they were often poorly printed. He therefore felt that students should be taught to draw their *own* geometric diagrams as soon as they arrived at the hydrographic schools; constructing these figures themselves would help them understand and remember the details. The English surveyor Adam Martindale felt that "draw[ing] a rude Draught" would "much help both the understanding and memory"; such "ocular Demonstrations" would in fact improve on Euclidean diagrams and be "more easily understood." When it came to comprehending the basics of trigonometry, Jonas Moore recommended that his reader draw a *mental* image: "You are not only to consider Triangles drawn before your Eye on Paper, but also those that are conceived to be made on the Ground, in the Air, upon the Sea or Waters, or in the Heavens."[78]

Instructors often asked students to engage with the material in ways that moved beyond the two-dimensional. From its founding, the RMS stressed teaching with instruments. The school received an eclectic range of instruments as bequests from benefactors and regularly bought new ones, purchasing from a variety of sources, including several female instrument sellers.[79] In the early years of the RMS, Jonas Moore persuaded the Mathematical Committee to send students to the Royal Observatory to study instruments with Flamsteed, "for their being entred into ye Practicall Use of Instruments, for ye takeing of Observa[tion]s, the know[ledge] of ye Starrs &c before they goe to sea." When Dr. Wood was hired in 1677, he pledged to take the boys out onto a "platform of Lead prepared for the purpose" to observe sunset and moonrise — even though, when Colson examined Crockett and Long in 1683, the instruments had not yet been installed.[80] Their absence, in combination with Paget's unfamiliarity with nautical tools, hindered the students' practice in observations.

French instructors seem to have done a more thorough job of teaching students how to carry out the basic daily observations. The teacher, author, and instrument inventor Boissaye de Bocage would conclude a day's lesson at the royal hydrography school in Le Havre by "tak[ing] the cross-staff and show[ing] them how to take an altitude, or [the same] with the quadrant or the astrolabe."[81] Most teachers and authors held that it was safest to know multiple ways to complete each task. An instructor in Rochefort, after explaining the day's dictated textbook passages to his students, would "make them do the operations and rules, such as pointing the charts, observing latitude and the variations with those instruments that one ordinarily uses at sea, of which he explains their uses."[82] Not infrequently, teachers took the students on excursions to allow them to apply the lessons. In good weather Coubert took his students to the harbor; others used small boats to survey the shore.[83] They learned how to use not only tools like the cross-staff but also techniques for assessing their accuracy. (By this date it was easier to purchase one than to mark the graduations upon it oneself, but there were still problems with professionally made tools.)[84]

If the RMS students did not have extensive opportunities to make outdoor observations, they nonetheless acquired crucial skills by manipulating navigational equipment *within* the classroom. The blue-coat boys, like students elsewhere, pored over model ships to learn technical vocabulary, aspects of rigging, ship design, even demasting and dredging.[85] Perhaps the most essential classroom tool was the globe (figure 3.4). The RMS inventory notes that the school owned three pairs of celestial and

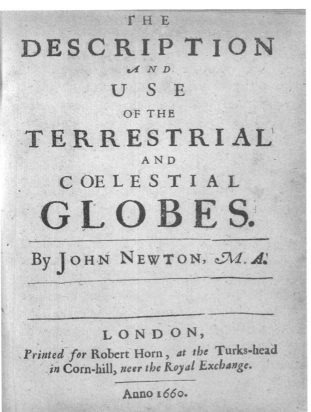
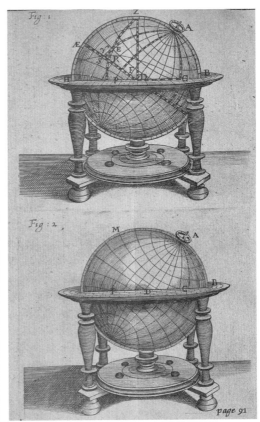

3.4 a and b.
A treatise explaining how to use globes for a wide range of calculations. John Newton, *Mathematical elements . . . in III parts* (London: R. Horn, 1660), pt 2. John Hay Library, Brown University

terrestrial globes. (Most institutions had at least one set, and students like Thomas Child, who was taking private lessons, were also expected to buy a basic pair.)[86] Early modern globes were not mere images of the Earth and heavens; they were scientific instruments. Not only were they a particularly intuitive tool for men who had already circled halfway around the world, but they were also extremely well suited for demonstrating astronomical concepts in a classroom.[87] The clearest instructions for using globes were penned by leading navigational authors like Martín Cortés and Willem Blaeu.[88] Globes could be positioned on a given angle, representing the observer's latitude. Once that was known, the student or navigator would use dividers to measure angles and distances. They could determine the positions of the Sun and stars, the ideal angle for a specific course ("great circle sailing"), and other conceptual details. Jan van den Broucke taught his students how to document a new or strange star (*nieu, vremde Sterre*) by triangulating from two familiar stars and then marking the location upon a blank paper-covered globe.[89] Robert Hues, writing in the 1630s, found globes an excellent alternative to the "very tedious and prolixe" trigonometry. Coubert explained how celestial globes did this representational work: "The sphere is a machine that represents the points, lines, and circles that one imagines in the heavens, in order to give reason to the movements of the heavenly bodies. One must get used to conceiving of circles as lines on the paper."[90]

This evidence that students worked with instruments alongside their mathematical exercises shows that their classroom education consisted of far more than rote memorization.[91] By taking up tools, marking positions on charts, identifying elements of the rigging, or carrying out observations outdoors, young sailors, while learning the physical operations required at sea, were also more likely to understand the concepts. Critics remained: when Isaac Newton was asked to assess the RMS curriculum in 1695, he was scornful. Not only did he deem "the whole scheme" that Paget had been teaching "soe confused & immethodical," but he also found it particularly "preposterous" that it should "comprehend little more than the use of Instruments, and the bare practise of Seamen."[92] He disdained the emphasis on instruments that Pepys found so illuminating. And yet, whether Newton approved or not, hands-on exercises had become an ingrained part of training.

It is worth revisiting for a moment Trinity House in the spring of 1683, as Long and Crockett's failed examination can now reveal something more subtle than a story about students left unprepared by an unsatisfactory teacher. A closer look at revised curricula and exam reports shows that, despite the seemingly inexorable shift toward more mathematics in English nautical education, there was a simultaneous push to retain manual dexterity. In the winter of 1682, Pepys, whose passion for pedagogy and high standards was matched by his attention to detail, noticed that when Edward Paget had conducted his preliminary assessment of the two youths, he had used an outdated curriculum. When Paget had been hired to replace the ineffective Robert Wood, he had contracted to teach five new topics atop the ten for which Wood had been responsible. The new curriculum settled on by the school board included more math — logarithms and arithmetic up to and beyond the rule of three — as well as drawing, astronomical observations, and a truly daunting list of no fewer than nineteen instruments.[93] Given his love for the latest technical tools, and his awareness of their utility at sea, Pepys felt very strongly that students be introduced to each of them. He made sure that Trinity House had a corrected copy of the curriculum so that future examiners could cover all the required topics.[94]

Alas, when Colson conducted his formal exam two months later, Paget's pupils were nowhere near competent on the hodgepodge of equipment. Both Long and Crockett failed to display mastery of at least a dozen instruments; Colson made no mention of basic tools like the sea compass or globes, and certainly not of the more elaborate "Semicircle & a Quadrant of Brass" or "Bond's dipping needle" (for measuring magnetic variation). Colson did note that the lack of equipment prevented examination of the boys' drawing abilities and their competency at astronomical observations.[95]

Pepys must have been frustrated. For all the Royal Mathematical School's cerebral emphasis on abstract mathematics, Pepys felt it was important that the students become familiar with instruments. The list of nineteen instruments was a straightforward inventory of all of Dr. Wood's equipment, made at Pepys's behest in January 1680 as part of his push to improve the school's reputation and repair its funding.[96] However, when the details of Paget's contract (his "articles") were compiled, no expert seems to have weighed in on which of these tools were essential. Pepys, in his drive toward efficiency and general competence, felt the students should make use of all the items the school already had. But mariners certainly did not need specialized and expensive inventions like a "tube or telliscope" for their daily observations, nor did they need to employ land surveyor's tools such as the "Staff & a Wyre-Chaine" or the mathematician's "Model for extracting the Cube-root."

Paget, never truly happy with his job, went on extended trips and left an assistant to run his courses. He drank heavily. Eventually he resigned, complaining of ill health.[97] Pepys's frustration with Paget's slapdash teaching may have been somewhat overblown, but his commitment to hands-on training was very much in keeping with contemporary trends in the maritime world. If the blue-coat boys had been introduced to the instruments on the list, they would have encountered a range of truly useful tools — not just the familiar backstaff and nocturnal but a full set of Gunter's sectors, and the "Double Horizontall Dyall & 2 Quadrants pasted on board," a type of sundial that showed the position of the Sun on the ecliptic and allowed users to calculate the altitude of the Sun at its azimuth.[98] In addition to introducing logarithms, arithmetic, and astronomical observations to the RMS coursework, Paget's new curriculum stipulated that the masters-in-training should learn to draw, an important step toward training their eye to be more adept at estimating distance, representing perspective, and understanding the optical distortions common at the horizon. These skills, long valued but deemed complex and elusive, could now be systematically taught. The updates to the curriculum demonstrate that, although the requisite skills in a navigator's daily practice had remained largely the same from the 1580s to the 1680s, the methods of carrying them out had multiplied. Consequently, the conventions of training also had to evolve in ways that effectively blended theory with practice.

Even though the RMS masters seldom introduced their students to all the navigational instruments on Pepys's list, they still used many instruments to ensure mathematical comprehension. In the classroom, a student would face a range of assignments: memorizing definitions, drawing geometric shapes, adjusting globes. These hands-on exercises would help cement concepts so that they would spring more readily to mind when students faced the examiners. These strategies appear to have been common in other schoolrooms, particularly in France. Despite the disdain of more cerebral mathematicians, teachers harnessed their classroom equipment to make abstract ideas concrete. As a result, far from being a bastion of theory, the classroom was in fact an important preparatory ground for learning applied techniques.

Ironically, relatively few of the early RMS graduates went on to serve in the navy. With their reputation for mathematical prowess, several were recruited abroad: as bookkeepers to Virginia, or mathematics teachers to Russia. Despite the institution's organizational challenges, the mathematics lessons seem to have been effective—albeit not necessarily in any of the ways that Pepys, Newton, or the Admiralty Board wished.[99] With varying success, the mathematical curriculum would be extended to other institutions in England.

As for the new mathematical style of nautical textbook, it too was widely adopted—but at different rates depending on local intellectual, social, and even political considerations. Certain maritime communities took longer to overcome the stigma of commerce associated with mixed mathematics.[100] In the Netherlands, as in England, financial computations were deemed part of the navigator's daily work, and business arithmetics were thus embraced as suitable and familiar models. The Dutch had a strong parallel tradition of *rekenboeken*—business arithmetic primers that formed the basis of primary mathematics education. Merchants, seamen, and young students alike studied these primers to learn how to handle money, measures and weights, and more complex financial situations. At the same time, navigation was an important enough topic for the Dutch that it had its own teachers, textbooks, and publishers, which complemented but flourished separately from the materials dedicated to commerce.[101] In seventeenth-century England, by contrast, where the maritime community was less influential, nautical material was initially a subsection of business mathematics rather than its own field. This somewhat different balance, along with the different educational options in each city, meant that when Dutchmen set out to study navigation, they already knew the rudiments of arithmetic, so these were not included in the *schatkamer*. English mariners, however, often needed to be introduced to basic mathematical concepts, and enterprising teachers were happy to oblige them.[102] The navy began formalizing internal lessons, paying midshipmen to teach "theory" and "seamanship." They also hired instructors to teach on ships, but men continued to take private classes as well.[103] English mariners who wished to improve their material standing seem to have been content to hire the same person to teach them to track their personal finances *and* their route.

In Spain, where the cosmographical style of navigation textbook was particularly dominant, it took longer to introduce a notably different model. It was not until the late eighteenth century that Spanish sailors regularly employed trigonometry.[104] French textbooks, drawing as they did first upon the Iberian tradition and then the Dutch one, never included business arithmetic. French mariners who wished to know about commercial interactions consulted separate treatises. Thus, their navigation books included trigonometry — thanks to Denys — but offered few lessons in the rudiments of arithmetic.

These curricula and teaching records show how mathematical masters and other experts continued to theorize about the best way to learn: by instruments, geometric calculations, or numerical tables. No matter which pedagogical methods were chosen, physical instruments could play a key role. Thus, where textbooks negotiated a balance between memory and math, schoolroom lessons added a third component: frequent hands-on opportunities to interact with instruments and visual material. This strategy, developed explicitly in some cases and unwittingly in others, fostered greater understanding of abstract concepts. The payoff of learning navigation in a classroom, rather than from books or on a ship, was that sailors gained the ability to convert these essential elements of the navigator's art into mathematical techniques — and vice versa.

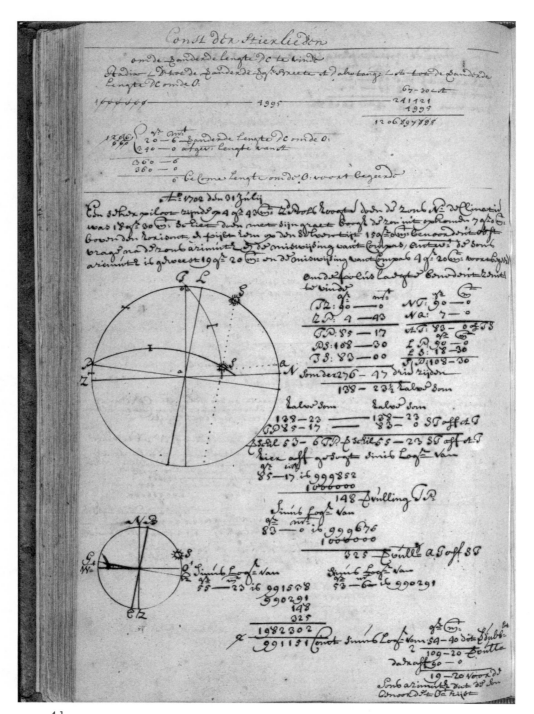

4.1.
Assuerus van Asson carefully worked through a typical textbook problem; he first needed to determine the Sun's azimuth in order to calculate the variation of the compass. Schatkamer (Netherlands [Maassluis?], ca. 1705), f. 71v. York University Libraries, Clara Thomas Archives and Special Collections, Toronto, ON

Chapter Four
PAPER SAILORS, CLASSROOM LESSONS
The Netherlands, ca. 1710

Shortly after the turn of the eighteenth century, a Dutchman named Assuerus van Asson sat in a classroom not far from the North Sea, diligently trying to calculate the Sun's azimuth (its distance from due north on the horizon plane) in order to determine magnetic variation (figure 4.1). Van Asson's teacher had drawn this practice question and dozens of others from three popular textbooks: a recent edition of Claas Gietermaker's *'t Vergulde Licht* (1660), Abraham de Graaf's *Kleene Schatkamer* (1680), and Jan Albertsz. van Dam's *Nieuwe Hoornse Schatkamer* (1702), just off the press. Van Asson was clearly a conscientious student. Over the course of his studies, he covered nearly 150 pages of a vellum-bound notebook with solutions to the set problems.[1]

Similar scenes were occurring elsewhere in the United Provinces. In 1709 another young man, a Swede by the name of Anders Wijkman, filled up his own notebook with the same circular diagrams and trigonometry. He had evidently paid for a series of lessons with a Dutch teacher, as many foreign sailors did, in order to prepare for his navigator's exam. In 1710 one C. Visscher, hailing from Sloterdijk in the vicinity of Amsterdam, worked his way through the latest edition of Gietermaker's *'t Vergulde*

Licht, which, having been published for the eleventh time earlier that year, dominated the market. Unlike most other students, Visscher did not write in a dedicated blank workbook but instead copied out Gietermaker's questions and diagrams on more than one hundred loose sheets of paper. After carefully solving the questions, he had the neatly written solutions bound into his copy of the textbook[2] (see figure 4.15).

For more than one and a half centuries, men with connections to the sea devoured textbooks like those of De Graaf and Gietermaker and produced manuscripts similar to those of Van Asson and Wijkman. Some of the men may have worked through navigational manuals on their own; indeed, authors like Jan van Dam promised that in his book, "this Art will be treated so clearly and orderly that a curious Pupil could learn it by himself."[3] However, in most cases these men would have completed their workbooks in a formal setting, either with a private tutor or, more likely, in a small school. These tall vellum-covered volumes were hardly unique to Dutch classrooms; contemporary examples survive from English, French, Scandinavian, and American sailors. However, Dutch mariners appear to have produced them in larger numbers and adhered more closely to published nautical texts, often citing specific titles and page numbers. At least three dozen examples of these workbooks survive from the Netherlands.[4] These navigation manuscripts are known in English as "cypher books" and in Dutch as "schatkamers"—treasuries—named after two pivotal early manuals, Lucas Waghenaer's *Thresoor* and Cornelis Lastman's *Schat-kamer*[5] (plate 4). Each of these documents tangibly commemorates a practitioner's expertise, often as the only surviving evidence of his life and mind. They are also one of the few windows into a navigator's professional education.

DUTCH NAUTICAL TEXTBOOKS AT THE TURN OF THE EIGHTEENTH CENTURY

Wigardus Winschooten, who published a nautical glossary in 1681, viewed his book as a friendly aide: "Here appears a Paper Sailor, which, having learned a great deal from old and experienced Seamen, is therefore ready to share the same with you."[6] Dutch textbooks developed characteristics quite different from the cosmographical emphasis of their Spanish predecessors but were not yet like those in England inspired by business arithmetic. Dutch schools, too, were different from the larger state institutions favored by the English, French, and Spanish. Both were particularly influential: foreigners, especially from German states and Scandinavia, flocked to Dutch schools, worked through textbooks, and carried their notebooks home with them. These volumes, printed and manuscript, enable us to delve into the curricula and classroom experience.

Like authorities everywhere, Dutch experts debated the optimum dose of theory, coming down resolutely in favor of significant mathematical foundations. How this theory was dispensed to mariners is uncertain, as no student voices survive from the first generation of Dutch navigational classes in the late sixteenth century.

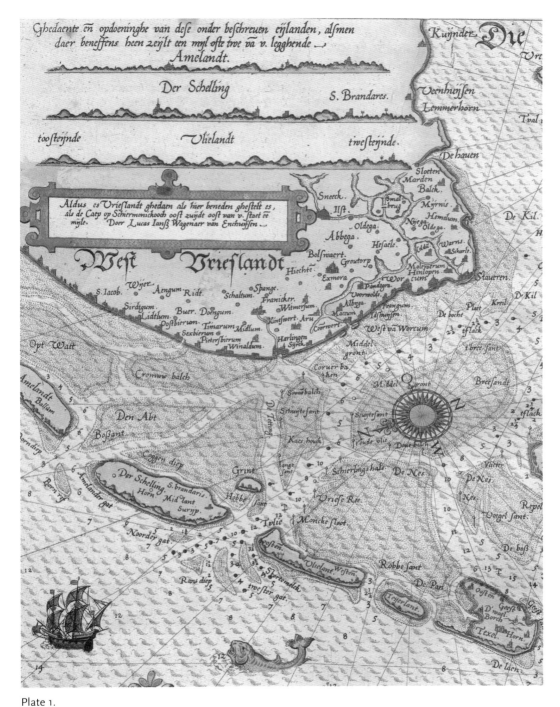

Plate 1.
Lucas Jansz. Waghenaer's new nautical charts are packed with information—showing sandbanks and channels, noting depths, and including small versions of traditional coastal profiles (*upper left*). Further textual directions are printed on the verso. This chart of the Zuyder Zee and the "famous Vlie"—the treacherous estuary of the River IJssel—shows why every Dutch mariner was pushed to master this tricky geography. Waghenaer, *[Teerste deel vande] Spieghel der Zeevaerdt* (Leiden, 1584), 79. Utrecht University Library, MAG : P fol 111 Lk (Rariora)

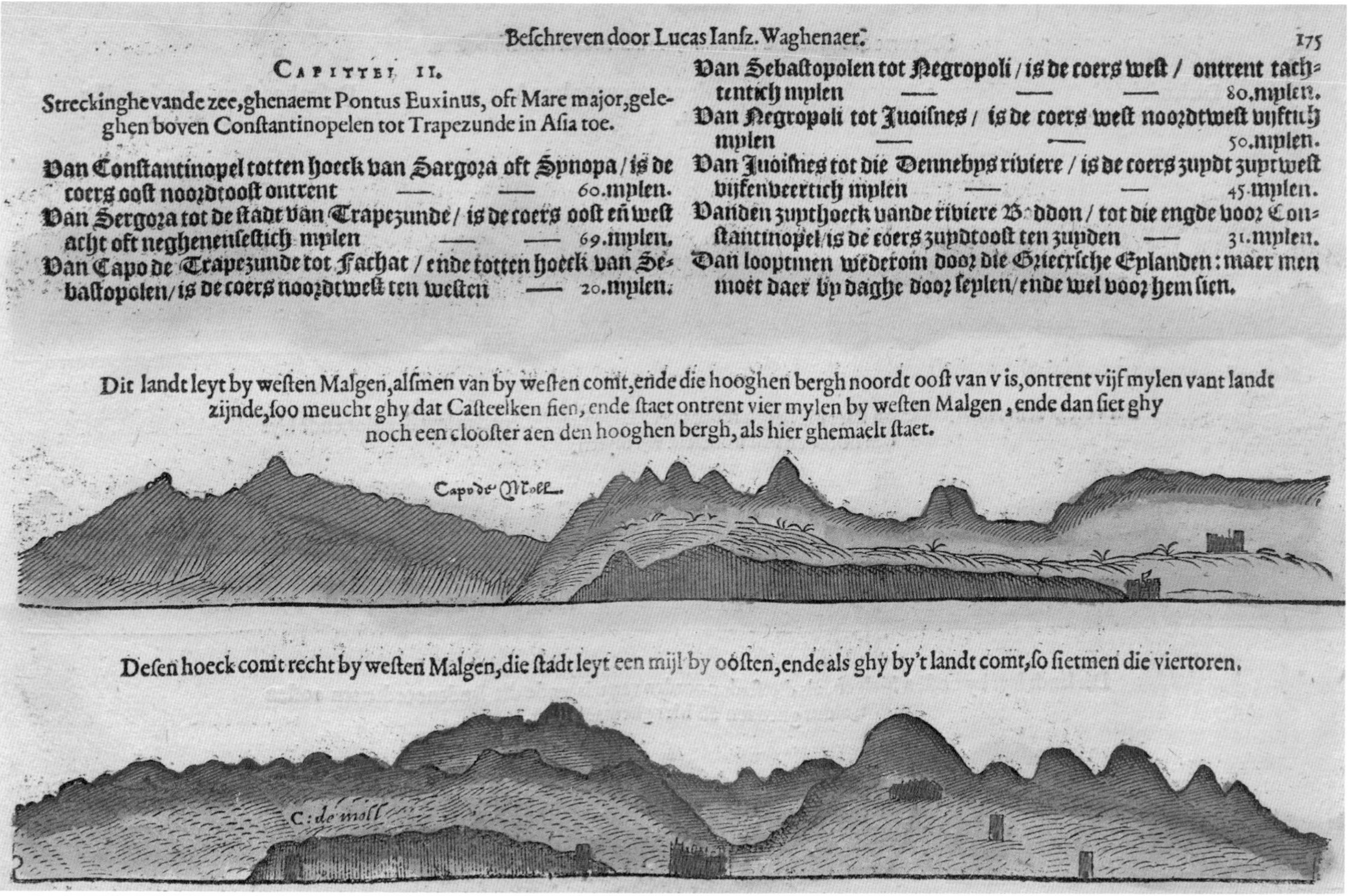

175

Beschreven door Lucas Ianfz. Waghenaer.

Capittel. 11.

Streckinghe vande zee, ghenaemt Pontus Euxinus, oft Mare major, geleghen boven Constantinopelen tot Trapezunde in Asia toe.

Van Constantinopel totten hoeck van Sargoza oft Synopa / is de
coers oost noordtoost ontrent — 60. mylen.
Van Sergoza tot de stadt van Trapezunde / is de coers oost ten west
acht oft negenenveertich mylen — 69. mylen.
Van Capo de Trapezunde tot Fassat / ende totten hoeck van Sebastopolen / is de coers noordtwest ten westen — 20. mylen.
Van Sebastopolen tot Negropoli / is de coers west / ontrent tachtentvijf mylen — 80. mylen.
Van Negropoli tot Iuosines / is de coers west noordtwest westrijsj mylen — 50. mylen.
Van Iuosines tot die Dennebys riviere / is de coers zuydt zuydtwest visfenveertich mylen — 45. mylen.
Van den zuydthoeck vande riviere Bodon / tot die enghe voor Constantinopel / is de coers zuydtoost ten zuyden — 31. mylen.
Van loopinnen wederom door die Gierische Eylanden : maer men moet daer by daghe door seylen / ende wel voor hen sien.

Dit landt leyt by westen Malgen, alsmen van by westen comt, ende die hooghen bergh noordt oost van u is, ontrent vijf mylen vant landt zijnde, soo meucht ghy dat Casteelken sien, ende staet ontrent vier mylen by westen Malgen, ende dan siet ghy noch een Cloofter aen den hooghen bergh, als hier ghemaeckt staet.

Capo de Mar.

Desen hoeck comt recht by westen Malgen, die stadt leyt een mijl by oosten, ende als ghy by 't landt comt, so sietmen die viertoren.

C. de mill.

Plate 2.
Waghenaer, facing resistance from sailors after publishing the *Spieghel*, then produced a smaller, more affordable volume. It contained just eighteen modest maps and returned to the traditional coastal profiles and extensive textual directions of the *roteiros*. Waghenaer, *Thresoor der zee-vaert* (Amsterdam, 1596), 175. Utrecht University Library, MAG : P qu 509 obl (Rariora)

Plate 3.
William Spink likely took his navigational lessons at an institution other than the RMS. He worked through mercantile problems as well as navigational ones. Here he solves a "plain sailing" problem trigonometrically but also notes the solution with Gunter's scale. Spink, Navigation workbook (ca. 1697–1705). © National Maritime Museum, Greenwich, London, NVT/47

Plate 4.
Navigation's "treasure chest" or "armoire de la navigation." Waghenaer, *Thresoor der zee-vaert* (Amsterdam, 1596), frontispiece.
Utrecht University Library, MAG : P qu 509 (Rariora)

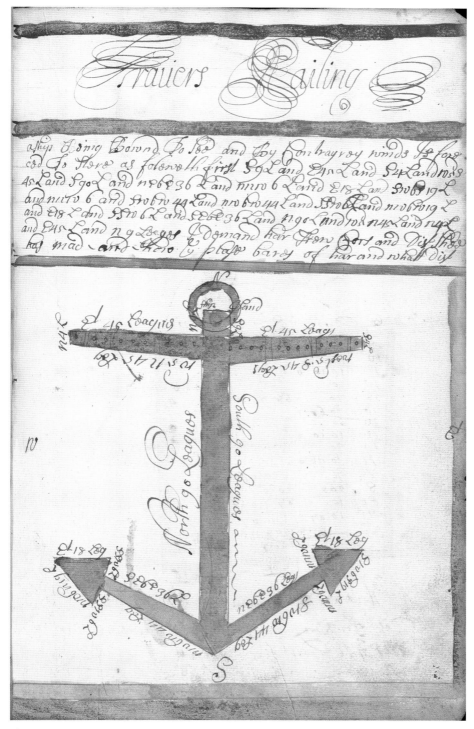

Plate 5.
William Spink enlivened his homework by computing traverse courses in the shape of geometric stars and cloverleafs, hearts and castles. This particular problem, which required him to trace a voyage comprised of twenty-three legs, resulted in a neat anchor. Spink, Navigation workbook (England, 1697–1705). © National Maritime Museum, Greenwich, London, NVT/47

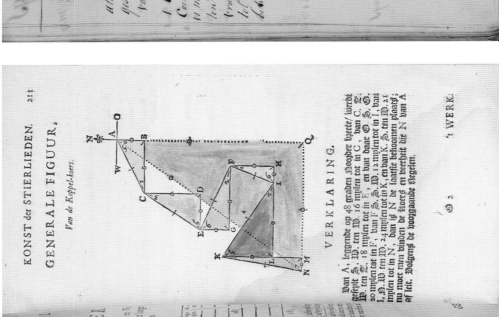

Plate 6.
K. de Vries, *Schat-kamer ofte Konst der Stier-lieden* (Amsterdam, 1702), 211. Het Scheepvaartmuseum, Amsterdam

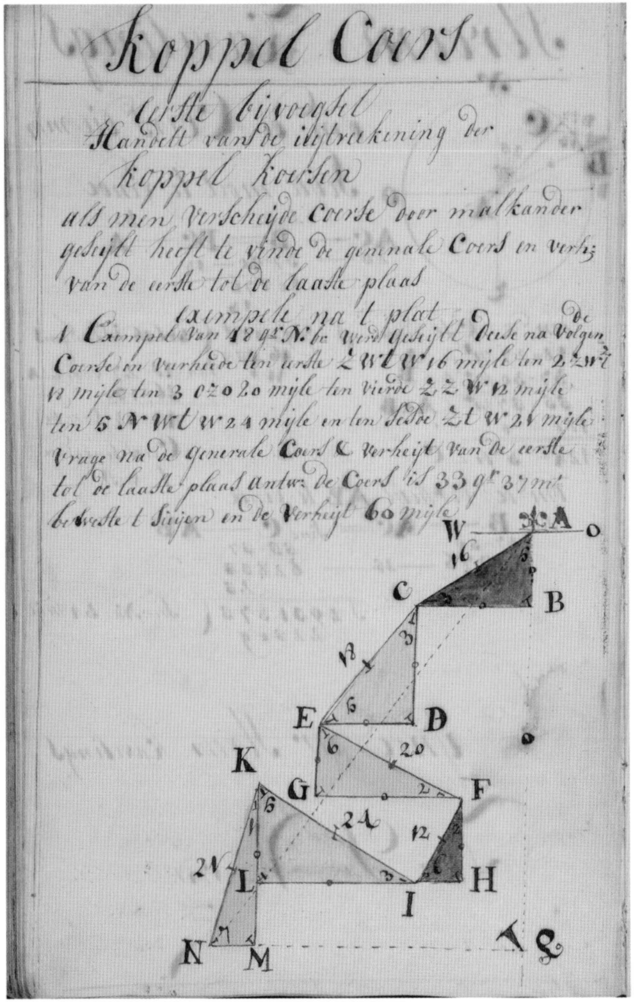

Plate 7.
H. de Vos, *Schatkamer* (Netherlands, 1748–53), based on De Vries. Collection Maritiem Museum, Rotterdam

Readers (or their instructors) augmented a handful of eighteenth-century Dutch printed texts with color and imitated these in manuscript workbooks, probably in hopes of clarifying the diagrams.

Plate 8.
The *Guardian* strikes "a floating Island of Ice." *Distressing Situation of the Guardian Sloop* (London: T. Tegg, 1809).
© National Maritime Museum, Greenwich, London, PAD6024

Plate 9.
Portrait of Midshipman Edward Riou, age 14 (1776), painted by Daniel Gardner.
State Library of NSW, ML 1263

(France and England provide only glimpses into the classroom when certain individuals — like Doublet, Long, or Crockett — get written into the archival record due to irregular behavior.) Dutch students a century or two later, however, produced a wealth of material. By analyzing these manuscripts along with published manuals, it is possible to recover unique details about the classroom experience. By assessing where manuscripts differed from textbooks (as well as voyage logbooks), and how they related to conventional professional exams, we gain a clearer picture of quotidian expectations and routines.

Dutch educators emphasized specific topics: students spent hours working on altitude formulas and analyzing courses both geometrically and trigonometrically. Their curriculum was closely aligned with the professional examinations of the Dutch East India Company (VOC), suggesting that even those men who never took an exam would be exposed to the core topics; in many ways these technical exams shaped the general curriculum and the broader cultural understanding of the science of navigation. Finally, these documents also demonstrate that certain traditional skills could no longer be taken for granted. Authors worked to break down the challenging art of estimating a boat's speed so that students would be familiar with at least the bare rudiments. Thus, the patterns set by nautical experts in the early seventeenth century would last for more than one hundred years. Then, in the mid-eighteenth century we see a shift to a more arithmetical approach to the constellation of navigational skills. This shift will ultimately narrow the audience for this topic, which had held national importance for so long. In the meantime, however, in the heyday of Dutch navigational education — when Wijkman and Visscher labored over their lessons — we can understand why mariners found it worthwhile to pay for lessons.

The textbooks that served as the basis of these navigators' studies were very different from the first generation of Dutch manuals. Many authors of important late seventeenth-century manuals also taught classes and tested candidates for the merchant companies' professional examinations, so they had a clear idea of the competencies that merchant companies sought in their employees and knew where students were most likely to stumble. These authors moved away from the Spanish cosmographical model *and* from Jan van den Broucke's emphasis on memorizing astronomical details and adopted something more like the English nautical business arithmetic. Yet as a group, these books retained certain recognizably Dutch features: a decided emphasis on the tides and an elevated level of mathematics.

Claas Gietermaker, for instance, published the extremely popular *Golden Light of Navigation* (*'t Vergulde Licht*) in 1660, blending traditional aspects with the new. Gietermaker, who spent half a dozen years as an examiner for the VOC, and who also took on the same responsibilities for the Dutch West India Company (WIC), valued comprehensiveness[7] (figure 4.2). He opened his book with elements that would have been familiar to Pedro de Medina and Michiel Coignet: an engraved compass rose faces page 1, and his first chapter treats the calendar, with its mnemonic "rule of thumb." He included two volvelles to help with tide calculations and no fewer than three "zodiac songs." The standard cosmographical definitions did appear but

4.2.
C. Gietermaker's *'t Vergulde[n] Licht* (Amsterdam, 1710) prominently displays an engraved compass rose. (Compare the vignette of the navigation lesson, *bottom right*, with figure 4.4.) Collection Maritiem Museum Rotterdam

not until fifty pages in.⁸ However, alongside this traditional content, Gietermaker also introduced many new features. He much expanded the trigonometric and astronomical tables so that his readers would have them all in one portable volume. He expected his readers to solve problems using trigonometry, providing them with 116 practice questions at the end of the book. He was also the first to include a model exam. The VOC assured Gietermaker's legacy by sending copies of *'t Vergulde Licht* on each of its vessels. As a result, between 1660 and 1774 the textbook saw no fewer than twenty-one editions, plus a French translation and a pirated edition.⁹

A generation later Abraham de Graaf, also an examiner, wrote the *Kleene Schatkamer* (Small Treasury). He placed still less emphasis on things cosmographical. His book comprised five sections: geometry was followed by trigonometry, both plane and spherical. The celestial sphere was relegated to the middle pages, ahead of assorted reference tables—sines, "wassende graden" (meridional parts) for the Mercator projection—and a set of practice questions. De Graaf explained the rationale behind this arrangement: "The first [topic], Geometry, is beneficial in order to understand all the others."¹⁰ As a classroom instructor and an examiner, De Graaf val-

4.3.
Some manuscript workbooks featured elaborate calligraphy. P. R. Komis, Schatkamer based on De Vries (1708–13). Het Scheepvaartmuseum, Amsterdam

ued brevity: he wanted the *Kleene Schatkamer* "to describe only the most important things, and [to have] as few pages as possible, so that one can easily carry it, or can use it as a handbook." To that end, De Graaf sacrificed accuracy and completeness. He omitted logarithm tables and included only every fifth number in another table for determining mid-ocean position, economies he felt would not be detrimental: "Although this description is short, we truly believe that it is clear enough."[11] In each of the maritime communities under discussions, we find adherents to these two opposite approaches to mathematics — comprehensive or selective. And yet, while other countries shared De Graaf's preference for ease of use, the Netherlands favored textbooks and classroom manuscripts that were like Gietermaker's, focusing on the benefits that would come from mastering mathematics.[12]

Some of these manuscripts were created by *liefhebbers*, enthusiasts who found maritime mathematics to be a fascinating mental exercise, a way to gain valuable insights into the technical underpinnings of their commercial enterprise.[13] These men had the education and time to devote to elaborate calligraphy and carefully colored figures (figure 4.3). Other dog-eared volumes are filled with the rough scrawls of men who had little opportunity to improve their literacy; such books were preserved not in estate libraries but by happenstance on merchant ships.

"THE BEST NAVIGATORS ARE ON SHORE": DUTCH NAUTICAL SCHOOLS

The manuscripts that Van Asson, Visscher, Wijkman, and others crammed full of numbers indicate that Dutch navigation instructors ran their classrooms in markedly different fashion from the hands-on approach that Samuel Pepys and others envisioned for aspiring mariners in England. No doubt Dutch sailors did value hands-on training.[14] In their uncontrovertibly nautical society, they had ample opportunities to perfect their manual skills aboard ship. (Even after the English Navigation Acts encroached on Dutch shipping activity, more than 15 percent of the workforce at the beginning of the eighteenth century was employed in maritime trades.)[15] But practitioners knew better than the others that to master the whole of a mariner's entire daily activity, theoretical and practical training must balance.

Dutch authors were not shy about touting theory. Gietermaker "freely confessed" that experience, for all its benefits, was still "rather rough, mysterious, and blind" knowledge. Consequently, it was "useful and necessary for those who make long voyages by sea, [to] join the experience of navigation they have gained by frequenting the sea, to science and the knowledge of it; for these two things should always progress together."[16] A generation later, Govert M. Oostwoud, a private instructor of mathematics in Hoorn, echoed Gietermaker's decree that "experience and art must go together." Even once one mastered the experience-based components—having memorized coastlines, understood how the ship responded in all weathers and currents, and gained the skill to decently estimate its speed—"still a man is not yet ready to sail as a Navigator." He also needed theoretical training, for it was necessary to have a thorough knowledge of altitude and magnetic variation, and the technicalities of charting, "not only for what one estimated but also to correct what one got wrong."[17]

For these authors, there was one natural place to grapple with these concepts: the classroom. Jan van Dam pithily intoned that "Theory can be learned on Land, Practice at Sea."[18] Joost van Breen, the Zeeland examiner, was more specific about this dichotomy. The topics acquired through experience remained those Coignet had itemized in the 1580s: the knowledge of local geography, harbors, hazards to avoid, and the direction and speed of currents in every place. To these Van Breen added tides—the time of high tide at the new Moon and full Moon—which were possible to find in a table (*Gradboek*), "but without experience [they] cannot be followed, at least not without great hesitancy and anxiety." In addition, estimation, speed, and the handling of instruments are "all proven easily on the sea." By contrast, mathematics, conclusions regarding latitude from solar observations, compass variation, and charts were reserved for the classroom.[19] Rhetoric in this vein seems to have been remarkably effective at persuading Dutch mariners in the late seventeenth century to sign up for lessons, lessons quite different from those taught to the blue-coat boys or the *gardes de la marine*.

Unlike neighboring polities, Dutch naval authorities did not rush to develop educational structures to train future officers. For instance, the Admiralty of Amsterdam briefly engaged the cartographer Hessel Gerritsz. as an examiner in 1618 but did not hire another one until Dirk Makreel was appointed in 1672. It did not establish its own school, the Seminary for Navigation (Kweekschool voor de Zeevaart), until 1785.[20] In many respects the institutions of the Dutch maritime world mirrored the decentralized organization of the Dutch state. The republican government of the United Provinces was far from monolithic. The navy was similarly decentralized, comprising separate admiralties in five different provinces. Although these often operated in concert, they had independent jurisdictions and somewhat different concerns and policies.

To a considerable extent, the branches of the navy viewed the large chartered trading companies as convenient nurseries. Indeed, from their founding at the close of the sixteenth century, the VOC and WIC were substantial, quasi-governmental enterprises. These companies were themselves uninterested in formal training facilities. The central administrators seem to have assumed that men could gain adequate seamanship and experience in other maritime occupations — perhaps the herring fishery or the Baltic trade. If truly new to the sea, they would earn their sea legs over the course of risky months-long voyages to distant trading factories. Those who were ambitious and intelligent would seek out individual instruction on the new mathematical techniques and then present themselves for the examinations that would loft them up the career ladder. When in 1743 the Amsterdam chamber of the VOC did attempt to remedy a surge in shipwrecks by opening a school, they soon deemed the institution not worth the expense and withdrew their support within the decade.[21] Around the same time, the VOC ran a school in Batavia for twelve years, while one of the colony's governor generals operated a private academy on the north coast of Java at the close of the eighteenth century. Both institutions aimed to train local men as able seamen rather than to expand the officer ranks.[22]

Although public lectures on navigation were less widespread than in England, in the Netherlands nonmariners enjoyed opportunities to learn about the subject. Between 1600 and 1635, the mathematician Adriaen Metius taught navigation at the University of Franeker in Friesland (northern Netherlands), obtaining special permission to hold class in Dutch rather than Latin.[23] In 1711 the Athenaeum Illustre in Amsterdam began offering public navigation lectures to an educated audience; these Latin presentations continued for more than a century. However, mariners — and a good portion of the general public — turned to vernacular options in large and small ports.

Small independent schools were unquestionably the most common venue for this technical education. Here the Netherlands aligned with England rather than France or Iberia. In Amsterdam at least sixteen schools operated for various periods in the seventeenth century.[24] Given the importance of navigation to the local economy,

4.4.
This quintessential studious scene, on the title page of Christiaen M. Anhaltin's *Slot en Sleutel van de Navigation* (Amsterdam: Doncker, 1659), served as the inspiration for later editions of Gietermaker's *'t Vergulde Licht*. Collection Maritiem Museum Rotterdam

these instructors could charge pupils hefty fees to guide them through the basics or to help them grasp more advanced theory. The entrepreneurial teacher Pieter Holm charged thirty-six florins for six weeks of classes at his home in Amsterdam, four times the going rate for conventional mathematics lessons. Pieter Ruelle promised to teach men "who had experience" in just eight to fourteen days.[25] Whereas navigation teachers in other countries had relatively little social influence, in the Netherlands these instructors were well respected: many men published trend-setting textbooks, served on examining committees for the admiralties or trading companies, and attracted students from elsewhere.[26] This reputation was even more burnished abroad: foreign heads of state recruited Dutch teachers and navigators and bought their books. Famously, Tsar Peter I traveled to Amsterdam for hands-on lessons from local shipwrights and nautical experts.[27]

These Dutch classrooms were intimate rather than institutional. Generally, half a dozen men sat at a modest table, surrounded by a few of the same instruments that had cluttered the handsomely appointed chamber sketched by David Vinckboons a century before (figures 1.1, 4.2, 4.4). In the latter half of the seventeenth century and into the eighteenth, bare-headed youth rubbed elbows with more established men sporting hats and even swords. They might work together with a globe and dividing compass, with a handful of books and atlases on the table and a nearby bookshelf, perhaps under a map hanging on the wall. This model of collaborative and interactive learning made its way into the iconography of numerous nautical texts (figure 4.5).

In the Netherlands, as in England, hands-on instruction was a key to unlocking the theoretical questions of navigation. Many teachers seemed to have owned at least

4.5.
Idealized navigation lesson. Jacob Aertsz. Colom, *De Groote Lichtende ofte Vyerighe Colom* (Amsterdam: J. A. Colom, 1652), title page, detail. Rijksmuseum, Amsterdam

a globe and a cross-staff for demonstrations.[28] (By the mid-eighteenth century, several entrepreneurial men started selling their students sextants and other cutting-edge instruments and then, for a fee, teaching them how to use them.) Instructors occasionally conducted lessons beyond the classroom walls. In Hoorn, an important port town, the celebrated instructor Dirk Rembrantsz. van Nierop trooped along the North Sea with his more intellectual pupils to make observations of comets.[29] In the final decades of the eighteenth century, when the *Kweekschool* opened in Amsterdam, the navy erected a model ship in the courtyard for the ultimate experiential learning: this vessel was sizable enough for the young students to climb the rigging[30] (figure 4.6). Nonetheless, judging from the humble, salt-stained manuals and manuscripts that survive in libraries and archives, the *schatkamers* painstakingly produced by Van Asson and his peers were the most important element of these Dutch classes.

MEMORABLE MATHEMATICS: CLASSROOM TRAVERSE DIAGRAMS

These student treasuries contain two features worthy of note: traverse diagrams and practice exams. The first was an exercise that points to the utility of classroom lessons, whereas the second, a modest set of preprinted questions, had a profound effect on navigation pedagogy and practice. Manuscripts were the quintessential type of "pen in hand" learning. By creating them, students began to visualize concepts — particularly geometric relationships — and memorize formulas. The analytical Dirk Kruik discussed the role of "Figures" in his *Gronden der Navigatie* (1737):

4.6.
In the late eighteenth century, Amsterdam's new Seminary for Navigation had a training ship set up in the courtyard. Lessons and examinations could now take place partially aloft; note the two dozen trainees in the rigging. Anonymous, "Depiction of the Exam at the Training School, Amsterdam" (1784–99). Rijksmuseum, Amsterdam

for him they served both as mnemonics and as an essential step in the process of problem solving. "The figure or illustration of the proposition . . . serves to refresh and strengthen the memory, so as not to lose that idea; rather, the same figure [allows one] to begin and ultimately to complete the solution."[31] According to the evidence in the many extant manuscript workbooks based on *'t Vergulde Licht*, students did find illustrations intrinsically necessary for their trigonometric computations. For instance, when Van Asson and Visscher set out to answer their practice exams, Van

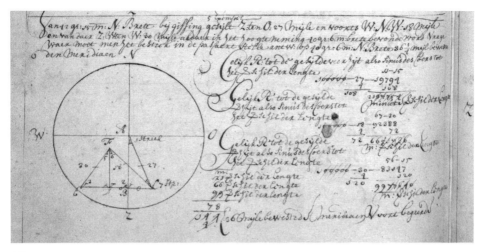

4.7.
Van Asson and Visscher answered the same question from Gietermaker's exam. A. van Asson, Schatkamer (Maassluis?, ca. 1705), f. [45]v. York University Libraries, Clara Thomas Archives and Special Collections, Toronto, ON

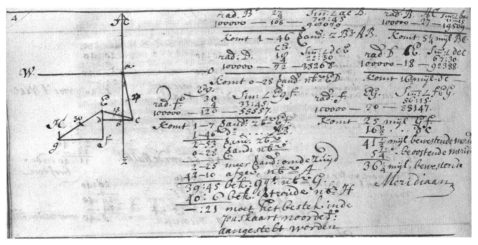

4.8.
C. Visscher's interleaved copy of Gietermaker, 't Vergulde Licht (Amsterdam, 1710), facing p. 151. Collection Maritiem Museum Rotterdam

Asson always used his drawing compasses to circumscribe a circle, a frame for each problem. Visscher, however, saw no need for this (figures 4.7, 4.8). When both men tackled the fourth question on Gietermaker's test, Visscher was the more careful, computing trigonometrically each leg of the vessel's course.[32] Van Asson relied on geometry more than trigonometry and made an error in his diagram that distorted the latitude. Remarkably, despite these different approaches, both men obtained the same answer: "36 miles west" of the original position (Visscher specified: "36 ¼"!).

PAPER SAILORS, CLASSROOM LESSONS ❖ **125**

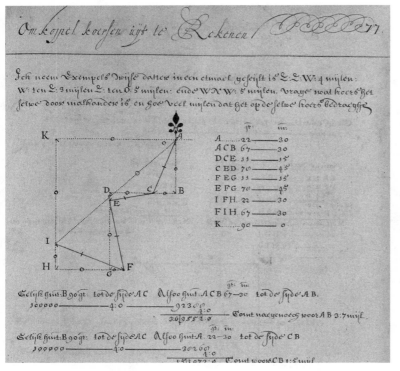

4.9.
To work a basic traverse course, Jan Sleutel drew two diagrams in his workbook. After carefully computing the four legs of the voyage, Sleutel constructed a larger triangle (AIK) that spanned the entire course. J. Sleutel, Schatkamer (Netherlands, ca. 1675), 77. Collection Maritiem Museum Rotterdam

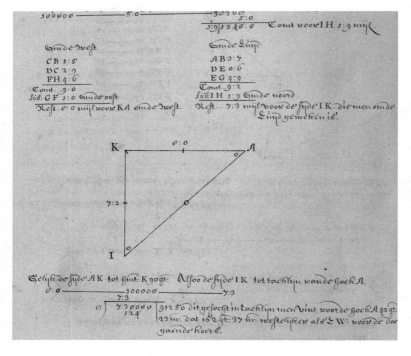

4.10.
On the next page of his workbook, Sleutel recopied triangle AIK, a representation of the entire day's displacement without any distractions caused by wind or current. Rather than computing all the components separately, he could then solve for one simple shape. J. Sleutel, Schatkamer (Netherlands, ca. 1675), 78. Collection Maritiem Museum Rotterdam

Van Asson's more basic geometry worked well enough and saved him a few minutes of tedious calculation.

Student navigators used diagrams to solve a range of problems. Not only could they draw a small circle and a few lines to quickly figure out the right rule for ascertaining the Sun's altitude—a graphic solution dating back to Medina's *Arte de navegar*—but they also turned to them first to resolve another problem, the traverse course. Once a day, at least, the navigator added up the numerous legs from the preceding twenty-four hours to determine a vessel's current position. Navigators were accustomed to tracking their progress on traverse boards, but these pegboards could document only the barest essentials of the course. As expectations increased about the accuracy of the day's records, sailors were taught how to record their motion in geometric terms and then to analyze the resultant shapes with trigonometry. Textbook authors devoted whole chapters to the "traverse course" (*koppel koers*), explaining the fairly straightforward process of converting a day's travel into a form compatible with a tabular logbook.[33] Instructors walked students through the construction of a right-angled triangle that represented the "difference of latitude" and "course" for a single trajectory and then demonstrated how to add together multiple triangles (figures 4.9, 4.10).

Relatively few textbooks included printed traverse diagrams. The earliest published example appeared in John Tapp's *Seamans Kalendar* (London, 1602). They remained scarce through the seventeenth century, with Dutch books featuring them more often than English or French works.[34] Gietermaker's *'t Vergulde Licht* did not contain any traverse diagrams but did include a single sample "koppel koers" in tabular form (figure 4.11). Sleutel, in fact, worked through this "little table," which turns out to produce a double-page diagram of a twelve-leg voyage[35] (figure 4.12). Teachers believed in the pedagogical value of drawing such courses and included set examples in their teaching. Although publishers chose to conserve paper rather than print these sprawling diagrams, students frequently produced them in class. Over the ensuing century, virtually every student who worked through Gietermaker's textbook drew and analyzed traverse courses in his manuscripts. In the 1760s, nearly a century after Sleutel completed his *schatkamer*, an anonymous student produced a very delicately calligraphed version of the same Gietermaker course[36] (figure 4.13). This century-long stability shows how standardized certain aspects of textbooks had become, but it also offers heartening proof of how lessons were taken up in the classroom.

Students regularly pushed beyond the printed text. When it came to traverse courses, teachers took opportunities to make the exercises more interesting and memorable. In certain texts, authors attempted to make their sample questions realistic—by naming real places, even if the distances were completely inaccurate.[37] Others took the entirely opposite approach, devising questions that produced fantastical courses. Far from workmanlike assemblages of triangles, or even complex courses that could plausibly be caused by adverse weather, instructors set up problems where, in the vein of children's "connect the dots" games, the ship often ended

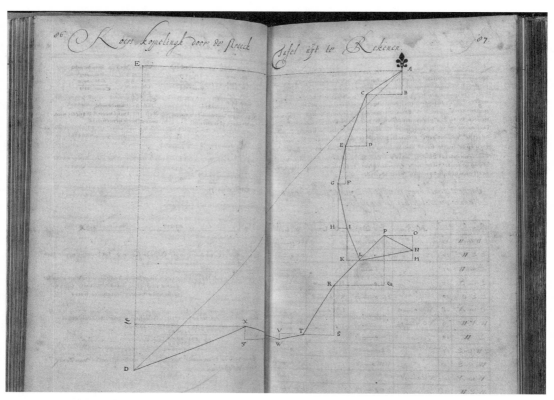

4.11.
"Koppel-koers Tafelken"—the table that served as the basis for so many student diagrams. C. H. Gietermaker, 't Vergulde Licht der Zee-vaert (Amsterdam, 1660), 79. Het Scheepvaartmuseum, Amsterdam

4.12.
Elaborate twelve-leg course based on Gietermaker's "tafelken." J. Sleutel, Schatkamer (Netherlands, ca. 1675), 86–87. Collection Maritiem Museum Rotterdam

up back where it started. Students completed courses in witty shapes: hearts, right-angled spirals, dodecahedrons, fanciful fortifications, and even anchors[38] (plate 5). These quixotic questions still used formulaic textbook language, and in most cases the students treated them seriously, carefully working out every element of the course and transferring the numbers to the adjacent tables.[39]

These fantastical courses shed light on the pedagogical experience of the early modern sailor. Students and instructors alike clearly found something valuable about the process of recreating courses from a series of sailing directions. From a pragmatic perspective, the fantastical questions may have been a way to make formulaic work more engaging. Wit was a potent pedagogical tool.[40] Even more crucially, drawing these diagrams was a way to practice visualizing intricate geometry, a vital skill for men out on the open water who needed to retain a mental picture of their vessel's course. Requiring far more paper than a published textbook could devote (the calculations that accompanied each leg of the traverse typically filled several additional pages), these exercises represent the supplemental teaching that could occur in a classroom but not when a sailor simply read on his own.[41] These inventive traverse

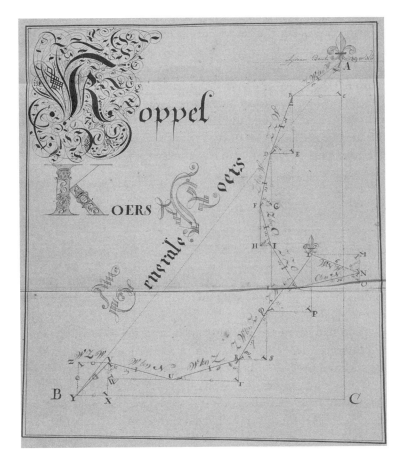

4.13.
The same traverse course, drawn nearly a century later. Schatkamer based on Gietermaker (Netherlands, ca. 1760). Collection Maritiem Museum Rotterdam H629

PAPER SAILORS, CLASSROOM LESSONS ✣ **129**

exercises did not contradict the more traditional means of disseminating knowledge; instead, they made it memorable.

No matter what choices they made in the classroom, few navigators drew out traverse courses at sea.[42] This disparity between classroom exercises and day-to-day practice need not undermine the relevance of textbook lessons. Far from signaling that teachers were out of touch with the realities of ocean navigation, or that their students disliked the graphical method enough to abandon it upon hitting the high seas, this is an instance of a two-stage approach to nautical problems. The graphical method helped students to visualize complex geometry and three-dimensional movement but was never intended to be done aboard ship. The classroom assignments imparted an understanding of the principles behind the numerical solution, at which point navigators could then employ tables or an instrument of their choice.[43]

If over the years Gietermaker's publishers expected students to be able to grasp the complexities of compound courses by reading his dense prose, classroom instructors using 't Vergulde Licht saw the pedagogical advantages of being able to convert words into images. Once distances, proportions, and bearings had been parsed, it took relatively little effort to transfer them into a table and find a numerical answer. Teachers recognized that aspiring sailors, as well as men with practical experience but little mathematical training, benefited from an intermediate visual step in the process of converting narrative—or the real event—to numbers.[44] Like those instructors who used globes and other three-dimensional demonstration aids in class, these teachers were not constrained by the limitations of the printing press (plates 6 and 7). Whether through dialogues or diagrams, practice questions or poems, instructors worked to facilitate comprehension in important ways. In order to persuade students to pay for the lessons, classes had to offer something beyond the basics in the textbook. Dutch sailors in the latter half of the seventeenth century and well into the eighteenth had serious motivation to return to school: they wanted to pass their navigator's examination.

EXAMS: "TO COMFORTABLY TEST THE KNOWLEDGE OF TRAINEES"

Not every man who sat in a navigation classroom and created a *schatkamer* would take a formal examination.[45] Certainly the elite merchants, and even some of the mariners, would have skipped the ritual. But by the first years of the eighteenth century, the navigator's exam had become a key part not just of the professionalizing process but of the educational one.[46] Once the standard components became widely known, they formed the core of classroom lessons. Even if some Dutch mariners somehow evaded these career milestones, they would certainly have been exposed to the test's main topics, which, from the 1660s onward, served as a standardized curriculum in navigational schools throughout the United Provinces.

The earliest record of Dutch navigators' exams dates to 1619, when the Amsterdam chamber of the VOC appointed Cornelis Lastman as an "examinator." He spent more than thirty years in the role, joined by Willem Janszoon Blaeu in 1633.[47] These two men were followed by their sons, keeping lucrative positions in the family. Claas Gietermaker became Joan Blaeu's assistant in 1661, shortly after he published his first two textbooks.[48] The WIC was slower to institute examinations, selecting Gietermaker as its first examiner only in 1663. However, in 1675 it implemented a striking rule: not only must the company's navigators pass an exam, but so must its captains.[49] This disregard for conventional nautical hierarchies, and for the respect traditionally accorded captains, hints that the caliber of the most senior men may have been declining, or at least that they warranted closer monitoring. The men charged with carrying out examinations for the Dutch merchant companies, as well as the branches of the navy, wore many hats. They often worked for several different organizations simultaneously, ensuring that candidates for the merchant companies and local admiralties faced very similar tests. Most are remembered for their textbooks, which seem to have been a prerequisite for taking up the post of examiner.[50]

Claas Gietermaker was the VOC's most influential examiner. Although he held the post for only six years, rather than decades, he introduced a simple feature in his textbook that would noticeably change nautical pedagogy: the model exam. He may have first included these few pages of questions to attract more students to his classes or to convince them to buy his textbook. Surrounded by rivals in the competitive Dutch maritime publishing industry, Gietermaker wanted 't Vergulde Licht to stand out. At the same time, mock exams were a natural combination of two popular pedagogical strategies, namely dialogues and practice questions. Gietermaker's all-important addition appeared toward the end of 't Vergulde Licht: chapter 57 consisted of "questions and answers between a captain and a navigator"—the oral exam that navigators had to pass in order to be promoted within the VOC. Then, in chapter 58, he provided a thirty-six question "Navigation Test, or Exam of Navigators"[51] (figures 4.14, 4.15). In the model of so many Dutch textbooks, Gietermaker began his exam with a question about the tides. Candidates were expected to compute the time of high tide in several key harbors, a multipart question that became a fixture of such tests.[52] Next followed a handful of questions on standard topics: courses and currents, plain charts, the core altitude observations (using the Sun and stars, and including the special case when the Sun was not due east or west), two methods of determining magnetic variation, and a concluding section on Mercator charts.

Whether Gietermaker chose to present familiar material in exam format as a way to capture a greater share of the maritime market or did so in response to students' demands, it must also have seemed to be a productive way to prepare candidates for the official rite of passage. With the slight, three-page "Toetse" (test), Gietermaker effectively codified the format of the written examination for the ensuing century. Readers seem to have clamored for these model exams. In the next fifty years, at least half a dozen textbook authors (many of whom also held examiner appointments)

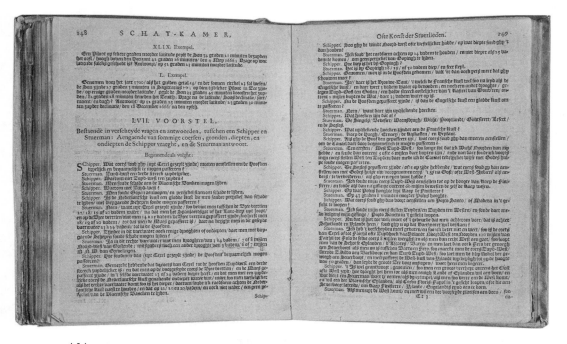

4.14.
The Dutch East India Company oral exam, published for the first time in Gietermaker, *'t Vergul de Licht der Zeevaert* (Amsterdam, 1660), 148–49. Het Scheepvaartmuseum, Amsterdam

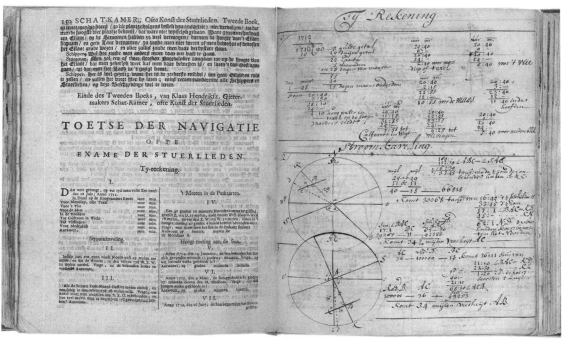

4.15.
C. Visscher answered Gietermaker's written exam in *'t Vergulde Licht der Zeevaert* (Amsterdam, 1710), 150, on a bound-in page of answers. Collection Maritiem Museum Rotterdam

rushed to include similar sets of questions in their manuals. Most chose to include answers so readers could correct their practice test, effectively converting the mock exam to a study guide.[53]

The topics covered by Gietermaker's written exam, and later ones based on his template, echo the competencies tested in Spain and England. Several questions were added on currents and magnetic variation (the Dutch emphasized the latter through the mid-eighteenth century for its potential to solve the longitude problem).[54] These mathematical questions typically required test-takers to derive a single number. As shown by the manuscript *schatkamers* of Wijkman and Van Asson, this process could take as many as a dozen steps and often relied on geometric diagrams to visualize the respective positions of celestial features and the observer. During their exam, navigators seem to have been permitted to consult mathematical or astronomical tables.[55] Certain numerical questions would have required either logarithmic and trigonometric tables or mathematical instruments like Gunter's scale. Similarly, ephemerides, almanacs, and tables of rhumbs would have been essential for many celestial, calendrical, and navigational problems. The goal of these exams, after all, was not for mariners to demonstrate a perfect grasp of mental calculation but to prove that they knew how to work with these tools in their daily problem-solving.

The oral examination that Gietermaker included in his 1660 textbook also made an impact. This was the first published version of the VOC's "Mond Examen" (verbal exam). Only those candidates whose written test demonstrated to the examiner that they were "good in numbers and calculating" could undergo the oral exam. The equipment master would pose the questions in his office, in the company of two members of the company board. As Gietermaker's instructional dialogue makes clear, these exams focused on local geographic knowledge. The nineteen questions "relating to some courses, ocean bottoms, depths and shallows: The *Schipper* asks and the *Stuurman* answers."[56] Once again, numerous textbook authors followed Gietermaker's lead, publishing similar documents, condensed into six or seven questions. Candidates ostensibly needed to know how to sail "from Texel [North Holland] to the East Indies," but were in fact quizzed on the immediately adjacent European waters: Scotland, the English Channel, and south to Spain. Sailors must have obtained this specific geographic knowledge — about the careful maneuvers necessary to get safely out to deeper waters — outside the classroom, for this was another topic about which the textbooks were silent.[57] The test-takers would have gleaned this crucial local knowledge from shipboard interactions, charts provided by the company, and personal experience.

The VOC authorities did not seem to object to their examiners publicizing the specifics of their examinations. If anything, the consistency of form and order of the questions in different books suggests that the VOC itself may have distributed transcriptions of the exam for publication to increase the likelihood of candidates passing. In the 1730s, Pieter Warius, examiner for the Hoorn chamber of the VOC, began publishing blank copies of the written exam, leaving spaces for test-takers to complete their answers (figures 4.16 a and b). Other authors less tightly connected

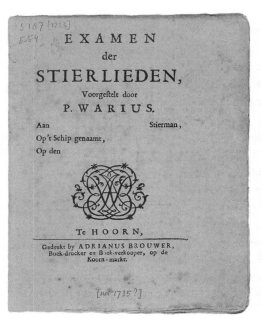 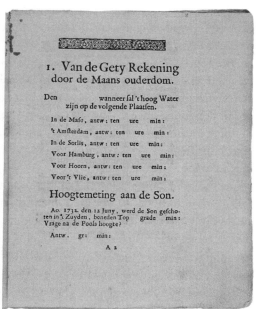

4.16 a and b.
In the eighteenth century, the examination process had grown routine. P. Warius printed up fill-in-the-blank booklets for exam candidates. *Examen der Stierlieden* (Hoorn, 1731). Het Scheepvaartmuseum, Amsterdam

to the companies began publishing creative takes on the exam. For instance, around the year 1701 the minor Amsterdam publisher Jacobus Robyn issued a small volume of tables and calendrical information that somewhat incongruously included an "Examination of Navigators" in dialogue form.[58] The nine-page text is unique within the genre: it is a discursive version of the written exam, where the interlocutors discussed navigational skills rather than local geographic knowledge. A "Jongman" chatted with an examiner about his experience and his aspirations before answering questions on altitude, cosmographical definitions, tides, magnetic variation, and the use of both types of charts.

Robyn's exam placed a premium on comprehension over calculation. For simplicity's sake, the questions used only right angles. When the candidate was quizzed on his knowledge of instruments, the examiner was interested more in how he would manipulate the observed number than in the technicalities of wielding the instrument; the manual side of things is once again assumed to be familiar from frequent use. This fictitious youth recited all the possible cases for each type of problem, deftly demonstrating his understanding of navigation's most important theories. This stands in sharp contrast to Gietermaker's test-takers, who were more likely to be required to calculate specific latitudes rather than to be probed on conceptual underpinnings.

This anomalous exam is perplexing. Dutch navigators had been required to write formal exams since at least 1660. It is difficult to imagine circumstances that would require them to address verbally the traditional topics from the written exam. This may have been intended as a guide for classroom recitations, although Robyn is not known to have been associated with a school. Regardless of whether anyone followed such a script in the exam room, this stylized conversation — where a candidate *talks* his way through the technical details of the *written* exam — illuminates what an experienced Dutch navigator might have been expected to understand and how he could have demonstrated his knowledge. It can arguably also provide insights into the inscrutable halls of Trinity House and the Navy Office on the other side of the Channel. In the Royal Navy's masters' and lieutenants' exams — where Charles Hadsell and others were plied with "moderate questions about navigation," and George Gage grilled about amplitudes and compass variation — English navigators might have responded along the lines of the "Jongman" to verbal questions about observations, instrument use, and other familiar subjects. Robyn's dialogue mirrors not only the topics of Gietermaker's written exam but those on the English lieutenants' certificates as well. However, whereas a Dutch *stuurman* facing a VOC examiner would be given a set of observations and required to compute magnetic variation down to degrees and minutes, using diagrams, logarithmic tables, and trigonometry, the English lieutenant would have faced a different task. Rather than expecting their candidates to carry out trigonometric computations mentally, British Admiralty examiners would have been more likely to ask for a description of *how* to obtain the relevant observations, or to pose questions about common cases, using round numbers, just as Robyn's imaginary examiner did.[59]

There is no question, however, that Dutch examinations resembled Gietermaker's more than Robyn's. Surviving *schatkamers* prove how closely classroom lessons adhered to the textbooks and, in turn, to the published exams. The men we met at the turn of the century — Van Asson, Wijkman, Visscher — diligently worked through extensive lessons, all of which included practice exams. Van Asson and Wijkman started at the beginning, most likely of Gietermaker's textbook; Wijkman even drew the ubiquitous thumb diagram that appeared on page 4 of *'t Vergulde Licht*. Van Asson came to this project already possessing a fine cursive hand and familiarity with standard abbreviations, whereas Wijkman's cruder hand suggests not only a weaker schooling (his spelling was irregular and often phonetic) but also that, if he was indeed Swedish, he had already spent considerable time speaking Dutch.

Van Asson's teacher, like many others, stressed mastering all the possible permutations of observing the Sun's altitude; he completed eighty questions on that topic before moving on to trigonometry. He worked though the standard topics, including currents and plain charts, always guided by the powerful conceptual tools of triangles and circles. Judging by his inked-out answers, Van Asson often made mistakes. Yet over the course of his lessons he gained confidence, progressing to more advanced topics such as Euclidean geometry. He triumphantly pointed out "errors" in the textbooks — but, more often than not, sheepishly reinstated them.[60] Van Asson, or more likely his instructor, also noted when certain questions were better solved geometrically; not every problem needed trigonometry. He was aware of the ramifications of minor errors, at one point articulating precisely the problem that the Corrections aimed to avoid: even a mistake of "one minute more or less in the latitude... can lead to ten to twelve minutes difference in the longitude."[61]

Anders Wijkman, although his *schatkamer* contained almost as many pages as Van Asson's, did not progress to as many advanced questions. He worked through many more traverse courses but spent little time on spherical trigonometry. The questions he answered, typical of contemporary textbooks, are for the most part contrived quandaries involving "Two captains on boats A and B" whose paths invariably cross at a location to be determined. But some evoke more interesting situations: in one question, one pilot had a plain chart and the other a round (Mercator) chart. In another, the navigator "sat two sailors down to their lesson" about taking altitudes from a constellation — a hint that shipboard lessons were common enough to become textbook fodder. (From their two observations, could Wijkman deduce their latitude? With the help of a circular diagram and some simple arithmetic, he could.) Wijkman was clearly a working mariner: he carried his manuscript with him on his next voyage and implemented at least some of his lessons. Several years after Wijkman had finished his course, he took up his book and scratched out one final calculation on the inside back cover — turning to the familiar calendrical formula to figure out the epact and the Moon's age for that day in the spring of 1712.[62]

If Wijkman clearly went to sea, it is less certain whether C. Visscher did. His Gietermaker volume, carefully interleaved with solved practice problems, bore no signs of salt water or even idle doodling. Visscher's script was polished, and he embarked on these lessons already comfortable with basic math. Rather than beginning at the start of the book, he instead took up the geometric constructions in book II. He answered most of the practice exam, although he skipped those questions about compass variation and the Mercator chart. He could well have been bound for an officer's commission, needing to become familiar enough with the formulas to pass his exam but unlikely to navigate any vessel himself.

By including extra practice questions in certain areas, authors and examiners indicated which topics their students should study most thoroughly. Gietermaker emphasized the somewhat less common operations — such as latitude measurements from stars as opposed to the Sun, magnetic variation from one observation rather than two, and charting courses using Mercator charts rather than plain charts. By contrast, the Amsterdam Admiralty examiner Dirk Makreel felt the most familiar ones were sufficient. (Notably, Makreel's exam did *not* begin with the tides; he focused on the polar altitudes.)[63] Should sailors spend extra time on the harder concepts or avoid them in favor of the common ones? Once again, there was no consensus on which approach was best — but the influence of Gietermaker, and his interest in thoroughness rather than shortcuts, may have pushed curricula and examiners to emphasize slightly more complex computations.

Mariners in Dutch cities and towns produced *schatkamers* throughout the eighteenth century, and these continued to contain practice exams. Klaas de Vries gradually displaced Gietermaker from his dominant position (his twenty-five-question exam was popular in *schatkamers* produced in the 1750s).[64] The physical process of taking the written exam evolved as well. Pieter Warius's streamlined fill-in-the-blank booklets suggest less interest in the process of obtaining the answers, just as long as they were correct. And yet over the course of more than a century, the order of the questions remained constant: virtually every Dutch exam began with computing the tides.[65] The emphasis and questions on the exam — which hearkened back to the long-ago manuals of Van den Broucke and Waghenaer, and their particular local concerns of sweeping tides in shallow waters — continued to set the template for the classroom.

These manuscripts further demonstrate that by the turn of the eighteenth century, Dutch practitioners were capable of sophisticated mathematics. Although errors were not uncommon, these men carried out pages of trigonometric calculations, wielding logarithms, geometric constructions, and extended iterations to obtain correct answers. Yet even men from polished backgrounds, like Visscher, did not necessarily venture into the most abstruse problems. Visscher's 1710 edition of Gietermaker concluded with a set of far more advanced questions. Treating cubics and more complex Euclidean problems in a string of cryptic notations, these were "for the speculation of *liefhebbers*"[66] (figure 4.17). Visscher chose not to work through these harder questions. Other *schatkamer* authors did, however, particularly as the century progressed. For instance, Dirk de West, who created a delicately decorated notebook around 1760, skipped the easier questions in favor of the harder.[67] West was part of a growing group of *schatkamer* creators who were exclusively interested in the theoretical aspects of navigation. Rather than being carried to sea, where they would become dog-eared and water stained, these manuscripts would remain pristine, and their authors would never take the opportunity to apply their lessons.

BACK ON THE WATER

Student-mariners were getting more adept at mathematics — or at least at parroting back prepared answers. But there were certain areas where their nautical training lagged. These young men were no longer arriving in a classroom after five or ten years on the water, where they would have learned at their father's or uncle's knee how to tell their compass or what sails a ship should carry in squally weather. As a result, the skills that seemed so intuitive and traditional a century before now had to be described and painstakingly taught. Here we find another reason that these mariners found it worthwhile to pay for lessons.

Take, for example, the question of assessing a boat's speed. Determining how fast a ship was traveling in the open water was up there with the longitude conundrum — one of the most difficult daily competencies to master. The Dutch mathema-

4.17.
Gietermaker's more challenging questions, "for the *liefhebber*," experiment with unusual notations for cube roots and other advanced concepts. Gietermaker, *'t Vergulde Licht der Zeevaert* (Amsterdam, 1710), 113. Collection Maritiem Museum Rotterdam

tician Adrian Metius maintained that it could be neither described nor explained nor taught to anyone. It must be learned by long experience.[68] Guillaume Denys found "the route . . . the most difficult thing to estimate, because there is nothing but prudence and experience that can produce it, and even after all the precautions one can bring to it, one is often wrong." And yet, as we have seen, this was the essential first step toward calculating distance and one's daily position. As Peter Perkins declared to the Royal Mathematical School pupils, "There's hardly any thing more necessary than to be able to make a good Estimate of the Ships way with any wind, according to all Circumstances."[69]

Fortunately, two changes early in the seventeenth century had simplified the process of measuring speed: in 1617 Willebrord Snellius had arrived at a more accurate measure of a degree of the meridian, which set a new value for the Rijnland mile. It also became more common to refer to a vessel's speed in miles per hour rather than fathoms. These changes made it far more straightforward to convert a few seconds' worth of progress into units related to the task at hand. On the basis of Snellius's estimate, mariners made an adjustment to the ropes that they tossed behind their vessels; they began tying them approximately forty feet apart. This had a harmonious result: for each section of the log's knotted rope that slipped overboard, the boat was traveling one nautical mile per hour, or "one knot."[70] By the later seventeenth century, most navigators were firmly convinced, as Perkins was, that when it came to determining a vessel's speed, "the most approved way, and now most followed is by our English Log and Log-line." Navigators would send the log and line overboard every hour or two and mark the speed with a peg at the base of their traverse board. However, it could be tricky to handle the line so that the wooden chip did not veer into the boat's wake, and it could not be used after dark. Furthermore, it often required one man to throw the log and line and oversee its unspooling while a second monitored the sandglass.[71]

Therefore, in the seventeenth century some Dutch authors felt a navigator might sometimes need to determine his vessel's progress without having a log-line handy. However, mariners were no longer accustomed to assessing the appearance of the waves or a passing log and then basing their day's reckoning upon those observations. Writers like Cornelis Lastman, Joost van Breen, and Claas Vooght thus set out to instill this intuitive skill. Lastman was the first to describe this process of careful estimation in print, in 1621. He expected the navigator to be an active observer: "Thus someone that sees daily how far the ship travels through the water, and makes himself an image of that, and remembers that when a ship makes similar progress it will sail so many miles in a day . . . so can he, with time, come to good estimates."[72] A generation later, in his 1662 *Stiermans Gemack,* Van Breen devoted an extensive chapter to the "Observation of Distance" using estimation or a tool.[73] He no longer assumed that his readers would be able to convert their sensory experience into a meaningful number, no matter how carefully they studied the ship's progress.

Van Breen claimed to have an "easy" method, which required just two essential things: a definite measurement for a mile, and a specific time period. The first step for the latter was to calibrate one's internal metronome using a sandglass: "With the help of this little glass, you get used to a certain steady counting, so that every time the glass runs you count the same number. For this, one must at first frequently do and redo it, to get firmly used to it. And having gotten very accustomed to it, it will remain useful all your life."[74] Such metrical counting blurred the line between a skilled human and an instrument. Whereas sandglasses were notoriously inconsistent, humans could train themselves not to lose time. Van Breen was far more specific than Lastman on how to act as one's own timekeeper, from where to stand on deck to how to measure speed at night when it was harder to see the water.[75] He gave concrete numbers: when observing a fleck of foam shooting past the hull, if the navigator counted up to 409, the ship was sailing at one mile an hour, whereas reaching only 102 indicated a faster, four-mile-an-hour clip. He warned that "it can also happen that your 'standard count' could become too slow or fast, but if that happens, you should just go back to practicing with the sandglass and you will learn to stay regular." He also provided a helpful formula should the first hourglass break and its replacement run faster or slower. This was not haphazard, intuitive counting.[76]

Claas Vooght, writing in 1695, also tried to explain how to become an expert estimator. Unlike Van Breen, he provided no exact numbers but instead invoked the language of focused memorization: to gauge the appearance of something intangible, "one must daily observe the speed of the ship through the water, and impress that in one's thoughts and remember how many miles the ship travels in one day."[77] By appropriating the language of popular texts devoted to the memory arts, Vooght highlights the significance ascribed during the early modern period to sight and memory. If traditional coastal navigators had derived much of their professional reputation from their ability to memorize the geographic features and hazards of their sailing routes, the same capacity continued to be essential in the age of transoceanic navigating: a good memory was inescapably twinned with diligent observation.

Just as subtle distortions could derail a ship's course or stretch out a log and line, many factors could subvert a properly timed observation. The idea of an "average day's sail" was crucial yet subject to any number of variables. Only once the navigator knew what kind of progress a particular ship could make in a day, in familiar waters under particular weather conditions, should he sail on other headings and begin to take magnetic variation, current, and drift into account. Lastman recommended that his readers repeat their careful observations several days in a row so that they could average the results.[78] Van Breen was careful to underscore that the navigator needed to modify his mental gauge of progress if he switched to a different type of ship. It was essential to be mindful of the vessel's general construction, the amount of sail she was carrying, how recently she had been careened, and even her national origin.[79]

Even with all these detailed pointers, Dutch authors took a great deal for granted about the content of the mental images and the process of forming them. They did not discuss the appearance of the surface of the water or the sails, for instance, or provide any guidelines about how particular factors affected average speeds; this was all left to the navigator to judge for himself. Still, they were far more specific than texts in other languages. Contemporary French and English manuals contain no comparable strategies for honing one's senses or emulating experts (the French seemed particularly daunted by the "impossibility" of devising rules about how to judge speed). And in truth, despite their respect for traditional methods, most Dutch authors *also* promoted instruments as efficient, simple, and, of course, accurate "helps."[80]

Given how much was at stake for the men at the helm, it is understandable that apprehensive practitioners in the Netherlands and elsewhere would seek out mechanical solutions. And yet, the history of navigation was not a linear trajectory toward more sophisticated technologies; rather, as this example demonstrates, throughout the centuries seagoing experts continued to rely on their perception. As late as the mid-eighteenth century, the entrepreneurial navigation teacher Pieter Holm still preferred flecks of foam, or even the Sun, to the log and line. Holm produced a hybrid solution to the vexing problem of speed assessment, combining traditional techniques with a simple numerical tool. In 1729 he began selling a tobacco box engraved with tables. ("He who carries this box in his pocket," Holm promised, "will never need another almanac")[81] (figure 4.18). In his accompanying textbook, Holm described "sailors teaching themselves to estimate using a pendulum, or the sails and the Sun, or by walking over the deck with a log-line."[82] Having used one of these methods to get a count for a set distance—Holm suggested "74" for marks forty feet apart on the hull—one could simply consult the table on the bottom of the box to obtain the boat's speed. With the assistance of a student, Holm produced these containers until at least the 1790s. This extended popularity is remarkable, as is Holm's expectation that late eighteenth-century sailors would still have a perfectly calibrated mental metronome and even be able to judge speed from the "sails and sun." Unlike the log and line, his boxes did not replace the navigator as arbiter of distance; rather, they respected his expertise while offering a shortcut to let him avoid math.

SEA CHANGES, CA. 1750

After more than a century of classroom instruction, expert teachers and authors finally pushed the field in new directions. Young officers hoping to climb the ranks within the VOC or a similar maritime concern would therefore seek out instructors like Claas Vooght and Pieter Holm. Some men presumably felt that sitting in a classroom would help them comprehend the material: they could ask questions, work through practice problems, look at visual aids, or manipulate physical ones. They may have wanted to study systematically those tasks that had once come effortlessly to seagoing men. Still others may have sought shortcuts to improve their chances of passing their examinations.

Pieter Holm offered an abbreviated course of study for those who wished to prepare for the navigator's exam. Students could take classes at his school—which bore the confident name "Ship Straight through the Sea" (*Schip Recht door zee*).[83] Rather than signing up for six weeks of lessons at thirty-six florins, they could pay a mere six florins if they just wished to cram for the exam. No details sur-

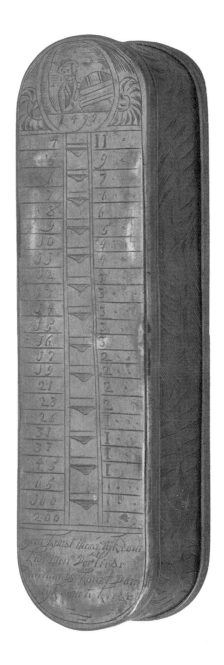

4.18.
Pieter Holm's "Sea Measurer" tobacco box (1754). The bottom of the box depicts a table that allows one to convert the speed of passing flecks of foam into knots. A sailor counts how long it takes a chip to float a distance of 40 Rhineland feet along the side of the boat, then cross-references from the left-hand column to that on the right. For instance, if he counts to 21, his boat is traveling at 2.25 knots (a fraction of a knot appears as a dot). The top of the box (not depicted here) features a perpetual calendar to determine the time of high tide: with the Sunday Letter and the epact known, one could find the Moon's age for any day of the year. Het Scheepvaartmuseum, Amsterdam

vive about these abbreviated sessions, although the men probably worked through only the exam questions. The accoutrements of Holm's lecture room do hint at the exercises that students would be expected to carry out: in addition to "six benches," there were "two desks at which to write [and] three tables at which to determine a position." Over nearly four decades, Holm taught five hundred men in his school. It was not uncommon for several members of one ship's crew to take classes together: in the winter of 1758, a gunner's mate, quartermaster, and third mate all took lessons before embarking on the VOC's *Kronenburg*. The following summer, another trio of officers enrolled together. Although they had little need to buy instruments from Holm—the VOC supplied all their equipment—each officer purchased one of the new octants that Holm was one of the first to market.[84]

In addition to selling instruments and writing materials, Holm naturally sold his own textbook. This 1748 manual, the *Stuurmans Zeemeeter* (Navigator's Sea-Ruler), was almost cryptically succinct, consisting of nothing more than a series of witty poems alongside dozens of practice questions. These students evidently already knew some trigonometry, for there were no introductory lessons here. The first page began with a poem about computing the "Golden Year and the Epact"—though he offers no explanation of how to obtain the epact once one has arrived at the Golden Number.[85] It seems unlikely that anyone not already enrolled in Holm's lessons could have followed the volume. Rather than deeming his textbook a cryptic failure, as some have done, we can in fact view it as a key component in a very successful educational empire. The *Zeemeeter* guided his disciples into his classroom.[86] After they had worked through their lessons, the book could serve as an aide-memoire—or as a review text for essential mathematical formulas. Despite its complete lack of explanatory text (and a tangential appendix focused on theology and chronology), Holm's volume conforms to main outlines of the Dutch textbook genre: it opens with a calendrical calculation and includes extensive practice calculations and the all-important tables (here, sine and rhumb tables). As his prefatory poem attests, Holm also clearly viewed mariners as members of the *liefhebber* community.[87]

This narrow audience—essentially just Holm's own students—indicates that, by the mid-eighteenth century, the Dutch maritime book market was shifting. Nautical publications continued to hold a critical place in Dutch society, even as the republic began to be unseated from its position of international maritime dominance. The market for nautical books had reached its peak by the turn of the eighteenth century. The energetic Dutch market could support more than one publishing house that specialized in maritime imprints. Firms such as Hendrick Doncker's and Johannes van Keulen's joined the fray, the latter eventually replacing the Blaeu family as the leading publisher of atlases and related cartographic works.[88] These publishers satisfied their audience by reprinting earlier popular textbooks with minor variations, updated questions, and new appendices.

In this market dominated by very successful titles, authors had to devise new strategies to attract readers. The moment had passed when teachers or examiners needed to write their own textbook in order to secure their position. Some chose to pen specialized texts rather than comprehensive introductions. Others took up the age-old strategy of critiquing their rivals — for printing errors, poor images, or an illogical approach. In 1755, for instance, Adriaan Teunisz. Veur published a slight work with the Van Keulen firm. To promote his book, he criticized one of the classics of the field, Lastman's *Schat-kamer* (first published in 1621). Veur took issue with how Lastman's first chapters were devoted to plane and spherical trigonometry, "the hardest work for a Student." Veur preferred Gietermaker's arrangement, which "first handled easy things, such as the Golden Number, tide calculations, altitude measurements of Sun and stars, etc., and this proceeds in a good order to the end, where at last the Spherical work is described."[89] (Gietermaker's text, incidentally, had recently been reissued — by Veur's publisher, Johannes van Keulen and sons.) Like De Graaf and many authors outside the Netherlands, Veur felt that navigational texts should be easy; it was more important to be concise than comprehensive.[90]

In the latter half of the eighteenth century, Veur's idea that certain math was too difficult for practitioners gained currency. So did his concerns about the best order in which to study navigation. Although Veur made no great dent in Lastman's reputation or the shape of the genre, another author-examiner soon did. Pybo Steenstra began conducting examinations for the VOC in 1763, and soon thereafter his new textbook replaced Gietermaker's on the *Lijste*.[91] He never ran his own school but lectured at the Athenaeum Illustre. He encouraged his audiences to seek out his small fleet of texts — on astronomy and geometry as well as navigation — along with a volume of tables and a printed version of his exam. He proudly advertised the *Grond-beginzels der Stuurmans-kunst* (Fundamentals of the Navigator's Art) as a clear improvement on the manuals of his predecessors, which he found dated and incomprehensible in places. One of most significant modifications appeared on the first page of the *Grond-beginzels*: for the first time in nearly two hundred years, a major Dutch textbook opened with something other than a lesson on tides. Instead, Steenstra featured logarithms and spherical as well as plane trigonometry. He did begin with definitions, but of an entirely different kind: the symbols for arithmetic (=, +, −, ×) and decimal fractions. He agreed with his predecessors about the importance of globes, treating them before cosmographical definitions. Tides appear much later in the book, with a small "Tafeltje" to help find the Golden Number, but no thumb diagram. Steenstra emphatically promoted mathematics. He viewed right-angled triangles as the cornerstone of the science of sailing: "If you do not understand these triangles, you are working blind"! He offered many examples to help his readers grasp the underlying concepts.[92] He, like many authors before him, was enamored with the power of print, proudly declaring that his book could make the young people who will steer our boats "perfectly experienced." He felt that these youth needed to go to sea (for "at least four or five years") before they should begin studying the intricacies of navigation, but even before going to sea they would need to learn geometry and astronomy.

Steenstra's views — about the relationship among print, math, and the water — confirm how dramatically the idea of technical training had shifted. Books, classrooms,

4.19.
Baron H. A. van Kinckel, Rough draft of *Luytenant's Exam* (Middelburg, May 15, 1767). Nationaal Archief, Den Haag, Adm. coll., 1.01.47.11, inv. nr. 13

and exams had become standard stages of a sailor's training. Men interested in a maritime career — or at least those aiming for the better-paying officer positions — were more likely to have attended school for a few years before going to sea, unlike earlier generations of Spanish or French men for whom navigational lessons were their first experience in the classroom.[93] Literacy rates for the general Dutch population had climbed steeply; so too had basic mathematical education. If would-be sailors came to navigation class without knowing how to estimate a vessel's speed, they could at least read the textbooks and carry out the necessary mathematical calculations. Then, upon completing their lessons, they could go to sea and apply the art they had learned in class.[94]

Navigation classes may initially have been intended to introduce abstract ideas and complex calculations to an audience unused to such things. But as those classes became linked to success in examinations, they increased in popularity.[95] As a side effect of this widening educational stream, people from all echelons of society grew more accustomed to turning to books for numerical information and abstract concepts — and also for technical and practical subjects. When Dutch readers needed

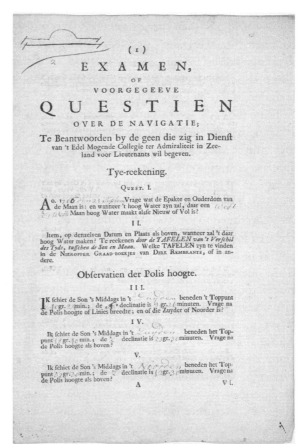

4.20.
Baron H. A. van Kinckel, *Luytenant's Exam* (Middelburg, May 15, 1767). Van Kinckel scored through the rough draft of the exam (Figure 4.19) after he copied the completed answers into the printed test booklet. The German admiral and diplomat also saved his ink-spattered classroom notes, which were based on De Vries's textbook. Nationaal Archief, Den Haag, Adm. coll., 1.01.47.11, inv. nr. 11A

help interpreting these volumes, with their tables, diagrams, and dense paragraphs, they sought an educator's expertise. In this age of print, navigation teachers had many nontextual techniques in their arsenal. From hands-on demonstrations of how to use instruments to subtle means of keeping students engaged in the material, it was these individualized aspects of classroom lessons that justified their cost.

Over the course of nearly two hundred years, nautical print culture, and its audiences, continued to evolve in the Netherlands. Instructors brought their own individual approach to the classroom, offering new interpretations of standard textbooks. While the most popular Dutch textbooks demonstrated remarkable hegemony, and a certain stasis, from the seventeenth into the eighteenth century, the period nonetheless saw significant technical advances. The tasks of small and large navigation had not changed, nor had the navigator's remit: to sail safely and efficiently from one place to another. And yet, in the century and a half since Medina and Cortés had codified the state-of-the-art Iberian knowledge, and men like Haeyen and Waghenaer began to do the same for the Dutch, new techniques and instruments had had their impact on practice. When it came to astronomical observation, the octant (invented in the 1730s) and then the sextant (1750s) would soon replace the cross-staff and the backstaff.[96] In the 1740s Cornelis Douwes, head of the Zeemanscollege in Amsterdam, devised an important new method for deducing latitude from a pair of solar observations. He never published a textbook, but his method would imminently be included in those of others, locally and abroad.[97]

From the early seventeenth century, the Dutch, like many European mariners, had taken up trigonometry enthusiastically. Men quickly invented tools — variants on the logarithmic scale, such as Gunter's sector and the universal quadrant — that took advantage of trigonometric or logarithmic principles to let users avoid onerous computations.[98] However, although these instruments were extremely popular elsewhere, particularly in France, Dutch mariners did not embrace them as exclusively. Rather, they continued to work with trigonometric tables; dozens of editions were issued and reissued for the Dutch market. Almost every standard textbook included these tables, bound in at the back for handy reference[99] (see figure 2.1). And men worked with them fluently, as every student manuscript attests. The emphasis in Dutch schoolrooms on solving problems mathematically — and working with trigonometric tables — seems to have had a lasting impact on daily practice.

By the eighteenth century, junior mariners born in the Netherlands were better educated and more comfortable with mathematics than their peers elsewhere. (Foreigners recruited to the VOC in its waning years, and to the navy during wartime, had lower levels of education.)[100] For quite some time, this higher level of academic preparedness allowed examiners to push students to new standards of precision. However, maritime gatekeepers gradually moved away from relatively advanced mathematics, choosing instead to focus, like Veur, on easier questions, in brief fill-in-the-blank exams similar to Warius's template. In 1767, when Hendrik August van Kinckel wrote his lieutenant's exam in Middelburg, he first worked through a draft (using the familiar diagrams to aid his calculations) and then transferred his answers into a fill-in-the-blank exam booklet[101] (figures 4.19, 4.20). At the same time, as Dutch nautical books were reissued with new appendices, filled with new and more complex practice questions, a subtle shift occurred. The idea that had prevailed a century before — that sailors themselves were intelligent *liefhebbers* who could make valuable contributions to the art of navigation — was fading. In the second half of the eighteenth century, authors stopped including working mariners in the audiences for the supplemental mathematical questions. Practitioners began to be warned away from overly theoretical material; a later edition of Gietermaker's *'t Vergulde Licht* added new formulas that were "not for navigators to put into practice" but only for "enthusiasts to solve."[102]

On the heels of Adriaan Veur's suggestion that sailors could handle only basic mathematics, a crack opened between *liefhebbers* and mariners. In the decades ahead, teachers increasingly wrote introductory manuals not for a general readership but for students alone. This pattern held true beyond the Netherlands, in locales where the status of sailors had always been a good deal lower. Unlike other technical fields that gained social cachet as they became more mathematized, sailors' work would not rise in status as their numeracy and capacity for theory did.[103] In keeping with the later eighteenth-century emphasis on precision and problem solving, Dutch nautical manuals evolved into dense volumes filled with homework problems — at which point the wider audience turned away. The disparate members of the Dutch maritime community gradually lost their shared cultural references, to say nothing of the mutual respect between those who carried out the minutiae of navigating and those who wrote about them. The skillful men who earned their living on the sea were no longer viewed as contributors to the science of sailing.

THE
Unfortunate Voyage
OF THE
Guardian Man of War.

YE sailors of Old England,
   Who boldly plough the main,
In hopes for to get riches,
   And honour for to gain;
Hear of the horrid danger
   The GUARDIAN underwent,
In a trice, on the ice,
   With a shocking force she went.

This ship she sail'd from England,
   A gallant one they say,
With convicts and provisions
   Was bound to Botany Bay;
When many leagues being out at sea,
   On an island she did drive,
Ice it's stock, like a rock,
   Yet they all were sav'd alive.

But happily they did get clear,
   Yet soon, alas! they found,
Thro' leaks the water pour'd in fast,
   When they the well did sound;
All hands to pumps, but still they found
   Endeavours were but vain,
They did stare with despair,
   For the leaks did on them gain.

5.1.
A contemporary ballad relating the *Guardian*'s misadventures: "The Unfortunate Voyage of the Guardian Man of War" (1790). State Library of NSW, Dixson Library drawer item 339

Chapter Five

LIEUTENANT RIOU IS PUT TO THE TEST

The Southern Indian Ocean, 1789

On December 24, 1789, Lieutenant Edward Riou and the crew of the HMS *Guardian* were sailing through rough seas some four hundred leagues south of the Cape of Good Hope when they found themselves under the looming shadow of two huge islands of ice. They had first sighted the icebergs two days earlier and judged them to be at least as long as their vessel and more than thirty feet tall.[1] As their last provisioning stop had been at Cape Town a long month previously, their water supplies were running low. The fifth-rate frigate, which had 120 men aboard, was taking convicts and provisions to the new penal colony at Botany Bay (Australia), and the 25 convicts had just been put on reduced water rations.[2] The young captain Riou sent a small ship's boat to collect some of the sizable chunks of ice floating nearby; however, the ice-clogged water made him nervous, so he set a double watch to warn if the *Guardian* came too close to the "floating islands." His caution was warranted but inadequate as a thick fog enshrouded the vessel. In the early evening the *Guardian* struck one of the icebergs, ripping a hole in the hull below the waterline and destroying the rudder. As Thomas Clements, the ship's master, vividly reported, "the ship shook from stem to stern in so violent a manner that we expected her to part in every joint, or that the shock would bring the over-hanging of the ice down on our heads, and at once bury us in its ruins"[3] (plate 8).

Riou kept admirably calm. He ordered all hands on deck to man the ship's four pumps at once and to jettison as much cargo as possible.[4] Clements's breathless report gives a sense of the immediate aftermath of the crash: at 8:15 p.m., the ship's carpenter reported two feet of water in the hold, "encreasing very fast"; by 9 p.m. the water was up to three and a half feet, but spirits were good. Only an hour later circumstances seemed hopeless, for two of the pumps had broken, and the water had risen to five feet. Fortunately, one of the pumps could be repaired, and the water slowed its ominous rush. At 10:30, less than three hours into the ordeal that would last nearly two months, Riou divided the crew in two to work in half-hour shifts. Careful even in a time of crisis, the "Captain ordered refreshments to each man, taking particular care that the grog should not be made too strong."[5]

The *Guardian* somehow stayed afloat all night in an "immense high sea." As Christmas morning dawned "uncommonly piercing," the crew struggled to "fother" the hull, wrapping first one sail and then another underneath the damaged vessel.[6] At first, the pumps seemed to be winning in the battle against the water in the hold — it decreased to only nineteen inches, but then rose inexorably back to four feet that evening, and up to seven feet on the morning of December 26. At this point, the desperate crew readied the *Guardian*'s four small boats so that at least some men could escape the drifting frigate. One, the launch, was large enough for fifteen, whereas the others — two cutters and a jolly boat — might each have held half that number.[7] In addition to water and provisions, each boat was outfitted with masts, sails, and a compass — a shockingly brief list of equipment for facing the open expanse of the Indian Ocean.[8]

Throughout all of this turmoil, Riou and his officers continued to track the daily observations as best they could. In his account of December 25, he identified a crucial dilemma: "We had now run — leagues from the Cape without having any lunar observation." The lacuna is telling: Riou had no ready number for their recent "distance run." He continued, "I therefore could not so well depend on my longitude: my researches made the nearest [land] somewhat between 50 and 100 leagues to the southward but in what particular direction 'twas impossible to say." To counteract this information gap, he "por[ed] over a chart of the world without being able to discover much, and . . . looked into Cook's *Introduction* for all the information I could get." The following day he guessed at a location of 44°10′ S. Long. 44°25′ E.[9] Riou realized that they were somewhere in the vicinity of the Prince Edward Islands — two remote specks of land which he had fortuitously sailed past with Captain Cook a decade before. Some of the crew wished to head south in that direction, but Riou felt strongly that there was a greater chance of rescue in the shipping lanes off the southern coast of Africa.[10]

The tireless lieutenant, who had injured one hand in the early hours of the disaster and thus could do less physical labor than his men, was "every where encouraging and animating the people as much as possible, for whose safety he shewed a concern that does him the highest honour." As the hour neared for the small ships to depart, Riou entrusted a letter to Clements, with what he believed would be his final words to the Admiralty. In a private speech to Clements and the other officers, Riou admitted that he felt the loss of the ship to be inevitable but staunchly refused to leave the *Guardian*.[11]

At nine o'clock on the evening of December 26, some three dozen souls set out in the inadequate boats, leaving the "scene of misery and distress."[12] A few men, inebriated and convinced of their imminent death, plunged directly into the waves. Only the largest of the vessels would ever be heard from again. Clements and his fellow passengers on this launch endured trying conditions, dividing up their paltry rations while blindfolded so as not to favor any one man, sharing thimbles of rum, trying to catch rain in handkerchiefs, and drinking their own urine. By January 3, 1790, they were near perishing.[13] Clements, who was splitting navigational duties with the *Guardian*'s gunner, reported that the night before they had "traverse[d], in safety, an ocean of such fierce and tremendous seas in different directions as is seldom met with." Imagine their euphoria when "at day-break . . . we discovered a ship at a little distance from us." The *Viscountess of Bentannie*, a French merchant ship, took the survivors aboard about ninety leagues east of Point Natal, and by January 18, 1790, Clements and company made it safely to Table Bay.[14]

What of the *Guardian* herself? Clements could only report that when he left the ship, her state was "hopeless," and "wholly unmanageable." Upon reaching the Cape, Clements forwarded Riou's "farewell letter" on to London. It took several months for the Admiralty to receive the news, but on April 24, those with family and friends aboard the *Guardian* were informed that their loved ones had perished.[15] Incredibly, however, just six days later, on April 29 the London papers published "extremely unexpected and miraculous" news: the *Guardian* had not in fact been lost at sea![16]

By some combination of good fortune and astute navigation, Lieutenant Riou had managed to keep the frigate headed in roughly the right direction for nearly two months. On February 21, Riou and his tattered vessel sailed into sight of Table Bay. How had he achieved this feat in the rough and unpredictable southern Indian Ocean, with no proper way of steering the waterlogged ship? Riou prevailed in part thanks to his skill as a persuasive commander, cajoling and chivvying the remaining crew and convicts to pump and bail for days on end despite failing equipment and flagging spirits. He was also resourceful and inventive, devising various "steering machines." And yet these skills would have been futile without one last essential quality: a deep familiarity with the art of navigating and all that that entailed. Although still young—he was only twenty-six when he received the commission for the *Guardian*—Riou was adept at making both accurate observations and reasonable dead-reckoning estimates under difficult circumstances. He knew enough seamanship to manipulate ropes and sails to bring the ship back around when she veered off course, taking into account variations in weather and ocean conditions. He also understood how to combat the effects of the relentlessly rising water on the crew's morale. Ultimately, Riou drew upon the practical and theoretical knowledge he had accrued over the preceding fourteen years—coupled with more than a little luck—to save sixty men's lives.

For most wrecks, there is no record of the final harrowing days or hours. However, historians are eminently well served in the case of the *Guardian*. The public avidly followed the news, first of the loss and then of the vessel's triumphal return. In addition to newspaper reports, Clements and an anonymous author quickly published narratives.[17] Within the year, there was a theatrical retelling of the disaster at the Sadler's Wells theater, where the stage could be flooded for nautical reenactments, and Lieutenant Riou and the *Guardian* featured in stirring tavern songs well into the nineteenth century[18] (figure 5.1). The clamor for details led Riou to revise his own extensive notes into a thorough account of the voyage, but he seems to have left this unpublished.[19]

Riou was a meticulous recordkeeper—it was a habit that he may well have learned in the navy, and one that certainly made him ideally suited for that career. Riou apparently had no issue with the increasingly onerous expectations that all naval officers keep hour-by-hour records of their voyages, and his family was no less punctilious. In a letter to his mother and sister, Riou asked them to "take great care" if they found one of his old logbooks, "as there are many little nautical remarks in it that may be of service in future!" They did indeed save it for him.[20] After his death at the Battle of Copenhagen at age forty, his relatives preserved many of his personal papers, including letters home, sketchbooks filled with rough doodles, journals from all his voyages, and even the manuscript workbooks from his student years. These rich documents, when analyzed in conjunction with contemporary navigation textbooks, not only tell us about the *Guardian*'s turbulent voyage but shed light on Riou's training and practice. They can help us understand how late eighteenth-century Britons conceived of theoretical training and how in moments of distress they applied many of those lessons. From Riou's rough notes even more than his polished writings, we can discern which of the skills and techniques described and analyzed in the preceding chapters Riou deployed. Was he an early adopter of any of the innovations of the late eighteenth century? What equipment did he have? How did tacit skills substitute for or supplement his classroom learning?

SHIFTING TIDES

Over the course of the eighteenth century a great deal had changed for sailors, including new instruments, revised techniques for solving classic problems, and updated institutions and forms of certification. The sextant and the octant, invented in the 1730s and 1750s, overtook the multitudes of quadrants and sectors to become the preferred tools for observing altitudes.[21] They used mirrors to reflect the image of the celestial object in the same field of view as the horizon, ensuring a more stable observation. The relative certainty of these readings made these new tools worth their higher price tag.

In a renewed effort to solve the perennial problem of longitude, in 1755 the German astronomer Tobias Mayer produced tables that made it possible to determine longitude using the "lunar distance" method.[22] In 1767 Britain's Astronomer Royal Nevil Maskelyne would update these tables in the *Nautical Almanac*. The process of "working a lunar" involved determining, from the ship, two different but simultaneous times — local time at the ship and Greenwich time — and then computing the vessel's longitude from the difference between those. The Sun's position, often obtained by double altitudes, indicated local time. Greenwich time required more computation: first the navigator had to find the altitudes of two different celestial bodies (Moon and Sun, or Moon and a star) and then the angular distance between them. After recalculating these values as if observed from the Earth's center and correcting them for the effects of refraction and parallax, the observer could obtain the equivalent time at Greenwich from the *Nautical Almanac* tables. This process was both complex and time-consuming: not only did the observations need to be repeated several times and averaged, but the computations alone could take the better part of an hour. The Royal Navy and the VOC soon expected all their navigators to be able to compute lunars to determine longitude, but most other mariners continued to rely on dead reckoning. Even with some type of geometric or trigonometric adjustment, they were resigned to a considerable degree of inaccuracy in their positions.[23]

These new approaches enabled — or required — mariners to aspire to more precise standards. Classes and examinations gradually incorporated these new technologies. At the same time, naval institutions were evolving in light of dramatic political and social shifts across Europe. After a series of intermediate reforms, the secretary of the marine, the marquis de Castries, restructured the French navy in 1786, finally moving beyond the Ordonnance of 1681. In the Netherlands, where new regulations had been introduced and then repealed in the 1740s, the VOC seemed to believe its declining fortunes could be reversed with enough attention to regulatory detail.[24] Both of these revised systems bore hallmarks of Enlightenment thoroughness. By contrast, the English — whose naval reputation was ever improving, despite losses to the French during the American Revolution — saw little need to change their formal system. Additional state-funded institutions in maritime centers offered naval trainees new educational opportunities: two state institutions for officers in Seville (1618) and Cádiz (1717), the Royal Naval College for officers at Portsmouth in 1733, the Kweekschool voor de Zeevaart (Seminary for Navigation) in Amsterdam in 1785, and half a dozen more schools in France.[25]

After nearly two centuries of stasis in the realm of navigational evaluation, in the 1780s and 1790s northern European authorities finally altered their expectations of navigators in meaningful ways. The Dutch, who had viewed examinations as a valuable part of their professionalization process for more than a century, published study guides to help prospective candidates prepare for their tests. For the first time since Claas Gietermaker had codified the questions in the 1660s, the order of topics on the written and oral exams changed. While questions pertaining to tides and compass variation continued to appear, VOC *stuurlieden* were expected to be up to speed with the newest instruments — sextants, octants, latitude tables, vernier scales — and methods, like fixing longitude by lunar observation.[26] The updated rules forbade the use of tables to solve longitude questions; the ban suggests that navigators had previously been allowed to consult reference works during their tests. Alongside lofty expectations about precision and frequency, the examiners also pressed the test-takers on their ability to use instruments.[27] This emphasis on manual skills had characterized the new curriculum at the Royal Mathematical School a century earlier and at the Casa de la Contratación a century before that. Rather than primarily assessing his knowledge of charts, the examiner quizzed the navigator about estimating speed, correcting instruments, and precision and could even send the candidate out into the field to make measurements. Examiners now sought out proof that the *stuurlieden* used the correct instruments (Did charts show compass prick-marks? Were any prohibited foreign charts present?) and squandered no resources. The VOC higher-ups also now demanded interim data in the mandatory logbooks, should they later wish to audit any shipboard calculations. Quick estimates no longer sufficed, as navigators were held to new standards of accuracy.[28]

In French exam rooms too, the emphasis was markedly different from the past. Instead of continuing Colbert's original preference for mathematics, aspiring officers were expected to focus first on military concerns and then on practical navigation. Each port had once been the province of local authorities, but the De Castries reforms of 1786 created a traveling corps of examiners to impart new national standards.[29] Three quarters of a century earlier, Isaac Newton had suggested holding rigorous public examinations to improve the reputation of the Royal Mathematical School. While the British did not adopt this model, the French did so from the earliest days of their exams. The candidates would face a room filled not just with maritime experts but with other authority figures such as the mayor and members of the local chamber of commerce. This process was intended to raise the stakes at the exams and thus inspire concerted studying.

The different French priorities are most evident in the examination for the prestigious position of commissioned ensign (*enseigne entretenu*). Bound for the navy rather than merchant careers, ensigns were a select group; in 1786 there were two hundred, and of these, only the top ten performers on a series of two tests would earn a commission. These young officers faced a first test that focused on "practical" matters, including rigging, maneuvers, naval gunnery, and naval tactics. Those who satisfied the examining panel advanced to the second stage of the competition. This time, a professor of hydrography joined the judges, testing the candidates on arithmetic, geometry, algebra, physics (statics and fluid mechanics), and—finally—the theory and practice of navigation.[30] (Like the English lieutenants of the previous century, French officers spent no time learning the local knowledge required for pilotage.) By contrast, those men aiming only for the lower rank of noncommissioned ensign faced modified exams, in which the rudiments of mathematics were covered before the more pertinent military tactics. Thus, at a minimum these candidates were familiar with the trigonometry that had become essential to basic sailing. (Those in noncommissioned positions would have little occasion to use algebra or physics.) Once again, though, this system equated time-served with experience: after two additional years of service, noncommissioned ensigns were automatically promoted to commissioned ensign, with no additional mathematical testing. The day-to-day practice on board ship would presumedly have given them sufficient calculating practice to reach the level of their more academically minded peers.

Whereas the French navy's exams can be characterized by their dual emphasis on military and mathematical prowess, the English navy a century after the time of Pepys and Newton had more prosaic expectations of its officers.[31] In 1791 John Hamilton Moore added a model exam to the ninth edition of his popular textbook *The Practical Navigator* (ed. princ. 1772). While familiar in Dutch publications, in British textbooks such mock exams were uncommon. Moore published a twenty-four-page chapter containing what he heralded as "The Substance of that Examination, every Candidate for a Commission in the Royal Navy, and Officer in the Honorable East-India Company's Service, must pass through, previous to their being appointed." This section in fact combined a series of previously published questions on pilotage, seamanship, naval techniques, and mathematical navigation. While it must have far exceeded the scope of the genuine exams, Moore presented it as a study tool, so that "young Gentlemen belonging to the Sea" might "refresh their Memories, previous to that Examination."[32]

Whereas the seventeenth-century English lieutenants' exam had omitted seamanship and problem solving in preference to a solid grounding in navigational tasks and physical skills, by the latter half of the eighteenth century the emphasis had completely reversed. Now the "Candidate for Commission" was expected to be conversant with the basics of seamanship, from loading cargo to mooring in particular harbors. The second part of the exam focused on working a ship "in all difficult cases that may probably happen." (Typical scenarios involved several stages: "Suppose it blows hard, you cannot carry your Courses, Night coming on and it is likely to blow harder, what will you do?")[33] The *pilotos* at the Casa de la Contratación had been concerned with just this sort of emergency preparedness — although they had not had such handy practice questions available to memorize.

In a departure from the assessments carried out at Trinity House or the Admiralty in the seventeenth century, Moore's exam began with questions that would have been familiar to navigators from more than a century earlier: the candidate was first quizzed on the calendar and the tides — those same components that opened every Dutch textbook. They then moved on to working simple answers about azimuths, for magnetic variation, and zenith angles, a new technique for finding latitude. Standard cosmographical terms were now defined through trigonometric ratios ("Q: How do you find the True Amplitude? A: As the Co-sine of the Latitude : is to the Radius : : so is the Sine of the Sun or Star's Declination : to the Sine of the True Amplitude"). Moore, like many navigational authors, encouraged sailors to learn several methods of handling each situation; he presented tables, trigonometry, and Mercator charts as equally useful ways of calculating a ship's course. He expected his readers to have knowledge of specific Royal Navy routines ("Proceed to unmoor Ship as it is done in the Navy") and emphasized the military uses of many of these actions. His exam included sections on sails and rigging and, for the first time, a narrative description of splicing rope.[34]

Although the relatively brief navigational sections included by Moore are highly mathematized — the Royal Navy had committed ardently to trigonometry and now expected officers to be comfortable with endless streams of sines and logarithms — the focus of the exam was overwhelmingly practical.[35] The officer needed to know the best places to anchor and the correct cables with which to do so. This exam, which tested pilotage and seamanship maneuvers, contrasted sharply with those faced by Charles II's lieutenants a century before, which had centered on classical celestial navigation and manual skills.[36] Virtually the only similarity was the component on rope work. The increasing focus on menial tasks reflects profound changes in how the Royal Navy viewed its officers.

These changes arose in part due to technological developments, but they also reflect an underlying shift in the attitude toward navigational professionals. Whereas authorities could assume that sixteenth-century mariners had certain competencies after spending considerable time at sea, by the close of the eighteenth century these competencies needed to be explicitly tested. Whereas it had once been taken on faith that English masters and Dutch *stuurlieden* quite literally "knew the ropes," now experienced men of all ranks were required to demonstrate quotidian capabilities. Similarly, the time had passed when to earn a commission English officers needed only to reef sails and carry out a few observations. Testimonies from captains were no longer sufficient to vouch for the more diverse pool of naval candidates, so even "gentlemen" were now required to submit copies of their birth certificates.[37] The English also began to insist that their officers be not only mathematically adept but well-grounded in the rudiments of pilotage, an art they had eschewed a century earlier. As for the French, the navy continued to rely on public pressure to instill good performance, turning displays of both naval and mathematical knowledge into landlocked competitions.

Over this same period, textbooks became more technical, addressing students rather than the curious amateurs (*liefhebbers*), and consequently losing much of their general appeal. The eighteenth century also saw the social rift between able seamen and officers grow more stark; the number of "tarpaulins" who worked their way up through the ranks to captaincies dropped sharply.[38] Instead, naval commissions were dependent on one's social connections. Riou sat at the intersection of these trends; while his social status opened doors, it was his mathematical training that allowed him to excel.

EN ROUTE

Edward Riou first went to sea at the age of twelve (plate 9). He was the younger son in a family that had ties to the army.[39] His father, Stephen Riou, whose Huguenot parents were naturalized in England in 1702, was a captain in the Grenadier Guards; his mother, Dorothy Dawson, was the daughter of Major George Dawson of Yorkshire. In the spring of 1774, the year he first enlisted, young Neddy began taking navigation classes with James Paterson and Thomas Adams. It is unclear how long he studied with these men, or whether he was in class with other students. Adams and Paterson have not otherwise been traced, but it is feasible to piece together some details of their lessons. On April 6, 1774, Riou inked the first decorative flourishes on a mathematical manuscript similar to those produced by studious *stuurlieden* across the Channel[40] (figure 5.2). Unlike Dutch teachers, Adams and Paterson focused on mathematical rudiments rather than navigational ones. They based their lessons on *The Elements of Navigation*, a textbook by John Robertson, the mathematical master at the Royal Mathematical School (1748–55) and the Royal Naval Academy at Portsmouth (1755–66).[41]

Robertson felt that his *Elements* made decided improvements over that of "other writers on this subject," who had "for the most part, cop[ied] from one another"; he felt that was "one reason why the art of Navigation has been so little improved within the last 150 years." By contrast, his six-hundred-page, two-volume opus aimed to make "some few improvements, if not in the art itself, at least in the manner of communicating it to learners."[42] Robertson intended his book for a wide readership, from the mathematically adept down to those uncurious souls who wanted only to learn "how to perform a day's work" (see figure E.1). He felt that the largest "class of readers" fell in between these extremes, comprising those "who want to learn both the elements and the art itself by rote, and never trouble themselves about the reason of the rules they work by." If he was dismayed that "most of our mariners" were more interested in rules than in the underlying explanations, he was also realistic, pointing each of the groups to different sections of the book on the basis of their degree of time and interest.[43]

Neddy Riou's lessons began as most English courses did, with the four arithmetical functions. In his first class, Riou also learned about Roman numerals, and then logarithms were introduced soon thereafter. A significant portion of the course was devoted to geometry — planar and solid — and Riou produced several fine perspectival diagrams (figure 5.3). Adams and Paterson then moved to the "Elements of Plane Trigonometry," a section drawn from a different textbook. The five cases of triangles are laid out in tabular form, treated in the same fashion Denys used a century before. At this point Riou's lessons finally touched on a specifically maritime practice. After learning how to solve each type of triangle by "construction" and "computation" — geometry and trigonometry — he copied a passage from Robertson: "There is another method called Instrumental, this method which is performed by the Gunter's scale being in great use at sea it will be proper to treat of it in this Place."[44] (In continuous use for the 150 years since its invention, Gunter's scale remained a simple way to avoid any calculations whatsoever.) Riou tackled only a small portion of Robertson's massive textbook. He continued with his studies, filling a second workbook that focused on more advanced spherical trigonometry, largely drawn from Robertson's book 4, "Of Spherics."[45]

In July 1776, the fourteen-year-old Riou joined Captain Cook's third voyage as a midshipman on the *Discovery*. A contemporary account of the *Guardian* disaster ex-

5.2.
Manuscript workbook from Riou's first navigation lessons (1774). © National Maritime Museum, Greenwich, London, RUSI/ER/3/1

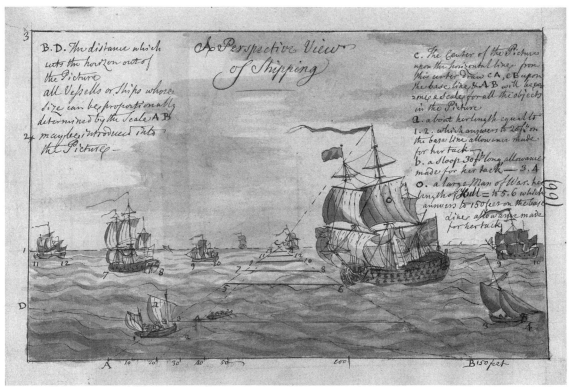

5.3.
"A Perspective View of Shipping." Riou, *Mathematical Lessons* (1774), p. 99. © National Maritime Museum, Greenwich, London, RUSI/ER/3/1

plained that Riou was "bred under Capt. Cook," noting that he "was a great favourite with that renowned Discoverer." Over the next four years, Riou had extensive opportunities to improve his grasp of "the theory and practice of maritime affairs."[46] Cook was an acclaimed navigator. It was a skill he had developed in a variety of settings, from the coal ships of northeast England to the naval fleet in Newfoundland and Nova Scotia during the Seven Years' War. He shared his enthusiasm for the latest theoretical approaches with a high proportion of his officers, many of whom had ample opportunity to apply those theories in remote corners of the globe.[47] On this voyage, Riou made surveys and drew charts, including one of Avacha Bay in Kamchatka, Russia. Two episodes in Cook's career would become particularly relevant for the younger navigator a decade later: early in the course of his third voyage, Cook sailed past the Prince Edward Islands in the southern Indian Ocean, only the third European to do so. During his second voyage (the published account of which Riou had on board the *Guardian*), Cook replenished water supplies from a floating iceberg, deeming that "the most expeditious way of watering I ever met with."[48]

Riou was promoted to lieutenant upon his return to England in October 1780. He

LIEUTENANT RIOU IS PUT TO THE TEST ✦ **159**

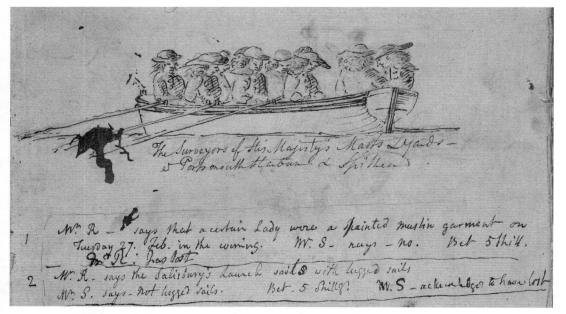

5.4.
"The Surveyors of his Majesty's Masts & Yards at Portsmouth Harbour & Spithead" plus records of various bets—about everything from lugged sails to maiden's dresses—almost all of which "Mr. R" won on account of his powers of observation and memory. *Mediator* and *Salisbury* rough logbook (1786–88). © National Maritime Museum, Greenwich, London, RUSI/ER/2/2

secured commissions in the West Indies, the Channel, and back in Newfoundland, where he was promoted again.[49] His journals from these voyages reveal a conscientious man who passed his days at sea playing backgammon, making minor bets with his fellow officers, sketching elements of the nautical world around him, and keeping his notebooks filled as best he could. Amid pages crammed with numbers, he added snippets of poetry, drawings of various ships, and—during one tedious voyage—a list of "Trifling Misfortunes" that vexed him that winter[50] (figure 5.4).

At the age of twenty-six, Riou received the commission to conduct the *Guardian* to Port Jackson (Sydney). Most of the frigate's guns were stored down in the hold, unmounted to make room for more than one thousand tons of supplies.[51] The *Guardian* was loaded not just with two dozen convicts, but with two years' worth of provisions for the fledgling colony (the voyage itself was expected to take at least six months). The cargo included everything from medical supplies to wheelbarrows, including wine from Tenerife, copious quantities of salted meat, 1,575 pairs of shoes and as many hats. Special pens were constructed for the livestock (horses, sheep, cattle, a variety of poultry, a dozen rabbits, and a pair of exotic deer from Mauritius). A collection of plants from the royal garden at Kew held pride of place. Joseph Banks, the wealthy naturalist on Cook's first voyage who then became president of the Royal

Society, provided careful instructions for their selection and care and drew up plans for a small wooden "plant cabin" constructed on the quarter deck.[52]

The *Guardian* shipped out from Spithead on September 8, 1789. Riou's log and computation book reveal a man with a keen eye and head for numbers (figure 5.5). He carefully followed the best practices of the late eighteenth-century Royal Navy, deploying a variety of instruments to carry out lunar observations. Considering the expedition's close ties to Joseph Banks and the Royal Society, Riou would certainly have had among his books an up-to-date copy of Maskelyne's *Nautical Almanac*. From his notebooks, we know that Riou had a sextant, the standard tool for altitude observations by the 1780s. Riou aimed to observe the Sun at noon—this conventional "meridian observation" involved taking several observations in a row in the late morning and assuming the highest of these was noon.[53] Due to cloud cover or other disruptions, it was not always possible to take readings at equal intervals of time, in which case one might miss the noon apex. To deal with this challenge, Riou also employed a new method of determining latitude. The "double altitude" technique had been developed in the 1740s by the Dutch instructor Cornelis Douwes. It involved making a pair of solar observations at equal times before and after noon. By employing spherical trigonometry, it was no longer necessary to catch the Sun at its zenith. Once these two altitude angles were determined, Riou could check his declination tables. He would need to carry out iterative sequences of approximations, which would eventually converge on his current latitude. On occasions when the horizon was obscured and the *Guardian* becalmed, Riou could use his sextant in conjunction with an artificial horizon.[54]

Riou's manuscripts show that he regularly applied many of the mathematical lessons he had learned in the classroom more than a decade earlier: the volume of "Day's Work" is filled with trigonometric calculations but also straightforward arithmetic. During his provisioning stop in Tenerife, Riou made some rough calculations inside the front cover of his computation book (he worked out how much time the ship's watch was losing, extrapolating from the average day's loss—two hours, thirty-eight minutes, and six seconds—to determine how much time it would lose over the course of a month, as well as in intervals as brief as fifteen minutes). Compared to navigators even half a century before him, Riou was more precise, noting when he corrected for refraction and parallax to obtain a more accurate longitude.[55] Throughout his logbook, Riou carefully named the manufacturers of most of his equipment—as if to acknowledge the individual peccadillos of each hand-crafted instrument.[56] Riou also made notes about the wind, sea birds, and other environmental details, items pertinent for practicing navigators but never discussed in published manuals or classroom notebooks.

To a large extent, Riou treated each day's activities as a scientific enterprise. The voyage of the *Guardian* was not one of the formal "scientific" expeditions that had become common in the eighteenth century. Rather, the questions that her commander pursued were in keeping with the instructions that had been issued to "all sailors" in the *Philosophical Transactions* a century before.[57] Each day he carried out

5.5.
Riou's record of his first visit to Table Bay, includes a coastal profile of Cape Town, noting the "Lion's Rump" and "Table Hill." Riou, Log Book of the *Guardian* (November 23, 1789). State Library of NSW, SAFE/MLMSS 5711/1

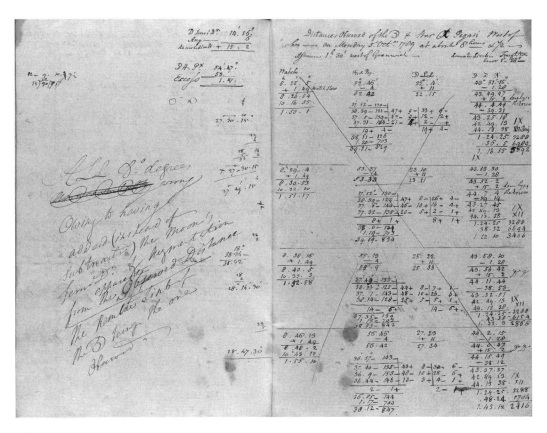

5.6.
On October 5, Riou attempted to carry out a lunar observation just south of Tenerife. Shortly after 8:00 p.m. he used his sextant to measure the height of the star alpha-Pegasi (Markab) above the horizon, followed by the height of the Moon, and then the distance between the star and Moon. Unfortunately, he could make no headway with his figures. Once he realized his error—"All degrees wrong owing to having added (instead of subtracted) the Moon's semi-D[iameter]"—he crossed out the whole page. *Guardian* "Computation book" (October 5, 1789). © National Maritime Museum, Greenwich, London, RUSI/ER/3/8

several careful observations, not all of them specifically nautical. The Admiralty had provided him with "a few barrels of Experimental Gun Powder," while the Victualling Board sent "a Cask & Bag of Bisket . . . each having scattered within it a pound of Whole black peppers as an Experiment to preserve the Bread." He recorded daily thermometer readings and observed a partial lunar eclipse on November 3 — a regrettably cloudy night. When it came to his daily position-finding observations, he might make upward of a dozen at a time in order to average them.[58]

During this portion of the journey, Riou was primarily preoccupied with the ship's watch and its exasperating inability to keep proper time. Riou had secured his "Time Keeper" before leaving London, most likely from Count Bruhl, a German diplomat and amateur astronomer. The Longitude Board—which had only just given John Harrison his long-awaited reward for devising an accurate marine timekeeper that did not lose time at sea—reportedly had none to spare for Riou's voyage. (Marine timekeepers remained expensive, and few had yet been made following Harrison's design. In the nineteenth century, the British Admiralty would equip every ship with three timekeepers to ensure their accuracy.)[59] Riou filled his computation book with pages and pages of rough calculations from the day's altitude observations that aimed to assess the accuracy of the watch. The instrument seems to have been running consistently slow out of Spithead, but then it abruptly sped up on September 26. (He had done these type of time calculations before, following Maskelyne's instructions in the early 1780s.)[60] Due to the complexity and sheer number of the calculations, Riou made occasional errors—adding when he should have subtracted, for instance, or making mistakes in reading his instruments—and was known to strike out entire pages in frustration[61] (figure 5.6).

At several points in the early stages of the *Guardian*'s voyage, Riou credited the young midshipman John Gore with the day's observations. Gore would prove so helpful in the final months of the journey that Riou would promote him to master's mate. This training process, where the most expert navigator provided personal shipboard instruction to promising young men, remained the best way to impart a feel for the equipment. Although Gore was just fourteen years old, he was well connected to the network of English "scientifics" who had gained their maritime experience in the company of Cook's great adventures.[62] All of these painstaking activities—the careful timekeeping, the assessment of various mathematical instruments, training an assistant—were interspersed between occasional religious services and regular rations of rum and tea. Yet few of these quotidian tasks would seem as important after the *Guardian* sailed into the path of the gigantic "mountains of ice" in late December.

AFTER THE CRASH

The future looked dire for the *Guardian* immediately after she was holed—a gale arose, leaving "the main topsail . . . hanging in a thousand ribbons to the yards" and as much as seven feet of water pouring into the hold (figure 5.7). Riou felt it his duty as captain to put on a brave face so that the sixty remaining men would not lose heart, although he personally "entertained not the least prospect of success." He "resolved to grasp at activity to employ their minds and to preserve them from those irregularities and despair which [he] saw were commencing."[63] This strategy seemed to work for Riou himself; as the days and weeks wore on, and the ship's fortunes oscillated between incrementally better and dramatically worse, Riou attempted one technical project after another in hopes of reaching safety.

From the earliest hours of the accident, the young captain had two main concerns: "My whole thoughts were on fothering and steering." The first technique, of using sails as barriers against the in-rush of water through the compromised hull, seemed to show modest success. However, it was very hard to drag the sails into position

5.7.
[Robert Dighton?], *The Guardian Frigate, Commanded by Lieutenant Riou, Surrounded by Islands of Ice in the South Seas, on Which She Struck, 24th December 1789, in Her Passage to Botany Bay, with the Departure of the Crew in the Jolly Boat* (London: for Carington Bowles, September 1790). State Library of NSW, ML 1112 (b)

underneath the vessel. Riou had his crew do this twice, on one occasion reinforcing the sail with blankets, and the next month with a mixture of rags, oakum, and animal dung to help it last longer. (This was another technique that Riou may have learned of from Cook, who resorted to it on his first voyage.)[64] If these were periodic efforts, pumping was constant. Every able hand on board took turns working — and repairing — the four pumps. But the men soon grew less willing to shoulder such Sisyphean labor. Some men drank away their woes, others threatened mutiny. Early on, a number ransacked Riou's private quarters, making off with his uniform, silver buckles, a case of rum.[65] There were also purely physical irritants: there was virtually no dry place on the entire ship. More and more men fell ill, so Banks's "garden house" became an infirmary. The ship was so well provisioned that Riou did not begin to worry about rations until early February. (To ward off scurvy, Riou opened two vats of sauerkraut, and had the men drink malt vinegar wort, another tactic he had picked up from Cook.)[66]

If Riou had his hands full managing the "many melancholy scenes" on his crippled ship, that did not distract him from his second major concern: how to steer the *Guardian*. The iceberg had torn off the rudder, so Riou urgently needed some other means of "getting her to head to the Northward and push onto the tracks of ships." There was little point in fothering the hull or constantly pumping if he could not control the vessel's path. Over the course of the eight weeks, he struggled to create three different steering machines. Riou vaguely remembered reading about some sort of emergency rudder, although he could find no details in his shipboard library. Inspired, he assigned half a dozen men to help the ship's carpenter to rig a makeshift rudder from the spare jib boom and six-foot-long sections of the gangways. The heavy contraption was mounted off the stern to little effect.[67] On New Year's Day, the seaman James Ross suggested a second option. Many years previously, he had sailed home from Newfoundland on a damaged ship that the captain had managed to steer with some kind of cable. Riou devoted several days to devising a similar device, tying short lengths of rope onto a log-line in hopes of trailing it astern and thus controlling the hull. Alas, by January 30, he concluded that the cable had "very little utility . . . in steering." Undaunted, Riou and the ship's carpenter began constructing a third apparatus, an adaptation of "Packenham's substitute rudder" that made use of spare topmasts and an anchor stock — but this too proved unwieldy.[68]

All the while that Riou was frantically working to shepherd the *Guardian* to safer waters, he drew upon myriad tacit skills that he had learned outside the classroom. He was fluent in the impenetrable nautical terminology that stymied so many young officers. He knew how each of the sails could affect the ship's progress, how shifting the cargo fore or aft could make the vessel more responsive in the wind. He estimated distances and speed with equanimity, paid careful attention to the unpredictable weather, and recorded any natural signs — such as "bonitos and flying fish" — that might provide clues about the current position.[69]

Riou never entirely abandoned his more theoretical daily tasks. He evidently retained the ship's full complement of instruments and charts. Even in this time of crisis, Riou would retreat to his quarters to carry out the requisite calculations, although his notes grew more perfunctory and all extraneous scientific observations disappeared. He was rarely able to make observations beyond basic altitude readings — the region's changeable weather and rough seas continued to prevent stable observations of the noon Sun, Moon, and stars. Even on a relatively good day, he "could not suppose [his] longitude to be right. It might be out ten degrees or more [considering] the manner in which we had been sailing about."[70] An error of such magnitude would translate into nearly six hundred nautical miles. His computation book skips from December 7, the date the *Guardian* originally sailed south from Table Bay, until the first week of February, when he was finally able to make two observations.[71] If he might previously have been bothered by such irregular position finding, in these conditions Riou found himself recalibrating his idea of adequate accuracy. On February 7, he exuberantly recalled the past week's favorable course: "We had a wind really like a trade and had been particularly fortunate in keeping her head in the right quarter, i.e. between N and SW." He went on to marvel at the fact that a range of five or six compass points now seemed "very fortunate."[72] Previously, he had aimed for a maximum of two or three points off the desired course, but this misadventure had relaxed his expectations.

In the second week of February, although Riou did not yet realize it, the *Guardian*'s fortunes turned. With the steady trade wind, they were finally heading to the southeast coast of Africa, rather than the less familiar coast of Madagascar.[73] The prospect of approaching land terrified Riou. Only two men on board had ever "doubled the Cape," which was known as a challenging place to navigate due to notorious winter storms and rogue waves. After being buffeted across the open water for two months, they faced "the danger of now falling in with an unknown coast without boats or even without the possibility left of making a raft." This fear spurred Riou back to a standard practice he had evidently neglected: "This day [February 8] I began to mark my track on the chart to see how I approached to what I most dreaded," the coast of "Natalia" (Natal). Concurrently with this geographic uncertainty, the pumps appeared to be once again losing against the rising water, the temporary rudder made scant difference, and Riou feared that the ship would roll over and lose her masts.[74] Although he admitted none of these anxieties in his retrospective narrative, Riou's deteriorating handwriting in his log reveals his precarious physical and mental state[75] (figure 5.8). The events of February 19 were particularly draining: sail stays kept breaking, men stole wine and fell asleep while on watch, and "the people had from negligence let the ship swing round to the wind on the starboard tack. . . . It was another instance of the folly and vice I had to deal with." Riou and Gore were able to make observations, a relief in light of their assumed proximity to land ("latitude was now a very consequential thing to know"), but Riou was "fatigued," and felt he could not trust his crew.[76]

The following day, February 20, Riou's tone brightened: three of the pumps were working, he spotted "gannets in great numbers"—land must be close—and the ship was "steering very well with[in] 4 and 5 points." Although the "dark unsettled weather and rain" hindered his visibility, "luckily [he] got a good meridian observation between the flying clouds and haze."[77] On February 21, a terse note belied his relief: "At 11, saw the land. At noon, Cape of Good Hope." This immensely happy news—that they had reached the one stretch of the southern African coast with viable harbors—required yet another change in navigational tactics. Could he make his way to False Bay? If he was already further west, he needed to avoid the submerged hazards and significant waves around the Cape and aim for Table Bay. Riou turned to another standard tool from the coastal navigator's arsenal: sounding. "I now thought it necessary to sound at all events, and as we were in the track of ships I wished to have every proof of where I was."[78] He armed the sounding lead, filling it with tallow, but determining the bottom proved difficult. He was able to compute a lunar that evening and was suddenly much more confident in his calculations.

The emotional and technical roller coaster continued into the next day: the steering cable was working again ("yet we were never certain of [the accuracy of] our machine") and because Riou was "perfectly unacquainted with the coast, [he] dreaded

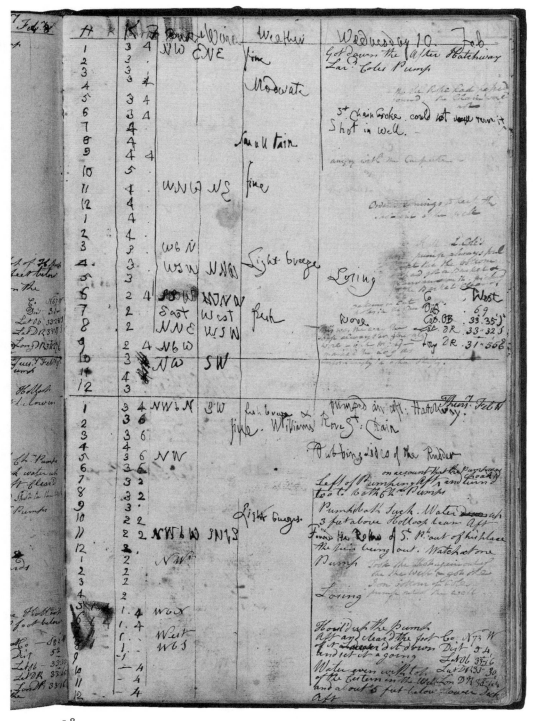

5.8.
Six weeks into the ordeal, Riou's far-from-meticulous records betray mounting strain. Riou, Log Book of the *Guardian* (February 8, 1790). State Library of NSW, SAFE/MLMSS 5711/1

the consequences of having to [deal] with the land in [his] helpless situation." He proceeded with caution, bypassing the more treacherous False Bay in favor of Table Bay, which he had left three months before. A small group of Dutch whale boats approached the *Guardian* and then towed her to safe anchor. With great relief, Riou and the remnants of his crew turned their back on the "deplorable situation of his Majesty's ship."[79]

Although Riou dearly wished to continue on to Australia, the *Guardian* could not be salvaged. Riou arranged for those convicts who had survived the ordeal to be pardoned; fourteen of them went on to enjoy freedom in New South Wales. He sold or forwarded on any stores that remained unspoiled, and returned to Britain in the spring of 1791, bearing with him the *Guardian*'s figurehead.[80] He faced a court martial for having lost his ship but was honorably acquitted and promoted, first to commander and then to captain.[81] Although Riou struggled with ill health, he held several subsequent naval commissions to the West Indies. In the spring of 1801 he sailed to the Baltic with the *Amazon* and there took part in the bloody Battle of Copenhagen under Rear-Admiral Hyde Parker and Vice-Admiral Lord Nelson. On the afternoon of April 2, he was wounded in the head by shrapnel. Unable to stand, he continued to direct the *Amazon*'s gunners from the deck until a cannon ball ended his life.[82]

PRECISION AND JUDGMENT

How does Lieutenant Riou fit into the late eighteenth-century scene of changing navigational expertise? In many respects, he was one of the gentlemen who increasingly filled out the Royal Navy's officer ranks. He enjoyed better than average social status, education, and intellectual abilities. Lieutenant Riou was lionized by contemporaries — both at the conclusion of the *Guardian*'s misadventures and after his untimely death. Contemporaries viewed him as an ideal combination of "skill, conscientiousness, application, and vigour." They praised his "calm intrepidity and presence of mind in facing danger" and his "sound and comprehensive understanding, which he has applied chiefly to the acquisition of professional knowledge."[83] Only Clements, the ship's master who saw him at the height of the crisis, acknowledged his somewhat volatile temper — but even this was a noble quality. And Riou's physical presence matched his moral compass: according to Clements, he was "one of the most elegant men; tall, well-made, with a face of much dignity," while a naval friend from his youth described him as an "interesting and delightful character," noting that "a pleasing gloom hung over his manly countenance.... His eye was particularly striking, beaming with intelligence."[84] In the debate over whether officers, navigators, or mariners must possess certain innate qualities, Riou left little doubt that he was a born leader.

However, Riou shared facets of his career with less elite sailors, the aspiring navigators who also leaped the various professional development hurdles discussed in this book: he went to sea as a youth and studied the standard textbook used at the Royal Mathematical School. He got through the *Guardian*'s crisis by leaning on his own years of experience and on those of his crew. In his desperate efforts to craft a steering machine, he remembered stories of similar circumstances and was able to turn them into three-dimensional solutions. The ship's carpenter and other men with deep experience also contributed. Steering the vessel became a group effort. Riou employed time-honored practices alongside newer tools and techniques. Although still a relatively young man, he had enough experience at sea to have mastered the quintessential skills of estimating speed and distance.[85] And yet even the meticulous Riou could not consistently make observations adequate enough to determine longitude — before the accident, weeks went by when the shifting weather and "flying clouds" obstructed his view and thwarted his best intentions. A navigator must strike a balance between precision and pragmatism.

The late eighteenth century is known as a time when scientists and practitioners of all bents strove toward ever greater precision and ever sharper data.[86] However, on a standard voyage, such exactitude was not always necessary. Astronomical observations could be taken with a backstaff rather than the more exact octant; tides could be calculated according to general guidelines rather than the most recent tables; and a vessel's route could be aggregated rather than charting each individual leg — as Riou did with his handwritten traverse tables. Sometimes the difference between the theoretically ideal method and an approximation was virtually imperceptible, and the challenging calculations or maneuvers canceled out any marginal benefits. Thus, the final decades of the eighteenth century tempered the old admiration for navigational rules. After a century of seeking out techniques that could be codified, simplified, and rolled out to larger groups, maritime communities were realizing that — for anything more than basic competence — authorities needed to find a way to impart just those ineluctable qualities that saved Lieutenant Riou and his ship and men: "sound and comprehensive understanding" and "presence of mind."

The French author V. F. N. Dulague made this move from rules and rote learning. In his *Leçons de navigation* (Rouen, 1768), he revisited the so-called Corrections, that trio of well-established rules, and argued for pragmatism and expert judgment. Dulague's approach is notable, given the prevailing French drive to simplify practices. Like Jarichs van der Ley 150 years before, he was concerned about intuition and guessing at sea. Despite the efforts to systematize navigation and improve longitude readings, Dulague felt that "the daily estimates of a vessel's route [were] founded on nothing but conjectures made with the help of a large number of measurements, all subject to more or less considerable errors."[87] Rather than seek out more rules or better tables, a more solid mathematical foundation or more precise instruments, Dulague offered a different solution. He sought far more active participation on the part of the navigator, calling for the pilot's "continuous attention, to render these errors as small as possible, an extreme vigilance to note them and remedy them at the earliest point; exquisite discernment or judgment clarified by deep theory about the art, and long experience to perceive the effects of those [mistakes] that one had noticed without being able to avoid."[88] Unlike earlier French authors such as Père Fournier, who was not sure whether practitioners could be trusted to share their knowledge without distorting it, Dulague respected these men, recognizing the theory that underlay their actions. In Dulague's updated system, the Corrections were no longer simply applied automatically according to the direction of the course but were chosen by the navigator after reviewing a host of factors: "The pilot should at least be able to decide on which side the error could have occurred, whether it was the fault of the rhumb or of the distance."[89] Whereas Jan Hendrik Jarichs van der Ley and Guillaume Denys had recommended that navigators follow a clear set of rules, Dulague required a navigator to possess intuition as well as training. In the era of rationality, rules were insufficient; the expert was needed more than ever.

To offset these new demands, Dulague accepted a degree of imprecision. In certain other areas — which he deemed less urgent than longitude — the "best practices" took far too much time, with little payoff. As traditional methods of tide calculation could produce as much as an entire day's discrepancy, Dulague felt that tables would "give much more precision." He found the typical deviation of fifteen or twenty minutes acceptable ("one could have pushed the exactitude of this number down to 7 or 8 minutes, but that would require overly complicated calculation").[90] He never recommended skipping observations — the best way to avoid error was to compute the position diligently every day — but he did feel that it made sense to simplify route calculations.[91]

Lieutenant Riou embodies this moderate approach, where discernment meets routine. In his formal education he learned the cutting-edge techniques — but in day-to-day practice he did not always implement them. Riou would have fit in well with the Dutch merchant companies, where navigators were expected to record minute observations in fastidious logbooks and to manipulate the latest instruments under the gaze of ever-more-exacting examiners. The examination process in England was radically different — with cosmographical definitions that were ubiquitous two hundred years earlier and a focus on local geographic knowledge and on preparing for disaster. In John Hamilton Moore's model exam, there were even questions about the exact predicaments in which Riou found himself, from losing one's rudder to fothering a leaky hull with a spare topsail. The 1791 edition of *Elements of Navigation*, the first to include this exam, saw publication just after Riou returned to London — one wonders if he was one of the "experienced Naval Officers" who weighed in on the questions.[92] In fact, it is more likely that many navigators would have known how to fother a hull and craft a replacement rudder. Although Riou was unusually lucky to survive on the *Guardian,* he was not at all unusual in his mathematical and practical knowledge. He, and any number of experienced navigators across Europe, fall not into the mathematical master John Robertson's third group of uncurious readers but into the smaller but critical first and second types of readers: "Those who have made a proficiency in the maths," and "those learners, who are desirous of being instructed in the art of Navigation in a scientific manner, and would chuse to see the reason of the several steps they must take to acquire it."[93] It was these intellectually curious navigators, willing to commit the rules to memory, and yet also ready to move beyond them, who were best prepared to sail into any situation. They, like Riou, would need flexibility and ingenuity, calm and reasoned responses, good judgment — and also careful computations — to carry the day.

THE ELEMENTS OF NAVIGATION.

BOOK IX.
OF DAYS WORKS.

1. BY Days works are here meant the practical methods of finding a ship's place every day at noon, and settling the course and distance she has then to sail.

The navigator requires for this purpose the following elements, or necessary things, to be known.
   I. The means of measuring and correcting the run of the ship.
   II. The nature and use of the sea and azimuth compasses.
   III. How to find the Sun's amplitude and azimuth.
   IV. How to find the variation of the compass, and to correct for it the courses the ship has sailed.
   V. How to find the ship's lee-way, and thence to correct the courses.
   VI. The nature and use of the instruments proper for observing the altitudes and distances of celestial objects.
   VII. How to find the latitude and longitude by celestial observations.
   VIII. How to correct the daily account of latitude and longitude.

Epilogue
SAILING BY THE BOOK, CA. 1800

In the summer of 1774, a group of thirty-one young French mariners embarked upon two modest naval vessels: the *Hirondelle* (Swallow) and the *Espiègle* (Alert). These boats, the French navy's first "floating school," plied the coast between Le Havre and Dunkirk for four months. Unfortunately, at least half the marines in the cohort fell ill during their time aboard, one died, and almost all got seasick.[1] A generation later, in the wake of the French Revolution, a new group of naval administrators was again convinced of the importance of spending time on the water. In 1800, J. B. P. Willaumez, an officer in the post-Revolutionary French navy, assessed the process of overhauling the fleet and training new navigators at the dawn of the nineteenth century. Willaumez had a creative plan to give trainees

E.1.
Shipboard instruments, old and new, and an updated list of an eighteenth-century navigator's essential tasks, the "Day's Works." The vignette shows (*from left to right*) a logboard for recording the details of the course sailed on every watch; a man with an elegant hat—most likely the master—transferring those details into a logbook; one mariner smoking a pipe with a traditional backstaff (Davis quadrant) under his arm; and another taking an altitude reading with a modern octant. J. Robertson, *Elements of Navigation* (London, 1780), 504. John Hay Library, Brown University

hands-on experience; rather than send them out to sea, he suggested mooring a training vessel on the Seine, in the heart of Paris.[2] If the lessons on the *Hirondelle* and *Espiègle* had mixed success, and Willaumez's boat was never launched in the Seine, these late eighteenth-century naval administrators were firmly convinced of the benefits of hands-on training. After all, as another French naval educator noted in 1761, "it is not in a [class]room that one learns the craft of the sea."[3]

This position is ironic, for by the end of the eighteenth century, virtually every navigator did indeed spend time reading books in a classroom. From Pedro de Medina's 1545 proposal to King Philip II that books and lessons could prevent shipwrecks, to the Royal Mathematical School's curriculum in the 1680s that required its boys spend at least eight years studying before even setting foot on a boat, it was ever more common for sailors to be introduced to definitions and theoretical foundations — of cosmography, geometry, trigonometry. Examinations, too, became an established part of navigational training. Across Europe, the navies and merchant companies valued the ability to calculate quickly positions, times, and courses. In the quest to smooth the path to expertise, maritime communities across Europe concluded that schoolroom lessons were essential. What for sixteenth-century mariners had been an eccentric detour was now de rigueur.

Over the early modern period schools had evolved significantly. Unlike the classes in private homes that had been the original model in England, France and the Netherlands, by the second half of the eighteenth century most classes took place in large, state-funded institutions. While literacy rates were climbing, and more of society had education, these navigational schools moved away from the resolutely intellectual curricula that had guided the English and French institutions around the year 1700. Over the centuries, the incoming students had changed: where schools had once tried to teach theory to fishermen or merchants' sons, already familiar with seafaring, they now more often taught raw recruits as unused to the deck as to the desk. These youth, perhaps drafted during wartime, or from an inland community, tended to be less capable or willing to tackle complex mathematics.[4] Thus, late eighteenth-century instructors were expected to do far more than just teach mathematics or how to sail between two points. They worked to enforce minimum standards and to ensure that enough new candidates passed their exams. They were also responsible for the discipline and character of their pupils. (During the same period instructors themselves became more closely regulated.)[5] At least one French professor of hydrography viewed himself as defending order — and squelching charlatanism — in a science "of the greatest importance."[6] From the early days, administrators had debated how much nautical experience educators required, ruling time and again in favor of university men who had spent little if any time at sea. Such choices had a lasting impact on how navigation was taught — for those who favored theory over practice turned naturally to lectures, notebooks, and practice problems.

Each of these countries faced political and economic pressures that affected their navies and merchant fleets. The eighteenth century, marked by mercantilist policies that championed trade, was also an age of many wars, often waged at sea. From the early eighteenth century, the Dutch Republic declined from its "golden age." It struggled to man its naval ships, and the merchant companies faced declining profits and had to recruit from further away. Yet when it came to professional benchmarks, expectations remained high: in the examinations and model logbooks, the administrators demanded an exactitude that had little utility at sea.[7] Across the Channel, the opposite was true. In spite of its cerebral curricula, designed by Flamsteed and Newton, the Royal Mathematical School produced very few naval officers. Academies for officers resisted the theory-heavy curriculum designed for the RMS boys; meanwhile, the lieutenants-in-training gained a reputation for "indiscipline and vice."[8] However, the Royal Navy had played a decisive part in England's victory over France in the Seven Years' War — so it felt largely free to continue with the program that had brought success. The French made the most dramatic shift in strategy. Even before their loss in the Seven Years' War, the navy saw signs that their trainees needed more practical experience. This observation led to educational reforms in the 1780s.[9] One early nineteenth-century proposal for a *collège de la marine* at Port-Louis emphasized the manual skills that had been completely overlooked by the *Ordonnances* of the 1680s. For instance, students were expected to study "the diverse types of ropes, knots, and strong strops, and mooring cables used in the navy," and in a special room in the college they would also learn "the manner of tying knots" and how to make rope. The proposal was among the first to include swimming lessons.[10] This move away from theory matched the navy's decision to launch the *Hirondelle* and *Espiègle*, with their queasy crews.

In some ways, this movement in the late eighteenth century was just another swing of the pedagogical pendulum, from theory back toward experience. There had been other times when mariners were expected to study on board. Besides the time-honored informal study at sea with more experienced crewmen (encouraged by Waghenaer), again and again, over the past two centuries, marines had advocated for shipboard training, from the Dutch lessons copied by Colbert to the Royal Navy's eighteenth-century cadre of certified instructors. Countless maritime documents prove that there had never been the least doubt about the need for hands-on training.[11] Many viewed it as the only way to learn the nuances of small navigation, while others averred that it taught better understanding of instruments or how a boat might handle in assorted weathers. Thus, the debate was in fact not between theory and practice but over their best balance and good order. Should practical training precede the classroom or follow? What theory was too hard to grasp with no experience, and what things could one learn only at sea?

Nautical experts perennially debated the value and minimum viable amount of shipboard experience (twenty days was too short for the mathematical master Edward Paget; nine months seemed sufficient for French naval officers; but English officers needed at least three years). They also tried to determine how much theory a naval man should have: Could he learn enough in one winter, six weeks, or even fourteen days?[12] Some felt that these new mathematical calculations slowed mariners down. In the 1790s, the Dutch naval captain Jan Stavorinus asserted that the VOC's calculation-intensive rules "impeded and fettered" the commanders of Dutch ships "in the improvement of navigation" so that "the English, the French, and others, so far outstrip us in the making improvements, new discoveries, &c."[13] Even on shore there was growing recognition that such extensive calculations could detract from navigators' performance. The French admiralty began to discourage instructors from teaching unnecessary theory. In the 1750s, the teacher at the *collège* in Croisic drew criticism for his "geometric stubbornness." He was spending time — and his employers' money — trying to teach algebra. (So high were his standards and so slow his pace that in three years he sent only two *pilotes* to exams.)[14]

However, by the eighteenth century, there was no question that navigators must know at least some theory. The emphasis had changed since Cortés's and Coignet's instrumental-cosmographical framework. Authors were now much more likely to identify mathematics, rather than astronomy, as the central theoretical component of the "science of navigation." The French mathematician Étienne Bezout reminded his readers "that almost all the methods in use in Navigation were founded on mathematical understanding."[15] While he held a healthy respect for hands-on experience, the VOC examiner Jan de Boer was equally clear: "Although Practice is the true basis for great navigation, nonetheless I would dearly like to state that the Navigator's Arithmetic is also necessary." When devising an educational program for trainees, he felt that "all Seamen should practice [both Practice and Arithmetic] in order to advance more and more."[16]

It was not that mathematics had somehow grown more relevant to navigational theory by the close of the eighteenth century; maritime practitioners had acknowledged its significance in the sixteenth century. Instead, navigational practice had evolved. Many more of the day's tasks required facility with numbers. Coastal pilots — small navigators — had long counted out the age of the Moon on their knuckles and divided the year by nineteen to obtain the Golden Number, kept time as they determined their boat's speed and then multiplied by the distance between two landmarks to quantify an average day's sail, to say nothing of settling personal bets and calculating their share of whatever goods they may have been trading in. Celestial navigation, however, required dozens more computations. Not only were navigators expected to write ever-more-substantial calculations, as the requirements for logbooks and other shipboard records stiffened, but these daily records demanded numerical fluency: navigators had to measure the length and direction of the day's course and compute the angle between the Moon and another celestial body. They had to take into account the variation in compass directions, reconcile discrepancies in positional estimates, and reckon a course to a third of a mile.[17] These newer tasks were anchored not by concrete metaphors like the clock or the human body. Instead, they were abstractions: trigonometric functions defined triangular wedges of various dimensions that fit on a globe or extended over one's head. By the close of the eighteenth century, navigators had added new instruments to the traditional ones — a sextant with a vernier scale, a marine timekeeper, still more extensive tables[18] (figure E.1). In addition to this daily numeracy, navigators had to prepare more assiduously for their professional examinations, completing mathematical calculations on the spot. Ultimately, as a French instructor phrased it in his midcentury textbook, *Les Principes de la Navigation*, "A good navigator possesses theory and practice equally."[19] Aspiring mariners now strove to learn not only from their more experienced peers but from teachers—in school and on board—and from books.

Wars and commerce as well as national policies and educational standards all shaped the maritime world—and the training of navigators. Yet, as we can see from this tour of bookshop, classroom, and ship's deck, one cultural phenomenon had an outsize impact on the practice of navigation in early modern Europe: the printed book. The nautical book market and the navigational manual had evolved greatly since Cornelis Claesz. ran his publishing house in the heart of Amsterdam. Books had come down in cost and could now be had even in towns and small cities. Some constants remained, such as inspiring rhetoric about the importance of navigation and claims to authorial innovation. However, while the content had stayed remarkably similar for more than two centuries, the audience had grown and changed. So too had the place of the written word in the navigator's daily work.

At the close of the eighteenth century, maritime communities still embraced textual authority but differently from how those of the early sixteenth did so, when Spain tried to teach barely literate pilots from university textbooks. Back then the cosmographers in charge wished to elevate their craft. They did so by appropriating the signifiers of learned professions: lectures, examinations, and writings. This strategy worked, for it persuaded the Spanish king to fund navigational training, and foreign administrators soon clamored for details—templates for schools, texts to translate—which they disseminated to generations of mariners. Two and three centuries later, the audience for navigational fundamentals had expanded dramatically, due to wider education, more literacy, and far more publications. By the Enlightenment, like many other urban workers, navigators took naturally to books.

In this volume, we have met many men who worked their way through navigation lessons. First, the anonymous merchants and mates who rubbed elbows in Cornelis Claesz.'s crowded store, purchasing books to satisfy their curiosity but also to gather tips on how to cross the ocean. Then, Jean-François Doublet, who forsook the wild corsair's life to become a star student. Nathaniel Long and Richard Crockett were Doublet's opposite: they had neither sufficient experience on the water nor the smarts to pass exams. Assuerus van Asson and Anders Wijkman fell somewhere in between, working assiduously on hundreds of math problems—though we will never know how their studies affected their careers. We have a better sense about Lieutenant Riou's trajectory but cannot tease apart the influences in his life. How much of his later resourcefulness was due to his two years under the tutelage of Captain Cook, one of the most intellectual of navigators? Certainly, that circumnavigation let Riou apply many of the new techniques then in vogue, while he must have honed somewhat different skills just patrolling off Newfoundland. What, in the end, did he owe to his natural conscientiousness, and what originated from his trigonometry lessons?

These men used books in their lessons, copied them out, studied from them, critiqued them, and carried them to sea. It is eminently clear that for us now books are valuable windows onto the maritime realm. They offer a record of expertise, education, and practice, real and idealized. Most directly, books preserve the perspectives of experts — or at least those deemed expert enough to transmit theory in print and in classrooms. Much of our knowledge of early modern navigational practice comes to us filtered through the voices of such teachers, who produced textbooks on commission or to promote their own careers. Books show us practitioners and practice, but imperfectly. Few authors, unless old sailors themselves, respected traditional practitioners, and most idealized the actual practices. Where working mariners once garnered respect for their deep knowledge of geography and technique, eighteenth-century authors no longer treated them as allies in the project to refine the science of navigation. Skilled navigators were increasingly barred from the ranks of curious enthusiasts and were instead demoted to mere students, presumed capable of only slow study of the rudiments of the craft.[20]

Publishers, editors, and translators also helped shape nautical knowledge. By reprinting certain texts — with updated or corrected tables, and new practice questions — publishing firms like the Blaeus, Donckers, and the prominent London cartographic house Mount and Page had long-term influence on techniques and goals. While Guillaume Denys was a pioneer in promoting trigonometry and the Corrections, and Claas Gietermaker the first to publicize the VOC's exam, both of their missions were aided by the publishers who repeatedly reissued their texts.[21] Between them, writers and printers created and publicized habits of mind that would inform classroom lessons and shipboard practices for many decades.

Books also reveal the elusive process by which learning happened. What strategies did authors and teachers provide in textbooks to help their students become "handy and circumspect Pilots"?[22] How did they try to supplement the print materials? Although the conceptual underpinnings of the science of navigation shifted — from cosmography to mathematics — and handbooks moved in step, many academic exercises and tools remained the same. The pedagogy of dialogues and definitions, medieval in origins, remained. These were supplemented by newer exercises: practice questions, and before long, mock exams.[23] Other technologies of print, from tables to volvelles, helped students to comprehend, recall, and solve. These two-dimensional devices had initially been secondary to hands-on manipulation of instruments — as we saw in the Royal Mathematical School curriculum, as well as in French classrooms like that of Boissaye de Bocage, who began his lessons by pointing out cosmographical definitions on globes — yet paper tools gradually became the default choice. Late eighteenth-century educators had to remind their pupils of the benefits that came from manipulating objects, once again encouraging them to get their hands on globes, instruments, and the ship's helm.[24] Close attention to these techniques allows us to picture how nonelite early modern individuals learned. The wealth of surviving nautical material can help show us how advanced mathematical concepts were acquired.

These documents also make evident the myriad ways in which illustrated material could aid cognition. Images were powerful crutches for audiences unaccustomed to reading and also helped any reader with more sophisticated concepts. Sailors benefited from the act of drawing geometric figures in introductory math lessons, plotting complex or even fantastical courses before adding up their components, and using simple circles to determine the correct formula for a given altitude observation. Drawing would become a key component of curricula in the years to come, as administrators recognized the benefits of training the sailor's eye as an "expert witness"—imparting the ability to view landscapes in perspective in order to gauge distance.[25] There are lessons here for the story of numeracy and visual culture in the wider history of education.

The ready availability of printed resources changed the larger arc of a sailor's career. Before the late sixteenth century, a mariner could not simply acquire a manual and work his way through it, alone or in a classroom. Now that the precepts of navigation were codified in increasingly affordable publications, navigators and those aspiring to such positions could approach their training differently. They could proceed linearly through a textbook or delve in to read about challenging topics. Books (and careful manuscript copies of them) were particularly useful for preparing for an examination, not just for presenting and explicating the exam questions but for allowing men to look back at the lessons well after they had passed the test. Mariners who climbed the ladder in the merchant companies or the navy would likely find themselves taking several exams a few years apart—so they could refresh their memories by rereading the textbooks or by returning to the classroom and working through more practice questions in their *schatkamers*.

Books also modified how navigators handled their daily tasks. If they were at sea and ran into difficulty making an observation, they could now look in their textbooks for instructions about alternative methods. Should foreign skies show an unfamiliar constellation or if weather conditions worsened, almanacs might hold the answers. If they were blown off course or, worse, shipwrecked, as Riou was, they could consult atlases (and perhaps published travel narratives or even blueprints for emergency steering machines). Although navigators might not always reach for a book at the height of a crisis, books were never far away in quieter moments. Furthermore, those men who had begun preliminary lessons on shore were often thrust into positions of greater authority when misfortune befell the official navigator. If a pilot's mate had paid careful attention during the daily observations at the binnacle, he might well be ready to step in to a navigator's role on his next voyage, if not immediately.[26]

Thus, we see how books and education could transform the lives of early modern seamen, changing the course of their careers and often their social standing as well. From the earliest wave of navigational lessons, certain experts presumed that everyone could benefit from training and that navigation could be transmitted in classrooms and through the pages of books. The Spanish cosmographers were the first to endorse this academic route, but soon the small entrepreneurial teachers joined the bandwagon. In some regions it took time to convince higher authorities that this was true, and it often took still longer to convince the practitioners themselves. And yet, by the turn of the eighteenth century, the Dutch, English, French, and Spanish maritime worlds were operating under another shared assumption: if one selected the right trainees (and instructors) and taught in a clear and logical order, one might cut a shorter path to expertise. Navigators were not born but made.

This view of early modern mariners, itself revolutionary, teaches us how technical education changed. Many of the men who, for self-betterment, paid for classes and took exams, would never set foot on a boat. They chose instead to put their newfound maritime knowledge to work in a shipyard or naval office. Others would take the lessons they had learned about keeping a daily logbook and become bookkeepers or perhaps merchants. There certainly were those who did end up at sea — and they often ascended quickly up the career ladder thanks to keen eyes, formidable memories, and numerical fluency. Many of these men began passing on tacit lessons to their children — no longer about the luff of the wind in the sails or about how to load barrels, but about logarithms and trigonometry. Expertise passed from one generation to the next, one social circle to another, through published and unpublished texts. Whether or not navigational schools and books sped the process of becoming an expert, they altered it irrevocably. They disseminated a different set of intangible skills from those imparted to earlier generations of apprentices: new modes of quantitative observation, precision, objectivity. Thus, technical and conceptual changes rippled across Europe and around the globe.

Although early modern mariners left few autobiographical records, they — and their instructors, examiners, and administrators — nonetheless produced a mass of documents. This enables us finally to draw conclusions not just about navigators as a group but also about the diverse maritime communities that flourished around European coasts and beyond. First, the documents associated with this training — from crown legislation to entrepreneurs' advertisements — preserve evidence of significant participation in scientific practices. They support an increasingly broad, vibrant account of the Scientific Revolution, where educated natural philosophers seek out information from practitioners, who in turn disseminate the same once it has been codified in print or classroom. The written documents left behind by sailors offer unexpected hints about unwritten, tacit knowledge; how they learned to train their internal metronome, their sense of precision, or their ability to gauge the three-dimensional world around them. In addition to such intuitive competencies, there were more overt ones, demonstrating the type of active learning that suited this challenging material. As pupils worked through their assignments, either alone or with assistance, they left evidence of priorities: drawing examples from certain sections of a manual or devoting time to extra practice questions when they found a specific topic particularly important or challenging.[27] We find traces of this in the messy, irregular, creative moments when practitioners took matters into their own hands — adding new notches to their personal cross-staff or, like Van Asson, boldly correcting the authors of his textbooks. Until this study, we have understood little about the precise pathways of transmission of this mathematized technical knowledge or about which skills were gradually codified for effective textual transmission and which remained in the realm of tacit knowledge.

At the same time, we find this history of navigators is not a teleological story of inexorable mathematization, where traditional techniques were sloughed aside to make room for rational modernity. One of the most traditional hallmarks of the professional navigator maintained its paramount importance: his memory. There was simply a sea change in the nature of what he was expected to memorize. Instead of geographic and calendrical details, which were increasingly easy to look up in reference works (especially as tables proliferated), the eighteenth-century navigator had to master mathematical calculations and remember formulas — and comprehend them as well.[28] This he did. Yet, even when offered a panoply of labor-saving inventions, he still continued to use a range of older techniques. Students, for instance, never stopped including familiar features in their manuscripts. Late eighteenth-century *schatkamers* and mathematics workbooks include both thumb diagrams and volvelles (figures E.2, E.3, E.4). Men would get their hands on outdated textbooks and take time to carefully work through century-old practice questions.[29] The longevity of early modern navigational textbooks had its own particular effect on practice. In many ways, the moment of putting something into print captures the past — preserving a set of practices known to that particular author, in his time and place. But it is also forward-looking: teachers intend to transmit their lessons to future generations. Navigational manuals prioritized preparedness, offering multiple solutions to nearly every problem. Thus, rather than splitting practice into old and new, traditional and innovative, such textbooks allowed the two to commingle — a leveling that preserved the older methods alongside the new.

The final salient characteristic of these maritime communities is their international and multilingual nature. Navigators were key players in an interconnected history of information: the records mariners made in their logbooks, to say nothing of the lessons they taught each other, show how information flowed across international borders. Useful knowledge — applied skills such as navigation — were objects of international competition and emulation among emerging empires. In his time as a Royal Navy administrator, Samuel Pepys inquired into the Spanish school in Seville and kept a close eye on French publications. Colbert, as Louis XIV's *ministre de la marine*, solicited reports of navigators' examinations in the Netherlands. Rather than secrecy, we see some cooperation and more opportunism: authors sometimes translated foreign works to benefit their compatriots or to increase their market share. (The translators in Dieppe fell into the first group, whereas the Dutch publishers who prepared English and French translations of popular texts hoped to profit by selling them abroad.) In other instances, mariners simply drew upon foreign texts, as when the Dutch turned to the French ephemerides, and the French to the English, reluctantly admitting the superiority of their rival's computational efforts.[30]

We see governments and other administrative bodies selecting suitable candidates for naval careers, funding schools, commissioning textbooks, and imposing professional standards by way of qualifying examinations, as well as offering rewards for

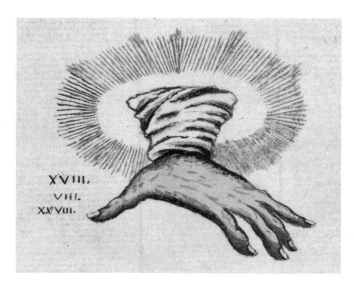

E.2.
Traditional paper tools continued to appear in student workbooks into the eighteenth and nineteenth centuries. Epact thumb diagram in H. M. Hoffman's Schatkamer (Amsterdam, 1818). Collection Maritiem Museum Rotterdam

E.3.
Compass rose in H. M. Hoffman's Schatkamer (Amsterdam, 1818). Collection Maritiem Museum Rotterdam

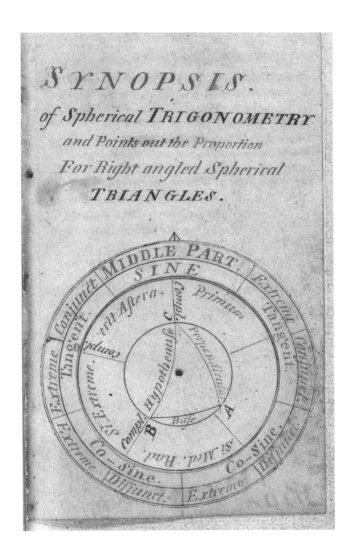

E.4.
Volvelle illustrating the "proportions" of various right-angled spherical triangles. G. Scargill, Manuscript volume containing instructional texts on spherical trigonometry and astrology, ([England], 1787). John Hay Library, Brown University

solving the longitude problem. In light of these attempts to "perfect the Art of Navigation," governments can be seen as "impresarios of knowledge." Their involvement could not be entirely for strategic advantage over rivals — in this age of international translations and informants, information flowed too easily to achieve that. Instead, they sought to find ways to generate competence at the demanding tasks inherent to navigating. Long before the bureaucratic accomplishments of the nineteenth century, here we find governments succeeding — more often than not — at improving and expanding a body of knowledge and raising the caliber of its practitioners. It was attention and funding from government authorities that made navigation cutting-edge research, the "big science" of its day.[31]

This book has highlighted similarities among academic traditions and training practices across Europe. The transnational approach is critical here — because so much maritime history focuses only on a single nation or navy at a time, we have lost sight of the shared tradition that made this system work so seamlessly. Far from a static world where ships, equipment, or skills did not evolve for centuries, we see an environment where practitioners were constantly adopting new techniques and modifying practices to suit their needs. By contemplating a history of the book for a group not known for its literacy, a scientific history of a subject long dismissed as a craft or an intuitive art, and an account of common origins and shared lessons rather than a tale of naval conflict, we learn that it matters for whom one sailed and from whom one learned. We have also seen that textbooks were read not only by gentlemen. Rather, they were pored over, annotated, and scribbled in by the modest, often anonymous men who enabled the expansion of early modern Europe.

With this clearer picture of the complexity of a navigator's tasks and his training, his intellectual life and his interactions with the wider world, we have learned something about practice, communication, and science in early modern society. Sailors were key links in global history *and* lively informants in local ones. Scholars today take the Scientific Revolution beyond the cabinets and courts of Europe to other corners of the globe. Often and again these navigators were there, stitching together port towns and factories, bustling metropoles and fledgling colonies. And their voices — of working men rather than the governing elites — offer a vital new chapter to the story of how science develops in the wider world.

Glossary

A term that appears only once in the book is generally defined in a note at that location in the text. Compiled with the aid of *OED Online* and Bennett, *Navigation: A Very Short Introduction* (2017).

altitude — angular distance of a celestial body above the horizon

amplitude — horizontal angular distance between where a heavenly body rises or sets and the eastern or western point of the horizon; complement of the **azimuth**; can be used to determine **magnetic variation**

armillary sphere — skeletal **celestial sphere** constructed of rings (fig. 4.5, lower right)

astrolabe, mariner's — circular brass instrument for measuring celestial **altitude** (fig. 4.5, upper left)

azimuth — the **bearing** of a heavenly body with respect to the observer, measured along the horizon; complement of the **amplitude**

backstaff — portable instrument for measuring solar **altitude** by the Sun's shadow (e.g., *hoekboog*, Nl., or Davis quadrant) (figs. 1.10 and 3.2)

bearing; heading (*rhumb de vent,* Fr.; *streek*, Nl.) — horizontal direction, the ship's course (angle made with the **meridian**)

bittacle (after 1750, binnacle) — housing for a magnetic **mariner's compass**

Casa de la Contratación de Indias (Sp.) — "House of Trade" in Seville (1503–1790)

185

celestial sphere (*de Sphaera*, Lat.; *hemelkloot*, Nl.) — model of the heavens; up through the seventeenth century, the geocentric Aristotelian model with concentric crystalline spheres bearing the planets (fig. P.2); more generally, sphere on which are projected the apparent positions of celestial bodies with the Earth at the center

compass — mathematical instrument for taking measurements and describing circles, consisting of two straight legs connected with a movable joint (fig. 1.11, man with atlas on left, and second example at globe on right)

compass, magnetic/mariner's — instrument indicating direction, using the north-seeking properties of a magnet that turns freely on a pivot (fig. 1.11, lower left corner; fig. 4.5, lower center)

compass rose — circular pattern indicating thirty-two directions (points or winds) (figs. 1.3, 4.2, E.3)

Corrections — set of three rules governing how to reconcile estimated and observed positions; invented by J. H. Jarichs van der Ley but most popular in France

corsair; privateer — privately owned armed vessel whose owners are commissioned by a nation (given a "letter of marque") to carry on naval warfare

cosmography — science that describes and maps the general features of the universe (both the heavens and the Earth)

cross-staff (*arbaleste/flèche/bâton de Jacob*, Fr.; *graadboog*, Nl.) — instrument for measuring angles, usually **altitudes** (fig. 1.11, right)

dead reckoning — finding position from a record of direction and distance, without astronomical observations

declination — the height of the Sun with respect to the celestial equator (ranges from 23.5° S to 23.5° N); the celestial equivalent of terrestrial latitude. Also an alternative term for **magnetic variation**

double altitude — refers to position-finding from two **altitudes** of the Sun or a star (or two stars simultaneously) that are on different **meridians** from the observer

ecliptic — apparent annual path of the Sun through the **celestial sphere**, caused by the Earth's approx. 23.5° tilt on its axis

epact — the surplus days of the solar over the lunar year, or the **Moon's age** in days on January 1

ephemeris, pl. ephemerides — set of tables of positions of heavenly bodies

fother — wrapping a sail (coated with dung, rope fibers, etc.) under a hull to patch a leak

gimbal mount — ring with pairs of supporting pivots, arranged to maintain a **magnetic compass** level

Golden Number; prime — related to each year's place within the nineteen-year "Metonic" lunar cycle. The Golden Number of any year may be found by adding 1 to the year and dividing by 19, the quotient showing the number of complete cycles elapsed since 1 B.C. and the remainder (or, if there is no remainder, 19) being the Golden Number of the year.

Guard Stars (*Wachters,* Nl.) — two stars in the constellation of Ursa Minor (Little Dipper), farthest from Polaris (North Star)

Gunter's scale/rule (*échelle anglaise,* Fr.) and sector — wooden ruler with logarithmic scales used in navigational calculations (a sector consists of a pair of two-foot-long rules joined with a hinge); conceived of by Edmund Gunter ca. 1620s. With a mathematical compass, users measure two distances along the rule to form the combined distance, the product. (fig. 2.2)

hydrography — discipline of surveying at sea and on the coast, and the making of charts. For the French, the term *hydrographie* was broader, often a synonym for navigation.

knot — unit of speed, one **nautical mile** per hour

latitude (*breedte,* Nl.) — angular distance of a position on Earth from the equator; can be determined by measuring the height of the Sun or a prominent star. Prior to the widespread use of instruments such as **cross-staffs** or **quadrants**, mariners would consult the **regiment** of the Sun or stars. Conventionally, one degree of latitude was sixty **nautical miles**.

lead and line (sounding lead, sink line) — instrument for measuring depth (fig. 2.9, upper left fig. S)

liefhebber (Nl.) — amateur, enthusiast

log and line (log-line, *loch,* Fr.) — instrument for measuring the speed of a ship, originally a loose chip of wood, then a small board attached to a knotted line (fig. 2.9, upper left fig. 14); the navigational journal kept on a voyage was named after this "log."

longitude (*leengde,* Nl.) — angular distance on Earth, measured parallel to the equator from a chosen **meridian**. The Earth's daily rotation makes it very difficult to determine how far east or west one has traveled without a reliable means of keeping time, which came only after the 1760s with timekeepers such as the device developed by John Harrison. Numerous proposals were ineffective (**magnetic variation**; the **Corrections**), or impossible without accurate astronomical tables (eclipses; **lunar distance** method, which did have some success).

loxodrome; rhumb line — line on a sphere cutting all the **meridians** at the same angle

lunar distance — angular distance from the Moon to another heavenly body, or a method of finding **longitude** based on such a measurement; popularized by N. Maskelyne in the 1760s

magnetic variation — difference between true and magnetic north. Leading seventeenth-century **longitude** proposal (because every spot on the globe was presumed to have a unique amount of magnetic flux, tracking this variation could determine longitude). Even after being debunked in the mid-eighteenth century, it was still necessary to correct for the deviation of the compass from true north, which was substantially higher in the early modern period than in the present.

mariner's compass — *see* **compass**

master — captain of a merchant vessel; or naval officer ranking next below lieutenant responsible for the navigation of a ship

Mercator chart (*carte réduite,* Fr.; *wassende graadkaart,* Nl.) — cylindrical projection where **meridians** and parallels cross each other at right angles to form a rectangular grid. This means all the parallels of latitude have the same length as the equator, and **loxodromes/rhumb lines** are straight segments.

meridian — great circle on the celestial or terrestrial sphere, passing through the poles and perpendicular to the equator; corresponding to a line of **longitude**

meridional parts (*croissantes largeurs,* Fr.; *wassende breedte,* Nl.) — the linear lengths of one minute of longitude on a **Mercator chart**, accounting for the increasing scale at higher latitudes

midshipman — naval officer of the junior-most rank

Moon's age — days past the new Moon (1–30), necessary to compute the time of high tide on any given day

nautical mile — measure of distance at sea intended as the length covering one minute of arc of a **meridian**; now fixed at 1,852 meters (6,076 feet)

navigation, large (blue-water; *grand,* Fr.; *groot,* Nl.) — sailing primarily in the open ocean; navigating with celestial observations

navigation, small (common; *commun,* Fr.; *kleyn,* Nl.) — sailing primarily within sight of land; coastal sailing

nocturnal — instrument for finding time from the orientation of the sky around the celestial pole; first described by Michiel Coignet of Antwerp in 1581

octant — 45° instrument for measuring angles up to 90° between distant bodies, using reflection to bring images into coincidence; invented in the 1730s (fig. E.1, right)

pilot — person who directs the course of a ship (helmsman, steersman). The English term is limited to the individual whose main task was to guide ships in and out of harbors; he was often brought on board as a vessel neared a port. However, in Spanish and French, *piloto/pilote* refers to someone with wayfinding responsibilities. The French used the terms *locman, lamaneur,* and later *pilote côtier/pratique* for small navigation. In Dutch, there was a distinction between the professions of *piloot/loodsman* (local harbors) and *stuurman* (large navigation), but the terms were used interchangeably.

pilotage — in English, navigation by observation of the coast and sounding depth; chiefly short-distance or **small navigation**; in French, general term for steering a vessel, used to refer to both **small** and **large navigation**

piloto mayor (Sp.) — "master navigator," appointed by Casa de la Contratación (from 1508) to teach and license navigators, and regulate their charts and nautical instruments

plain/plane chart — chart representing Earth as a flat surface, with the lines of

latitude and **longitude** straight and parallel; adequate for short voyages but introduces errors over longer distances

portolan chart — late-medieval and early modern style of navigational chart, centered on the Mediterranean and crossed by radiating **rhumb lines**

prime — *see* **Golden Number**

privateer — *see* **corsair**

quadrant — graphical instrument for measuring angles between distant bodies, or distance above horizon; a quarter circle of wood or brass with a graduated arc of 90° (fig. 2.2, lower right). The sinical quadrant (*quartier de réduction* or *proportion,* Fr.) was marked on each side with intersecting **sines** (fig. 2.3).

regiment of the Pole / North Star — tables or list of rules for eight positions of the **Guard Stars**; the correction to be applied to the **altitude** of the Pole Star to obtain the **latitude** of the observer

regiment of the Sun — table showing Sun's **declination** throughout the year

rhumb — compass **bearing**, or the course of a ship following a constant bearing; *see also* **loxodrome**

rule of three — mathematical formula that solves for one unknown when three other quantities are known; a standard tool of medieval merchants

rule of thumb — counting on joints of one's thumb to facilitate calendrical and tidal computations (figs. 1.7–1.9, 1.16)

rutter (*routier,* Fr.; *derrotero,* Sp.; all from *roteiro,* Port.: "route-book") — "pilot guide," textual sailing directions

sailing triangle — right triangle with the boat's point of origin and its point of arrival as the vertices and the course the hypotenuse (taking into account the direction of travel/**bearing**/**rhumb** and the distance covered/run/*chemin,* Fr.) (fig. 2.9). Less frequently a vertical triangle consisting of the observer's zenith, the celestial pole, and the Sun or other celestial body.

schatkamer (treasure-chamber/chest) — Dutch term for volumes of navigational knowledge; English manuscript workbooks often termed "cyphering books"

seamanship — handling a ship at sea, including the physical techniques of steering, mooring, etc.

sector — folding rule with pairs of lines used for calculation; *see* **Gunter's sector**

sextant — 60° instrument for measuring angles up to at least 120° between distant bodies, using reflection to bring images into coincidence; invented in the 1750s

sinical — constructed using **sines**

Sinus, **sine** — trigonometrical function, being the ratio of the opposite side over the hypotenuse; general term for set of tables of all trigonometrical functions

sounding line — *see* **lead and line**

spherical trigonometry — the study of triangles drawn on the surface of a sphere *stuurman*, pl. *stuurlieden* (Nl.) — navigator, mate

Sunday (Dominical) letter — the weekday upon which January 1 falls in a given year, enumerating backward from A (Sunday) through G (Monday)

Sun's right ascension — measured eastward (along the celestial equator) from the point where it crosses the celestial equator at the vernal equinox (0° on March 21); as it increases approximately four minutes a day, this observation can provide the date. The celestial equivalent of terrestrial longitude.

traverse (compound course; *cours composé,* Fr.; *koppel koers*, Nl.) — path taken by a ship, which may comprise a number of different **bearings**

traverse board — wooden board marked with the points of the **compass**, with holes and pegs by which to indicate the course of the ship (fig. 1.4). This information would be transferred to a log/traverse book (fig. E.1, left).

traverse table (table of rhumbs; *table de loch,* Fr.; *streek-tafel,* Nl.) — showing how far to sail on any given course (**rhumb/bearing**) to raise one degree of latitude (fig. 1.5, lower right)

VOC, Verenigde Oost-Indische Compagnie — Dutch East India Company (1602–1799)

volvelle — movable paper tool, often with a rotating disk and/or pointer or string

WIC, West-Indische Compagnie — Dutch West India Company (1621–74)

zenith distance (*topafstand,* Nl.) — the complement of the **altitude** (measured from the zenith point directly overhead); used to calculate the distance in **nautical miles** from the observer's position to the geographic position of the celestial body.

Notes

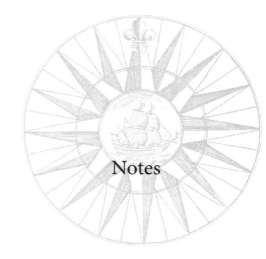

ABBREVIATIONS

AN-Paris	Archives nationales de France (Paris, France)
BMR	Bibliothèque municipale de Rouen (Rouen, France)
BnF	Bibliothèque nationale de France, Richelieu (Paris, France)
CUL	Cambridge University Library (Cambridge, UK)
HSM	Het Scheepvaartmuseum (Amsterdam, Netherlands)
LMA	London Metropolitan Archives (London, UK)
MALSC	Medway Archives (Strood, Rochester, UK)
MMR	Maritiem Museum Rotterdam (Rotterdam, Netherlands)
NL-HaNA	Nationaal Archief (Den Haag, Netherlands)
NMM	Caird Library, National Maritime Museum (Greenwich, London, UK)
OTYA	York University Libraries, Clara Thomas Archives and Special Collections (Toronto, ON, Canada)
PL	Pepys Library, Magdalene College, Cambridge University (Cambridge, UK)
PWDRO	Plymouth and West Devon Record Office (Plymouth, UK)
QQA	Bibliothèque et Archives nationales du Québec (BAnQ-Q) (Québec City, QC, Canada)
QQS	Musée de la civilisation, Bibliothèque du Séminaire de Québec, Fonds ancien (QMUC BSQ) (Québec City, QC, Canada)
SHD-V	Service historique de la défense (Vincennes, France)
SLNSW	State Library of NSW (Sydney, Australia); incl. Dixson and Mitchell Libraries
TNA	National Archives of the UK (Kew, Richmond, Surrey, UK)

INTRODUCTION

1. Van Breen, *Stiermans Gemack* (1662), frontispiece: "Piloot Siet Met Gemack, Matroos Dragen Taback." (See glossary on the equivalence between the terms "navigator" and "pilot.") On Van Breen's poetry, see K. ter Laan, *Letterkundig woordenboek voor Noord en Zuid* (1952).

2. Van Breen, *Stiermans Gemack* (1662), 111, ch. 8. On the navigator's method for estimating the boat's speed using the log and line, see chapters 1 and 4. The triangular appendage could also be a lead weight, a "lead and line" used for depth sounding, but those were typically more cylindrical in form; cf. figure 2.9.

3. Dunton, *Young Students-Library* (1692), preface. See similar patriotic rhetoric from the French, e.g., Louis XIV, *1681 Ordonnance* (1714), preface, and the Dutch, Philips, *Zeemans Onderwijzer in de Tekenkunst* (1786), [vii], "Our Republic owes her greatness, in reputation and power, primarily to Commerce, [and] . . . all Commerce, cannot be practiced, without help from Navigation."

4. There are monograph-length studies of French and British naval education, and the Spanish system has been well covered: Anthiaume, *Evolution et enseignement de la science nautique* (1920); Vergé-Franceschi, *Marine et éducation sous l'ancien régime* (1991); Dickinson, *Educating the Royal Navy* (2007); Sandman, "Cosmographers vs. Pilots" (2001); Garralón, *Taller de Mareantes (1681-1847)* (2008). However, most accounts are briefer case studies, see note 15 in this chapter.

5. Political and technical histories are difficult to disentwine. Portugal: Bensaúde, *Histoire de la science nautique portugaise* (1914); Boxer, *Portuguese Seaborne Empire* (1969); Bethencourt and Ramada Curto, *Portuguese Oceanic Expansion* (2007). Spain: Lamb, *Cosmographers and Pilots* (1995); Portuondo, *Secret Science* (2009). England: Taylor, *Haven-Finding Art* (1956); Waters, *Art of Navigation* (1958); Rodger, *Wooden World* (1986), *Safeguard of the Sea* (1997), and *Command of the Ocean* (2004); Levy-Eichel, "'Into the Mathematical Ocean'" (2015). See also Rasor, *English/British Naval History to 1815: A Guide to the Literature* (2004). Netherlands: Boxer, *Dutch Seaborne Empire* (1965); Feyfer, *Het Licht der Zeevaert* (1974); Koeman, *Flemish and Dutch Contributions* (1988); Israel, *Dutch Primacy in World Trade* (1989); Bruijn, *The Dutch Navy* (1993). France: Mémain, *La marine de guerre sous Louis XIV* (1937); Chapuis, *A la mer comme au ciel: Beautemps-Beaupré* (1999).

C. A. Davids is one of the few scholars to take a transnational approach to the development of navigational science in some of his works, e.g., "On the Diffusion of Nautical Knowledge" (1988), "Transfer of Technology between Britain and the Netherlands" (1991). See Dunn and Higgitt, *Navigational Enterprises* (2015), for a recent transnational treatment focused on the eighteenth and nineteenth centuries.

6. I follow Rob van Gent's distinction between the "practice (*praktijk,* routine) of navigation" and the "practice (*beoefening,* study) of the science of navigation"; Van Gent, "Het Sterrenlied in het Hollandse Zeevaartonderwijs" (2005), 217.

7. Mariners are omitted from most standard accounts of the "Scientific Revolution" published before 1990, e.g., Koyré, *From the Closed World to the Infinite Universe* (1958); Dijksterhuis, *Mechanization of the World Picture* (1961). Two important early exceptions are Merton, *Science, Technology and Society* (1938), and Zilsel, "Origins of William Gilbert's Scientific Method" (1941). Both promoted the mariner as a valuable contributor to the development of technology and society, the "scholar-craftsman" debate.

8. Zilsel, *Social Origins of Modern Science* (2000), and Cormack, Walton, and Schuster, *Mathematical Practitioners and the Transformation of Natural Knowledge* (2017), revisit Zilsel's questions. Inspired by Michael Polanyi, *Personal Knowledge* (1964), a range of scholars has

explored the idea of manual expertise—the artisan's unwritten or "tacit" knowledge—e.g., Latour, *Science in Action* (1987); Mukerji, "Tacit Knowledge and Classical Technique in Seventeenth-Century France" (2006). On bodily knowledge, Bennett, "Practical Geometry and Operative Knowledge" (1998); Smith, *Body of the Artisan* (2004). On material evidence of artisanal knowledge: Van Helden and Hankins, "Introduction: Instruments in the History of Science" (1994); Daston, *Things That Talk* (2004); Smith and Schmidt, *Making Knowledge in Early Modern Europe* (2007); Bertucci, *Artisanal Enlightenment* (2017).

9. Successful efforts have been made in nonmaritime domains to smooth the rifts between science and technology, head and hand, and to blur the categories of artisan and innovator—e.g., Roberts, Schaffer, and Dear, *The Mindful Hand* (2007); Long, *Artisan/Practitioners* (2011).

10. The anonymous author of *The Accomplish'd Sea Mans Delight* (1686) called for "the better breeding, continuance, and increase of Pilots amongst us." See chapter 3, note 31, on Samuel Pepys's patriotic worries. On parallel Spanish concerns, see Pérez-Mallaína Bueno, "Los libros de náutica españoles" (1989), 469. In the Netherlands, VOC and naval administrators complained about the declining quality and numbers of "boat people" (*bootsvolk*), particularly in the eighteenth century (although notably few such complaints were leveled against the merchant marine); see Bruijn, "De personeelsbehoefte" (1976), 246–47. Politics affected these cycles as much as sailing conditions. Following the revocation of the Edict of Nantes, for example, France saw a shortage of skilled mariners as Huguenot sailors left for Britain and the Netherlands; McNeill, *Atlantic Empires* (1985), 67. Diderot and d'Alembert expressed concern: "We have lost many sailors; therefore, there is nothing for it but to create more"; *Encyclopédie* (1765), 10.124, "Marine." On the need to "accelerate" training, see Bezout on the 1773 Ordonnance, cited in Crisenoy, "Les ecoles navales" (1864), 786.

11. Willaumez, *Projet pour former des élèves marins à Paris même* (1800), 4. See epilogue for his hopeful proposal.

12. Des Nos, "Mémoire sur l'escole du canon de Brest" (Feb. 2, 1682), AN-Paris MAR/G86, No. 2: He lists the "places that seem to me the most fertile [sources] of men suited to this trade," such as the diocese of Vannes in Brittany. Talon Letter to Colbert (Nov. 2, 1671), Recueil: Hydrographie sous Colbert, BnF N.A.Fr. 9479, f. 21, "une pépinière de navigateurs . . ."; cited in Chapais, *Jean Talon* (1904), 422–23.

13. Boteler, *Six Dialogues about Sea-Services* (1685), 61–62. See also *Considerations on the Nurseries for British Seamen* (1766).

14. This transition has been characterized as a shift from coastal to open-water sailing; lengthier voyages required more mathematical methods. See Taylor, *Haven-Finding Art* (1956); Davids, *Zeewezen en Wetenschap* (1986); and chapter 1. For Pepys, "nurseries" were related not to marine trades but to schools; see Pepys, *Naval Minutes* (1926), 414, 420, where he recorded half a dozen domestic examples of "Nurserys or Lectures for Navigacon" and made special notes to inquire into Dutch and French institutions.

15. In addition to the key authorities cited in note 4 in this chapter, see on Spain: Carlier, "Los colegios de pilotos" (1993); Arroyo, "Las ensenanzas de nautica" (1994); Jiménez, *El Real Colegio Seminario de San Telmo* (2002); Garralón, "La formación de los pilotos de la carrera de Indias en el siglo XVIII" (2009). England: Johnston, "Making Mathematical Practice" (1994). Netherlands: Crone, "Pieter Holm en zijn Zeevaartschool" (1930); Crone, *Cornelis Douwes . . . en Zeevaart-Onderwijs* (1941); Davids, "Het Zeevaartkundig Onderwijs [1795–1875]" (1984); Davids, "Het navigatieonderwijs aan personeel van de VOC" (1988). France: Neuville, *Établissements scientifiques de l'ancienne marine* (1882); Audet, "Hydrographes du Roi et cours d'hydrographie au collège de Québec" (1970); Fauque, "Les Écoles d'hydrographie en Bretagne" (2000); Geistdoerfer, "La formation des officiers de marine" (2005). For works

focusing on astronomy, see Kennerley and Seymour, "Aids to the Teaching of Nautical Astronomy" (2000); Boistel, "Training Seafarers in Astronomy" (2010).

16. On administrators minimizing theory, Russo, "L'Enseignement des sciences de la navigation" (1959), 192. For examples of the widespread concern that mathematical aspects of celestial navigation were too difficult for sailors, see García de Palacio, *Instrución Náutica* (1587; trans. 1988), 81; Sandman, "Cosmographers vs. Pilots" (2001), 102, for sixteenth-century Spanish complaints about would-be pilots; Waghenaer, *Spieghel der Zeevaerdt* (1584), xvi; Saltonstall, *The Navigator* (1636), 13; Vooght, *Zeemans Wegh-Wyser* (1695), 36. Van Breen was one of many who promised to teach navigation entirely without math; *Stiermans Gemack* (1662), poem on verso of frontispiece.

17. Roberts et al., *Mindful Hand* (2007); Long, "Power, Patronage, and the Authorship of Ars" (1997); Long, *Openness, Secrecy, Authorship* (2001).

18. Van Breen was the first examiner for the Zeeland chamber of the VOC (1670–79); Davids, *Zeewezen en Wetenschap* (1986), 399. He promised to teach mariners to use his own invention, the "mirror-staff" (*Spieghel-Boogh*), prominently depicted twice on the frontispiece.

19. Capt. Charles Saltonstall disdained such men who could not apply their knowledge outside their comfortable studies; *The Navigator* (1636), "To the Reader." See further discussion of Jacob Cats's motto in chapter 4, note 14.

20. See Pomata's chapter on physicians' preference for experience over theory in Daston and Lunbeck, *Histories of Scientific Observation* (2011); Wolfe and Gal, *The Body as Object and Instrument of Knowledge* (2010), 1–5, on empirical medicine over theory; Davis, "Sixteenth-Century French Arithmetics on the Business Life" (1960), 33. Warwick, *Masters of Theory* (2003), is an important study of the tight ties between theory and practice in the nineteenth-century classroom.

21. See the extensive bibliography of "studies in expertise and experience (SEE)" in the special volume edited by Ash, "Expertise and the Early Modern State," *Osiris* 25 (2010), esp. n. 17. On the mutually reinforcing relationship between theory and practice in various fields, see Cormack, *Charting an Empire: Geography* (1997), ch. 6 and conclusion; Mahoney, "Christiaan Huygens: Longitude" (1980); McGee, "From Craftsmanship to Draftsmanship (1999).

22. Numerous studies of early modern education are pertinent here, including those on more learned subjects (such as geography, mathematics, and astronomy), because many instructors took them as models for *all* subjects, regardless of the level of their pupils: Grendler, *Schooling in Renaissance Italy* (1989); Caspari, *Humanism and the Social Order in Tudor England* (1968); Hunter, *John Aubrey and the Realm of Learning* (1975); Feingold, *The Mathematicians' Apprenticeship* (1984); Cormack, *Charting an Empire* (1997). Taton, *Enseignement et diffusion des sciences* (1964); Chartier, *L'éducation en France du XVIe au XVIIIe siècle* (1976); Parias and Rémond, *Histoire générale de l'enseignement* (1981); Brockliss, *French Higher Education* (1987); Terrot, *Histoire de l'éducation des adultes* (1997); Burger, *Amsterdamsche Rekenmeesters* (1908); Davids, "Universiteiten, Illustre Scholen" (1990). Extensive work has been done on elementary education — the introduction to reading, writing, and some basic arithmetic for young children or older learners. These "little schools" were taught in the vernacular, while grammar schools were devoted to learning Latin. Relatively few examples regarding the details of artisanal or technical training exist, although Charlton, *Education in Renaissance England* (1965), includes navigators in pt. 3, ch. 9, "Informal Education."

23. Rodger, "Commissioned Officers' Careers in the Royal Navy, 1690–1815" (2001); Rodger, "Honour and Duty at Sea, 1660–1815" (2002); Geistdoerfer, "La formation des officiers de marine" (2005); Teitler, *Genesis of the Professional Officers' Corps* (1977). On the debate over the

class status of officers in the Royal Navy, see Davies, *Gentlemen and Tarpaulins: The Officers and Men of the Restoration Navy* (1991), but also Rodger, *The Command of the Ocean* (2004), 133.

24. Pérez-Mallaína Bueno, *Spain's Men of the Sea* (1998); Fury, *Social History of English Seamen* (2011); Howell and Twomey, *Jack Tar in History . . . Maritime Life and Labour* (1991); Van Royen, Bruijn, and Lucassen, "Those Emblems of Hell"? *European Sailors and the Maritime Labour Market, 1570-1870* (1997). Sailors have been framed as an anti-intellectual proletariat: Rediker, *Between the Devil and the Deep Blue Sea* (1989); Van Royen, "Personnel of the Dutch and English Mercantile Marine" (1991); Linebaugh and Rediker, *Many-Headed Hydra: Sailors, Slaves, Commoners* (2000). On their economic and political power, see, e.g., Dening, *Mr Bligh's Bad Language* (1992); Lane, *Pillaging the Empire: Piracy in the Americas* (1998); Lunsford, *Piracy and Privateering in the Golden Age Netherlands* (2005); Hanna, *Pirate Nests and the Rise of the British Empire* (2015); Perl-Rosenthal, *Citizen Sailors* (2015).

25. Ferreiro, *Ships and Science* (2006); Valleriani, "Knowledge of the Venetian Arsenal," ch. 4 in *Galileo Engineer* (2010). Historians of colonialism and technology have presented the ships themselves as powerful vectors, e.g., Cipolla, *Guns, Sails and Empires* (1965); Law, "On the Methods of Long-Distance Control" (1986); Sorrenson, "Ship as a Scientific Instrument" (1996); Janzen, "A World-Embracing Sea: Oceans as Highways" (2004), 103. Samuel Eliot Morison made even grander pronouncements: "Navigational methods in effect around 1500 lasted, with many refinements but no essential changes, until 1920–30"; *The Great Explorers* (1978), 32.

26. Davids, *Zeewezen en Wetenschap* (1986), 294–99; Crone, *Cornelis Douwes* (1941), 30; Dickinson, *Educating the Royal Navy* (2007), 11; Rodger, *Wooden World* (1986), 265; Anthiaume, *Evolution et enseignement* (1920), 1.92. The early Spanish exams have been more thoroughly treated: Navarro García, "Pilotos, maestres y señores de naos en la carrera de las Indias" (1967); Lamb, *Cosmographers and Pilots* (1995); Sandman, "Educating Pilots: Licensing Exams" (1999); Sandman, "Cosmographers vs. Pilots" (2001).

27. For typical materials, see the "Sailing Letters Project" (a.k.a. the "Prize Papers Consortium"), a research partnership between the National Archives (United Kingdom), and originally the Koninklijke Bibliotheek (Den Haag), now Carl von Ossietzky Universität (Oldenburg); Doe et al., *Buitgemaakt en teruggevonden* (2013).

28. Textbooks are attracting renewed scholarly interest, considering their longevity not simply as static and conservative but as dynamic and evolving; see, e.g., Campi et al., *Scholarly Knowledge: Textbooks in Early Modern Europe* (2008); Oosterhoff, *Making Mathematical Culture* (2018).

29. One of the most extensive national bibliographies is Adams, *English Maritime Books Printed before 1801* (1995); see also Adams, *The Non-cartographical Maritime Works Published by Mount and Page* (1985). For France, see Polak, *Bibliographie maritime française* (1976), and Polak and Polak, *Supplément* (1983). For Dutch works, see Peters, *Crone Library* (1989), and Hoogendoorn, *Bibliography of the Exact Sciences in the Low Countries* (2018).

30. Few studies of maritime books extend beyond bibliographic lists. Exceptions include Adams, "Beginnings of Maritime Publishing in England" (1992); Polak, "Les livres de marine en français au XVIe siècle" (1993), 41–54; Davids, "Van Anthonisz tot Lastman. Navigatieboeken" (1984); Davids, Van der Veen, and De Vries, "Van Lastman naar Gietermaker" (1989); Mörzer Bruyns, "Nederlands Zeevaartkundeboeken [1800–1945]" (1985); Schepper, "'Foreign' Books for English Readers" (2012); Netten, *Koopman in Kennis: De Uitgever Willem Jansz Blaeu* (2014). For an edited volume on French works, see Charon, Claerr, and Moureau, *Le livre maritime au siècle des Lumières* (2005). Gulizia, "Printing and Instrument Making"

(2016), offers a "cosmopolitan" publishing history of Medina's *Arte de navegar* that considers humanists and their patrons as well as maritime readers.

31. Eisenstein, *Printing Press as an Agent of Change* (1982), and numerous responses, e.g., Johns, "How to Acknowledge a Revolution" (2002).

32. This estimate includes works in Latin, Portuguese, Spanish, Dutch, English, French, German, and Swedish. I follow Peters, *Crone Library* (1989), XXXIv, in deeming the *Regimento do Estrolabia e do Quadrante e Tractado da Spera do Mundo* (Lisbon, 1509), written in the 1480s, to be the "oldest known textbook on the art of navigation." I do not include books that focus only on shipbuilding, maritime law, or travel narratives written by mariners. Nor I do include any books written *for* sailors that do not discuss technical practices—a book of prayers is omitted; however, books on trigonometry for sailors as well as ones on drawing are included. I do not include the "sphere" books of Apianus, Sacrobosco, or Gemma Frisius. I consider the content of certain general interest works in my analysis (e.g., Blome's *Gentlemans Recreation*, 1686) but do not include them in the census. These statistics are compiled from Adams, *Crone Library*, Hoogendoorn, Polak, etc.

33. In light of the very disparate geographic, economic, and technological issues at play, this study considers only merchant and naval branches of navigation rather than fisheries, whaling, or inland navigation.

34. Waters, *Art of Navigation* (1958), 497–98. See also Davids, *Zeewezen en Wetenschap* (1986).

35. In *Secret Science* (2009), María Portuondo has argued convincingly about the "secrecy" of the Spanish empire in the sixteenth century. The state was strongly invested in maintaining control over such valuable geopolitical information, keeping any new geographic details on a single, central chart, the *padrón real*. See prologue, notes 25 and 29.

36. E.g., Schepper, " 'Foreign' Books for English Readers" (2012); Collins, "Portuguese pilots at the Casa de la Contratación" (2014). Also, chapter 2, note 25, on foreign translations in Dieppe.

37. Long, "Power, Patronage, and the Authorship of *Ars*" (1997); Long, *Openness, Secrecy, Authorship* (2001); Frasca-Spada and Jardine, *Books and the Sciences in History* (2000). See Bell, *How to Do It* (1999) on the more general growth of pragmatic, instructional texts. An early maritime example is analyzed in Long et al., *Book of Michael of Rhodes* (2009), whose fifteenth-century textual production was likely motivated by his career aspirations rather than his navigational concerns.

38. Administrators and instructors embraced books before most sailors did. In each of the countries in question, the authorities commissioned new textbooks to teach their semiliterate mariners.

39. Darnton, "First Steps toward a History of Reading" (1986); Price, "Reading: The State of the Discipline" (2004); Yeo, *Notebooks, English Virtuosi, and Early Modern Science* (2014); Blair, *Too Much to Know* (2010).

40. For example, Captain John Smith (of Virginia fame) thought that publishing a book might help him find a position upon his return to England: Smith, *An Accidence or The Pathway to Experience Necessary for all Young Sea-men* (1626).

41. Oughtred, *Circles of Proportion* (1660), cited in Pepys, *Naval Minutes* (1926), 413, note written ca. 1695.

42. The key moment that it becomes possible to study a skill is often the moment that a practitioner's account appears in a text. Eamon, *Science and the Secrets of Nature* (1994); Pamela Smith, The Making and Knowing Project (Columbia University), www.makingandknowing.org.

43. For an introduction to early modern mnemonic practices, see Yates, *Art of Memory* (1966); Spence, *Memory Palace of Matteo Ricci* (1985). Although we now often associate diagrams with juvenile and pedagogical literature, in the early years of print, woodcuts were comparatively expensive and thus uncommon, except in works for more elite audiences. Even as they were increasingly sought out as conceptual aids, their presence or absence had more to do with the publisher's budget than authorial intent.

44. Campbell-Kelly et al., *History of Mathematical Tables* (2003). For instance, Nathaniel Bowditch accuses John Hamilton Moore of "criminal inattention . . . [for] by reckoning the year 1800 as leap year, he had made an error of 23 miles in some of the numbers. This error was the cause of losing two vessels to the northward of Turk's Island, and bringing others into serious difficulties." Bowditch, *Improved Practical Navigator* (London, 1802), xiii.

45. Mosley, "Objects, Texts and Images in the History of Science" (2007), provides a helpful overview. See Ferguson, *Engineering and the Mind's Eye* (1992); Shirley and Hoeniger, *Science and the Arts in the Renaissance* (1985); Baigrie, *Picturing Knowledge* (1996); Lefèvre, *Picturing Machines 1400-1700* (2004); Kusukawa and Maclean, *Transmitting Knowledge: Words, Images, and Instruments in Early Modern Europe* (2006); Baldasso, "The Role of Visual Representation in the Scientific Revolution" (2006); Dackerman et al., *Prints and the Pursuit of Knowledge in Early Modern Europe* (2011).

46. Franklin, "Diagrammatic Reasoning and Modelling" (2000); Vanden Broecke, "The Use of Visual Media in Renaissance Cosmography" (2000). In this category I include models, globes, instruments, and paper representations such as volvelles: Gingerich, "Astronomical Paper Instruments with Moving Parts," (1993); Schmidt, "Art—a User's Guide" (2006); Schmidt and Nichols, *Altered and Adorned* (2011).

47. Enenkel and Neuber, *Cognition and the Book* (2004); Tribble, " 'The Chain of Memory': Distributed Cognition in Early Modern England" (2005); Eddy, "The Shape of Knowledge: Children and the Visual Technologies of Paper Tools" (2013).

48. Bruijn, *Schippers van de VOC* (2008), 136, dismisses the exam questions as "predictable"; see also Davids's critiques in chapter 4. The manuscripts have been similarly overlooked; Crone, for one, doubts that mariners implemented many of the lessons copied into their textbooks; *Cornelis Douwes* (1941), 36.

PROLOGUE. A MODEL EDUCATION —SEVILLE, CA. 1552

1. In Spanish (and French), the term *piloto* (*pilote*) referred to men who had wayfinding responsibilities, unlike the English *pilot,* whose main task was to guide ships in harbors. See glossary.

2. The earliest details of these exams appear in AGI Patronato 251 R.22 (July 2, 1527), published in Pulido Rubio, *El Piloto mayor* (1950), 140–43. The examinations originally took place in the house of the *piloto mayor* but were soon relocated to the Casa de la Contratación.

3. The best entry points are Lamb, *Cosmographers and Pilots of the Spanish Maritime Empire* (1995); Sandman, "Cosmographers vs. Pilots" (2001); Portuondo, *Secret Science* (2009).

4. See *cédula* (royal decree) of July 2, 1527, in Pulido Rubio, *El Piloto mayor* (1950). Hakluyt, *Principal Navigations* (1600), 3.862–64, 866–68, printed account by Pedro Dias, taken by Richard [Grenville] in 1585/86; reprinted in Waters, *Art of Navigation* (1958), 555–57.

5. Hakluyt, *Principal Navigations* (1600), 866.

6. Most sixteenth-century mariners were illiterate. Cortés, *Arte of Navigation* (1561), f. iiii: "current ignorance of Pilots: at these dayes . . . fewe or none of the Pilotes can scarcely reade,

and are scarsely of capacitie to learne." Pérez-Mallaína Bueno, *Spain's Men of the Sea* (1998), 84. María Portuondo strongly doubts that "these books [were] within the average pilot's reach." *Secret Science* (2009), 54.

7. Cosmography was not distinguished from geography until the close of the sixteenth century; see Cormack, "Maps as Educational Tools" (2007), 3.622. Pedro de Medina defined cosmography as "the description of the world . . . includ[ing] geography and hydrography . . . so that we are engaged in describing heaven and the elements of which the world is made [*geos*, which is earth and *hidros* which is water]." Medina, *A Navigator's Universe* (1972), 165. Portuondo, *Secret Science* (2009), ch. 1, offers an "intellectual history of cosmography," from its classical origins (Ptolemy's *Geography*) through its fifteenth-century reinvention as a humanistic discipline, when it became part of the arts faculty of early modern universities.

8. Sacrobosco's *De Sphaera*, written in Paris in the early thirteenth century, was the most popular academic introduction to astronomy up to the seventeenth century. See Crowther et al., "The Book Everybody Read" (2015); Oosterhoff, "Reading Sacrobosco" (2015). Sacrobosco was very popular in Iberia, with more vernacular translations of the *Sphere* into Spanish or Portuguese (ten between 1510 and 1650) than in the other European vernaculars combined. Pulido Rubio, *El Piloto mayor* (1950), 73–75; Sandman, "Educating Pilots" (1999), 104. Apianus's *Cosmographia* (1524) contained four volvelles; Frisius added a fifth; see Dekker, "Globes" (2007), 3.150.

9. See Pérez-Mallaína Bueno, "Los libros de náutica españoles" (1989), 477–78, on the role of the *Sphere* in navigation texts, and Sandman, "Cosmographers vs. Pilots" (2001), on the numerous sphere treatises published by Spanish nautical authors in the 1530s and 1540s.

10. Medina, *Arte de navegar* (1545), Prologo (aii–aiiv). Although Medina aspired to an appointment at the Casa, he did not succeed; however, he was named an honorary royal hydrographer. Medina had assistance from other cosmographers in writing the *Arte*. Guillén, *Europa Aprendió a Navegar* (1943), 9; López Piñero, *El arte de navegar en la España del Renacimiento* (1986), 162; see also Cuesta Domingo, *La obra cosmográfica y náutica de Pedro de Medina* (1998).

11. Cortés, *Arte of Navigation* (1561), f. iiii, Dedicatory letter. See, e.g., ch. 9, cross staff; ch. 11, "an instrument for determining latitude at any time of the day."

12. The observer's latitude is essentially equivalent to the altitude of the celestial north pole. Sonar, " 'Regiments' of Sun and Pole Star" (2010), 9.

13. Loxodromes were first described by the Portuguese mathematician Pedro Nuñes in 1537. The Flemish cartographer Gerhard Mercator in 1569 invented a new projection where such paths appear as straight lines. On a plain sea chart, these paths appear as portions of a spiral-shaped curve. (Medina provided only a small square diagram to illustrate the concept, f. 31.) Randles, "Pedro Nuñes and the Discovery of the Loxodromic Curve" (1990).

14. Larger than most subsequent manuals, Medina's *Arte de navegar* was a 216-page folio (29 cm tall), with its title page printed in red and black. Medina included eighty-eight different images (with no movable parts), compared to Cortés, who had thirty-two (including five volvelles: three borrowed from Apianus and Sacrobosco, and two related to the Pole Star).

15. He began with a circle showing the four cardinal directions and then subdivided it until all thirty-two winds appeared in the fourth diagram. Medina, *Arte de navegar* (1545), ff. 19v–20v.

16. Medina, *Arte de navegar* (1545), f. 75v, ch. VII; the series spans ff. 72–74v. On the "sky man," Hicks, *Voyage to Jamestown* (2011), 83.

17. Medina, *Arte de navegar* (1545), rhumbs: 32v–34, declination: 39–41, 42v–49. On rhumb sailing, see chapter 1.

18. Medina's *Regimiento de navegación* (1552), which was intended for the newly estab-

lished school of navigation, omitted much of the theory of spherical geometry that Medina deemed inessential. The never-published *Suma de Cosmographia* (1561) presented information on astrology and navigation to a nonmaritime audience. López Piñero, *El arte de navegar* (1986), 162. Sandman describes this period as a moment when certain mid-level cosmographers ("theory proponents") made a bid to raise the status of the profession; they managed to persuade the king of its importance, but he then hired instructors who had university degrees. "Cosmographers vs. Pilots" (2001), 21–22, 213.

19. Portuondo, *Secret Science* (2009), 58, n. 98; Lamb, *Cosmographers and Pilots* (1995), 3.40–57.

20. García de Palacio, *Instrución Náutica* (1587); all citations from facsimile edition, trans. Bankston, *Nautical Instruction* (1988), 143.

21. García de Palacio, *Instrución Náutica* (1587; 1988). The captain's list (142) focused on personal qualities rather than on technical skills, while mariners (149) and ship's boys (150) were presumed to have neither.

22. García de Palacio, *Instrución Náutica* (1587; 1988), 143; Cuningham, *Tractaet des Tijdts. Almanach . . . byzonder voor alle Navigateurs ende Piloten* (1605), dedicatory letter.

23. Bourne quoted in Wright, *Errors in Navigation* (1599), "To the Reader"; Smith, *An Accidence or The Path-way to Experience Necessary for all Young Sea-men* (1626), 36.

24. The 1552 *cédula*, published in Pulido Rubio, *El Piloto mayor* (1950), 73–75, stipulated one year of lessons but this was scaled back to three months in 1555. See Sandman, "Cosmographers vs. Pilots" (2001), 288, n. 157. Dias reports two months; Hakluyt, *Principal Navigations* (1600).

25. The Spanish crown kept careful proprietary control over nautical charts, adding the latest geographic details to the crown's master map, the *padrón real*. In light of this, there were as yet no published atlases or nautical charts for the *pilotos* to consult. See Portuondo, *Secret Science* (2009). *Derroteros* would have circulated first in manuscript copies and from the 1520s began to be available in printed form.

26. Dias report in Hakluyt, *Principal Navigations* (1600), 3.867, and Waters, *Art of Navigation* (1958), 556; Pulido Rubio, *El Piloto mayor* (1950), 141; Sandman "Cosmographers vs. Pilots" (2001), 118.

27. Cortés, in designing his regiment of the North Star, sided with "Astronomers" rather than the tradition-bound "mariners"; *Arte of Navigation* (1561), 73v. Unfortunately, he followed a German astronomer who had published an incorrect value, making Cortés's tool dangerously inaccurate; figure P.1. Sonar, " 'Regiments' of Sun and Pole Star" (2010), 10; Waters, *Art of Navigation* (1958), 45–46. See further Sandman, "Educating Pilots" (1999), 103; Lamb, "Science by Litigation: A Cosmographic Feud" (1969).

28. Given the indoor location of the exam, it seems that *pilotos* were not expected to make astronomical observations, although regular inspections of their instruments before each voyage offered authorities repeated opportunities to verify that they knew how to use them.

29. In *Secret Science* (2009), María Portuondo shows convincingly that the Spanish crown made concerted efforts to maintain the secrecy of its geographic information (forbidding not only the circulation of charts but also the publication of certain manuals; see also Sandman, "Cosmographers vs. Pilots" (2001), 121). However, Portuondo's study ends at the close of the sixteenth century, at which point the printing press had already begun to override any efforts to block the flow of information.

30. The Portuguese crown named Pedro Nuñes royal cosmographer in 1529 and chief royal cosmographer in 1547. From 1580 until 1640, the Portuguese and Spanish courts were unified by marriage, so many of their institutions were administered similarly. Spain and the Low

Countries were linked by the Habsburg crown up through 1572. France had more limited diplomatic and commercial ties and thus may have obtained information through intermediaries. Davids, "Ondernemers in kennis" (1991), 39.

31. Waters, *Art of Navigation* (1958), 103–5, 516. Borough's commission, which was never signed, stipulated that his responsibilities would have been to "geve rules and Instructions towching the poyntes of navigacion and at all other tymes to be redye to enforme theim that seke knowleg at his handes."

32. Richard Eden produced the English-language *Arte of Navigation* (1561) at the behest of a private trading company, the Muscovy Company. López Piñero, *Ciencia y técnica en la sociedad española* (1979), 202, lists ten English editions, while Navarro Brotóns, *Bibliographia physico-mathematica hispanica* (1999), 118–120, finds just six.

33. Although the French route of transmission is obscure, Nicolai's translation was extremely popular, seeing editions in Paris and Lyon. Three Italian editions appeared in Venice in 1554, 1555, and 1609; and two English editions in 1581 and 1595. Waters, *The Iberian Bases of the English Art of Navigation in the Sixteenth Century* (1970).

34. Medina and Coignet, *De Zeevaert* (1580). See chapter 1, for Cornelis Claesz.'s Amsterdam editions. Meskens, *Practical Mathematics in a Commercial Metropolis* (2013), 16.

35. Grendler, *Schooling in Renaissance Italy* (1989); Willem Frijhoff, "Graduation and Careers" in Ridder-Symoens, *A History of the University in Europe* (1996), 2.361.

36. Hakluyt, who was not known to be in England during the period when Dias was in captivity, may have requested Dias's report by letter; see Quinn, *Hakluyt Handbook* (1974), 1.293. In his dedicatory letter to R. Cecil, Hakluyt included advice for the crown regarding "government in sea-matters"; *Principal Navigations* (1889) 12.9–10.

37. The English merchant marine briefly shared the impulse to educate and regulate sailors: in 1614 the English East India Company hired Edward Wright to teach and examine its mariners, but this arrangement lasted only two years. Waters, *Art of Navigation* (1958), 243, 294. On exams in France, see chapter 2; in England, chapter 3, and the Netherlands, chapter 4.

1. FROM THE WATER TO THE WRITING BOOK —AMSTERDAM, CA. 1600

1. Vinckboons, "[Nautische Gesellschaft/Nautical Company]" (ca. 1610), Inv. 13372, Albertina Museum, Vienna.

2. On Vinckboons (1576–1633), see Schapelhouman, "David Vinckboons" (1996); Bleyerveld and Veldman, *Netherlandish Drawings of the 16th Century* (2016).

3. Coignet, *Nieuwe Onderwijsinghe* (Antwerp: Hendrick Hendricksen, 1580), f. 3: "We generally call the art of navigation the knowledge [*wetenschap*] of rules [*regule*] to be able to safely steer a ship from one harbor to another." Using similar terms, the first English definition of the "art of navigation" appears in Dee, *Elements of Geometrie* (1570); see Waters, *Art of Navigation* (1958), 521.

4. The herring buss (*buis*) was a large two-masted fishing vessel with square sails; *fluyts* were oceangoing cargo ships. In the late sixteenth century, Holland and Zeeland held dominant positions in the herring fisheries and in the bulk trades in grain and timber with the Baltic. See Gelderblom, *Zuid-Nederlandse Kooplieden* (2000). The ubiquitous canal boats linking the River Rhine to the sea were an exception, needing little wayfinding along their route.

5. In the early 1590s companies were organized to trade in Asian spices. These were amalgamated into the famed VOC (f. 1602) — see Gaastra, *Dutch East India Company* (2003) — and WIC (f. 1621); Postma and Enthoven, *Riches from Atlantic Commerce* (2003).

6. See, e.g., Israel, *Dutch Primacy in World Trade* (1989); Israel, *The Dutch Republic: Its Rise, Greatness, and Fall* (1998); Brook, *Vermeer's Hat* (2008).

7. Bruijn, *The Dutch Navy* (1993). Burke, *Venice and Amsterdam* (1994), 66: commerce with the East Indies flourished from 1600 on, with up to 50 percent of Amsterdam's elite connected to this trade (33 percent to the WIC and VOC).

8. Lesger, *Rise of the Amsterdam Market and Information Exchange* (2006); Cook, *Matters of Exchange* (2007); Jardine, *Going Dutch* (2008). Many natural philosophers and artisans across the Netherlands concerned themselves with projects relevant to maritime interests, including lens grinders producing telescopes and other instruments, engineers and theoreticians devoting attention to shipbuilding and mechanics, cartographers developing new projections, and mathematical practitioners endeavoring to solve the longitude problem.

9. Akveld, Hart, and Van Hoboken, *Maritieme Geschiedenis* (1977), 2.200; they note that in 1670 the Dutch had more ships than England, France, and Scotland combined. Van Lottum estimates that in 1607, 18 percent of "the male labour force in the coastal provinces were employed aboard Dutch deep-sea vessels"; by 1635, this rose to "almost a quarter," then declined to 17 percent in 1680 and 15 percent in 1780. Van Lottum, "Maritime Work in the Dutch Republic" (2017), 841. Van Royen, "Mariners and Markets" (1994), 50: in 1610, there were thirty-three thousand sailors to man all the merchantmen, VOC and naval; this rose to fifty thousand in 1680, while the population remained essentially stable, at two million. De Vries and Van der Woude, *The First Modern Economy* (1997).

10. Akveld and van Hoboken, *Maritieme Geschiedenis* (1977), 2.126–30; four branches of shipping map to different geographic regions: trading in the North Sea, along the coast of France and Spain to the Mediterranean, to the Cape of Good Hope and beyond, and the navy.

11. Van Lottum, "Maritime Work in the Dutch Republic" (2017), 841, calls the Netherlands during this period "one of the first truly international labor markets." Scandinavia and Germany provided the largest numbers of men (see 843–45). Migrants exhibited relatively lower literacy rates and "skill" than Dutch workers (848).

12. On average, VOC captains and masters earned six to ten times what an able-bodied seaman did, and first mates (the navigators) four to seven times. In the Dutch navy, the master earned three times the able seaman, and the first mate outearned the master. Van Royen, "Mariners and Markets" (1994), 54. Akveld and van Hoboken, *Maritieme Geschiedenis* (1977), 2.141.

13. Although in Dutch there is a terminological distinction between *piloot* (pilot) and *stuurman* or *stuurluy* (navigator, mate), in the sixteenth and seventeenth century the terms were frequently used interchangeably; further, see glossary.

14. The organizational structure on early modern vessels appears hierarchical but emphasized redundancy and precautionary collaboration. Captains often deferred to men of modest rank who laid claim to a deeper knowledge of wayfinding, see descriptions of the collaborative process in *1681 Ordonnance,* cited in Justice, *Sea-Laws* ([1709?]), 292; García de Palacio, *Instrucción Náutica* (1587; 1988), 143; Boteler, *Six Dialogues* (1685), 6; Monson, *Naval Tracts [1624]*, 4.30. If agreement proved impossible, the navigator had the final say.

15. Crone, *Cornelis Douwes* (1941), 100; Bruijn, *Schippers van de VOC* (2008), 135. On the French trio of pilots, see chapter 2, note 21.

16. Davids contends that voyage length was the largest factor driving navigators to adopt new technologies; this explains why Dutch VOC navigators embraced "advanced" techniques. *Zeewezen en Wetenschap* (1986), 383 and ch. 11.

17. In Dutch and French, *kleyn* or *commun* and *groot* or *grand*. See, e.g., Blome, *Gentlemans Recreation* (1686), 9–10.

18. Saltonstall, *The Navigator* (1636), 14–15. See also Smith, *A Sea Grammar* (1627), 35, "The boyes . . . euery Munday . . . [are] to say their compasse." (This would become known as "boxing the compass.") Anthonisz., *Safegard of Sailers, or great Rutter* (1584), f. 76v, also provides an unlabeled diagram to facilitate memorization. In early books, these diagrams are referred to as "the winds," e.g., Cortés, *Breve compendio* (1551), f. lxi, "Demonstracion de los vientos."

19. On compasses: European mariners were familiar with magnetic lodestones from the thirteenth century on; see Jonkers, "North by Northwest" (2000), 151, 160, 164, and 180, on the discovery of the north-seeking needle and its development into the compass but also on the many problems in how the compass functioned. On tides, Howse, "Some Early Tidal Diagrams" (1985). See, e.g., Davis, *Seamans Secrets* (1595), f. A2.

20. Reyersz., *Stuurmans-Praetjen Tusschen Jaep en Veer* (1637), 22, "Why is it that a navigator . . . hits the ground sooner than the estimate [predicts?]" To guard against this common problem (which Reyersz. deemed to be caused by the distortions of plain charts; see note 43 in this chapter), navigators were advised to modify their estimates. When approaching an unfamiliar coast, the mathematician and Royal Navy gunner William Bourne (*Regiment for the Sea* [1574], f. 49v) declared it "twentie times better to be throughly persuaded that he knoweth it not, than to thinke he doth knowe it [when it was] not . . . that place." The French discussed this at length; see, e.g., Boissaye de Bocage II, "Memoire" (Havre de Grâce, 1682), AN-Paris MAR/G86, f. 22.

21. Cornelis Lastman was the first Dutch author to describe this process of careful estimation in his influential *Schat-kamer, Des Grooten See-vaerts-kunst* (Amsterdam, 1621). In *De Sectore & Radio. The description and use of the Sector* (1623), Edmund Gunter dubbed a free-floating chip of wood "the Dutchman's log." Waters, *Art of Navigation* (1958), 428, contends that Gunter expected navigators to use that simple tool in conjunction with his eponymous sector (see glossary), but it seems likely that sixteenth-century Dutch mariners carried out rough approximations without such a logarithmic device. See further chapter 4.

22. Bourne, *Regiment for the Sea* [1574], f. 43v, describes how to track "the swiftnesse or slownesse that the shippe goeth" by reciting "some number of wordes." Unfortunately, no examples of such phrases seem to survive in any English sources, but they could have been similar to the calendrical mnemonic "Thirty days hath September . . . ," which dates to the eighth century. White, "Calendar Mnemonics" (1942), and see chapter 3, note 22.

23. From at least the mid-sixteenth century, an English navigator would likely throw the "log and line" off the vessel's stern. See Hicks, "Navigating the *Mary Rose*" (2009), 357. Waters, *Art of Navigation* (1958), 434, notes that Richard Norwood was the first to describe adding knots to the rope, in his *Sea-Mans Practice* (1637). The Dutch learned of the log and line from Waghenaer's translation of William Bourne's *Regiment,* which Cornelis Claesz. published in 1594, 1599, and 1609. See also Lakeman, *Een Tractaet, seer dienstelijck voor alle Zee-varende Luyden* (1597), f. [6]v–[7]r.

24. See Mainwaring, *Sea-Mans Dictionary* (1644), 110.

25. See the diagrams in Medina (1545), Cortés (1551), and Waghenaer (1584). Blaeu (1608), ch. 20, has a tabular regiment, a "streek tafel"; Bourne (1574) provides the same information on a windrose. Cotter, "Early Tabular . . . Methods" (1978), 5, 11.

26. The small wooden log, free floating or tied to a line, soon gave its name both to the verb ("to log") and to the paper record of the speed and distance it tracked. On the evolution of formal logbooks, with mandatory daily observations stipulated by shore-based authorities, see Schotte, "Expert Records: Nautical Logbooks" (2013). Charts were not considered essential parts of the navigator's equipment until the late sixteenth century. Waters, *Rutters of the Sea* (1967); Taylor, *The Haven-Finding Art* (1956).

27. Tides shifted back forty-eight minutes each day. Tide tables suitable for use at sea were being produced in Brittany from the mid-fifteenth century. See Rose, "Mathematics and the Art of Navigation" (2002); *Sir Francis Drake's Nautical Almanack, 1546* (1980).

The systematic use of one's fingers as computational devices dates back at least to the elaborate monastic symbols described by the Venerable Bede in the eighth century, and examples appeared in navigational manuscripts by the 1300s. Cajori, "Comparison of Methods . . . Dates by Finger Reckoning" (1928); Yeldham, "Early Method of Determining Dates by Finger Reckoning" (1928); Murray, *Reason and Society in the Middle Ages* (1978); Sherman and Lukehart, *Writing on Hands* (2000).

28. From 1539, almanacs began to include calendrical information. Waters, *Art of Navigation* (1958), 17ff. On the Golden Number, Howse, "Some Early Tidal Diagrams" (1985), 374.

29. Van den Broucke, *Instructie der Zee-Vaert* (1609), 5: "Adam offered no prayer, and God's strength fed all the people" (translation changes the initial letters of each word, voiding the utility of the phrases); Seller, *Epitome* (1681), [penult f.]. A nonreligious Dutch example appears in Vooght, *Zeemans Wegh-Wyser* (1695), 320: "Als Dirk Die Gans Bout Eet, / Greep Claas Foort Al De Flees" (When Dirk ate the goose leg, Claas grabbed for all the meat). For additional examples of these mnemonic phrases, see Schotte, "A Calculated Course" (2014), 419, table 1.

30. Coignet, *Nieuwe Onderwijsinghe* (1580), f. 3.

31. Sir William Monson believed that those who chiefly sailed near coasts and harbors had more skills than those who "go to the East Indies and long voyages"; *Naval Tracts [1624]*, 30–31. For Coignet's curt list, see *Nieuwe Onderwijsinghe* (1580), f. 3.

32. See Garcie, *Rutter of the See* ([1536]), f. a.iiv; Rösler, " 'De Seekarte Ost' " (1998), 103–4; Waters, *Rutters of the Sea* (1967). The prominent maritime publisher Cornelis Claesz. printed several rutters in the 1590s (see note 73 in this chapter). On portolans, which were rarely used in northern Europe, see Randles, "From the Mediterranean Portolan Chart" (1988).

33. For an elaborate example of a mariner's personal notebook containing sailing directions, see Long et al., *Book of Michael of Rhodes* (2009).

34. On this transition (around the fifteenth century) to our modern grid, see Waters, *Art of Navigation* (1958), 47–48.

35. Coignet, *Nieuwe Onderwijsinghe* (1580), f. 3. He credits Apianus, Gemma Frisius [Phrysius], Cortés, and Medina for writing the "astronomical fundamentals" (*beghinselen*) that every pilot should know.

Jim Bennett describes the two types of sailing as "bearing and distance technique"—dead reckoning—and "Latitude sailing," noting that far from being mutually exclusive, they were "complementary and were used together for . . . centuries to come." Bennett, "Mathematics, Instruments and Navigation, 1600–1800" (2005), 44.

36. The English captain John Davis is credited with inventing the backstaff in 1594; see his *Seamans Secrets* (1595). The *hoekboog* (angle staff) was first depicted in Blaeu, *Zeespiegel* (1623).

37. The Sun's apparent path—the ecliptic—has a maximum distance of 23.5° from the celestial equator. Portuguese sailors had developed this "regiment of the sun" in the late fifteenth century, drawing upon medieval astronomical tables. On early modern tables, see note 88 in this chapter.

38. Gemma Frisius added a figure of a nocturnal to Apianus's *Cosmographia* in 1550. Coignet was the first to explain how to use the instrument.

39. The Little Dipper's position shifted back one minute a day, returning to the same place in the sky once a year.

40. See Waters, *Art of Navigation* (1958), 76. Mariners used rhumb tables for this type of sailing.

41. There were contests in Spain, the Netherlands, and England. The Dutch States General offered rewards for promising ideas, accepting forty-six submissions between 1600 and 1775; Galileo submitted one of the unsuccessful proposals. See Dunn and Higgitt, *Ships, Clocks, and Stars: The Quest for Longitude* (2014), 36; Davids, *Zeewezen en Wetenschap* (1986), 82; and chapter 5, note 59.

42. On magnetic variation, see Jonkers, "North by Northwest" (2000); Jonkers, *Earth's Magnetism in the Age of Sail* (2003). During this period the magnetic compass remained an object of fascination; see Gilbert, *De Magnete* (1600); Keteltas, *Ghebruyck der Naeld-Wiising* (1609); Le Telier, *Vray moyen de trouver la variation de l'aymant* (1631).

43. Coignet *Nieuwe Onderwijsinghe* (1580), f. 19v, discussed the dangers that could arise from "erroneous and fictitious" charts with inconsistent scales. The first English account of Mercator's system was published by Edward Wright in *Errors in Navigation* (1599). Mercator charts "became a matchless instrument for sailors because it enabled them to solve graphically and simply all the problems relating to loxodromic [rhumb line] navigation." However, they were "rarely used by seamen until well into the seventeenth century." Randles, "Pedro Nuñes and the Discovery of the Loxodromic Curve" (1990), 129.

44. E.g., Garcie, *Rutter of the See* ([1536]), "Maryners . . . with outte leaue of the mayster and drynke dranken and make noyse and stryfe . . . ," Fournier, *Hydrographie* (1667), 124, ch. 23, "Yvrognerie." *De Vroolyke Oost-Indies-Vaarder, of Klinkende en Drinkende Matroos* ([ca. 1780]) included songs about rowdy sailors.

45. Positive representations of navigators include Westerman, *Groote Christelijke Zee-Vaert* (1664), f. B3v, prefatory poem, "Ghy stoute kloeck Zee-man"; Witsen, *[Architectura navalis]* (1690), 474; Tollen, *Het Stuurmans en Lootsmans Hand-boek* (1723), ch. 1.

46. Decker, *Practyck vande Groote Zee-Vaert . . .* (1631), dedication. On trigonometry, see chapter 2.

47. The French described this daily group activity in Louis XIV, *1681 Ordonnance de la Marine* (1714), 142.

48. Waghenaer, *Mariners Mirrour* (1588), 3.

49. Waghenaer's sailing career stretched from 1550 to 1579, at which point he became a collector of maritime dues in the important port of Enkhuizen, his hometown. When he lost this job in 1582, he turned to cartography. His atlases would prove so successful that the whole genre was soon known by an anglicized version of his name, "waggoners." Molhuysen and Blok, *Nieuw Nederlandsch biografisch woordenboek* (1911), 7.1304–5; Koeman, "Lucas Janszoon Waghenaer" (1965).

A request appears in Waghenaer, *Spiegel der Seefahrt* (1589), "Der Autor zum Leser." Cornelis Claesz. issued similar appeals in other publications; see Van Hollesloot, *De Caerte vander Zee* (1587), cited in Schilder and Van Egmond, "Maritime Cartography in the Low Countries" (2007), 3.1390, n. 37. Waghenaer also had pilots attest to the quality of his charts in front of a notary, see Schilder, *Early Dutch Maritime Cartography* (2017), 74. A later example appears in Moore, *Practical Navigator* (1791), final page, advertisement. Such pleas can sound like ritual, but they were evidently quite practical; see Ann Blair's work on "Conrad Gesner's Paratexts," *Gesnerus* 73, no. 1 (2016): 73–122.

50. Waghenaer, *Thresoor der zee-vaert* (1592), ch. 1. Georges Fournier, S.J., held similar disdain for the "frank beast" who withheld information about his navigational decisions: Fournier, *Hydrographie* (1667), 124. This idea, of navigators willfully hiding the records of their calculations "to make themselves indispensable," could be corrected, the Sieur de Viviers suggested to Colbert, by "establish[ing] Schools of Navigators." See Viviers, "Memoire sur les Ecoles de Pilotes" (April 1681), AN-Paris MAR/G86, ff. 47, 49v.

51. Boteler, *Six Dialogues* (1685), 3.

52. In its English translation, *The Light of Navigation* (1612) is subtitled "A breef and short introdvction for the vnderstanding of the Celestiall Sphaere, as farre as it concerneth the Art of Seafaring." In 1595 Blaeu spent six months in Hven studying cutting-edge astronomy with Tycho Brahe. Upon returning to the Netherlands, he turned his back on his family's herring business, as well as the carpentry and clerical work he had been doing, and set up shop as a printer, globe and instrument maker, and cartographer. Van Netten, *Koopman in Kennis* (2014), ch. 2.

53. Authors of numerous early atlases invoked the imagery of a light shining, in reference both to the ways in which maps brought far-flung corners of the world into the light and to their authorial role in illuminating such caches of wisdom; e.g., Ortelius's pioneering terrestrial atlas, *Theatrvm orbis terrarvm* (Antwerp, 1570) was translated as the "mirror" or "theater" of the world (*Spieghel der werelt*, 1577; *Miroir du Monde*, 1579; *Theater of the World*, 1603). See Reeves, *Galileo's Glassworks: The Telescope and the Mirror* (2008). As books were accepted as repositories of knowledge, the metaphor of the shining light would be replaced by that of the "treasury"; see plate 4.

54. In the final decades of the sixteenth century, a few teachers were offering lessons in Amsterdam. The preface of Albert Haeyen's book of sea charts, *Amstelredamsche Zee-Caerten* (1585), indicates that he was running a school; Robbert Robbertsz. "Le Canu" taught in the city between 1592 and ca. 1611, and around 1600 the theologian, astronomer and cartographer Petrus Plancius offered occasional lessons, including to the crew of the merchant vessel about to set sail for China. Crone, *Cornelis Douwes* (1941), 95, 97; Davids, "Ondernemers in kennis" (1991), 37. See Davids, Everaert, and Parmentier, *Peper, Plancius en Porselein* (2003), for current theories about Plancius's role as an educator.

55. Burger, *Amsterdamsche Rekenmeesters* (1908). Kool, *Die conste vanden getale* (1999), on the introduction of "Italian bookkeeping." Meskens, *Practical Mathematics in a Commercial Metropolis* (2013), ch. 4, for Antwerp arithmetic schools.

56. During the sixteenth century, presses across Europe issued ever-more affordable documents for all sectors of society. The market was particularly healthy in the Low Countries, where presses put forth not just books but other printed material, including engraved prints, maps, and blank books for record keeping. See Suarez and Woudhuysen, *The Book: A Global History* (2014), 351–56; Lesger, *Rise of the Amsterdam Market* (2006).

57. Anthonisz., *Onderwijsinge vander zee* (1558) 42 pp.; ed. princ. 1544 (no known copies; included the first printed coastal profiles); Keuning, "Cornelis Anthonisz" (1950), 54; Meskens, *Practical Mathematics in a Commercial Metropolis* (2013), 141–42.

58. The Eighty Years' War (1568–1648) saw the Northern and Southern Netherlands split along religious lines; see Israel, *Dutch Republic* (1998). Suarez and Woudhuysen, *The Book* (2014), 353, notes that the war brought into the new Protestant country scores of skilled printers and booksellers as religious and economic refugees.

59. Waghenaer, *Mariners Mirrour* (1588), ch. 25: "Amongst manie Pilots there is an opinion that they had rather use written mappes, then such as are printed, esteeming the printed mappes to be imperfect." Coignet was similarly concerned about cartographic inaccuracies: *Nieuwe Onderwijsinghe* (1580), f. 29. Blaeu vouched for his superior printed productions, noting that not only had his predecessors made mistakes (see note 77 in this chapter) but the ocean bottom and width of channels had changed since they had published their works. Blaeu nonetheless included in *Licht der Zee-vaert* (1608), ch. 22, five pages of tide tables. Waters, *Art of Navigation* (1958), 326.

60. Anthonisz., *Onderwijsinge vander zee* (1558), includes a list of "harmful errors," caused in part by mates "through negligence, inattention, drunkenness and inexperience"; cited in

Keuning, "Cornelis Anthonisz" (1950), 54. Colbert's French naval correspondent, Sieur de Viviers, feared that most pilots could neither read nor write, "Memoire" (1681), f. 49. See also prologue, note 6, for Cortés's candid opinion. But the Dutchman Robert Cuningham was more positive, see prologue, note 22.

61. On the popularity of travel literature, Youngs and Hulme, *Cambridge Companion to Travel Writing* (2002); Lunsford, *Piracy and Privateering* (2005). On dialogue pamphlets, Dingemanse, "Rap van Tong, Scherp van Pen: Literaire Discussiecultuur in Nederlandse Praatjespamfletten (circa 1600–1750)" (2008).

62. The Blaeu firm published blank books and regulations, see, e.g., NL-HaNA VOC 5165.

63. Waghenaer, *Mariners Mirrour* (1588), preface, and see note 75 in this chapter.

64. The French term *amateur* appears only in works translated from the Dutch: Coignet, *Instruction nouvelle* (1581), dedication; Gietermaker, *Le Flambeau Reluisant, ou proprement Thresor de la navigation* (1667), XLV Proposition; the prefatory address to the readers is similarly glowing: "Bien-Aimé Lecteur." On the category of *liefhebber* and its influence in art and science, see Keller, "Art Lovers and Scientific Virtuosi?" (2016).

65. Reyersz., *De vaste Grondt der Loflycker Zee-vaert* (1622), 3; "Leer-Gierighen Lezer." In England, by contrast, Blagrave, *Mathematical Jewel* (1585), addressed the "curteous, gentle" reader, whom he assumed had a separate "chamber at home" to serve as a private study. Zamorano, *Compendio del arte de navegar* (1588), "Al Lector" addressed its elevated reader as "curioso y discreto."

66. Changing print technology made Waghenaer's atlases possible, greatly simplifying the problem of accurately replicating cartographic details. See Schilder, *Early Dutch Maritime Cartography* (2017); Van Netten, *Koopman in Kennis* (2014). On the benefits of tables for sailors, see Digges, *Prognostication everlasting* (1567), subtitle, and Thomas Blundeville, *Exercises* (1594), f. 341.

67. Davis, *Seamans Secrets* (1595), f. [C3]v. A Pole Star volvelle could be printed on just two sheets of paper, making it much less expensive than, for example, the six pages and sixteen woodblocks that Medina had needed to present the same information; see figure P.3.

68. Blaeu, *Light of Navigation* (1612), "To the Reader" expressed concerns about "unnecessarie things / wherewith some Seafaring men trouble themselves in vaine" and feared that "these newe found Lengths [longitude calculations] are built . . . without ground or true foundation." Waghenaer felt that geometry was too difficult: *Spieghel der Zeevaerdt* (1584), xvi. Conversely, Jarichs van der Ley, *'T Ghesicht des grooten Zeevaerts* (1619), "To the Art-loving Reader."

69. See, e.g., Van Dam, *Nieuwe Hoornse Schatkamer* (1751), "alle noodige zaken."

70. The business, "*opt Water int Schrijfboeck by de oude Brugghe*," was also known as the "Golden Notebook" (*Vergulden Schrijfboeck*). For an overview of the process and the personnel involved in printing in the sixteenth century, see Gerritsen, "Printing at Froben's" (1991).

71. Claesz. sold books in his hometown of Enkhuizen beginning in 1578 and ran a shop from 1593 to 1595 in the port city of Hoorn. He had a huge stock, sourced largely from the Frankfurt book fair and from fellow publishers. (Between 1604 and 1608 he acquired 1,273 new titles at Frankfurt.) See Van Selm, *Nederlandse boekhandelscatalogi* (1987), 176–79; Lesger, *Rise of the Amsterdam Market* (2006), 228–237. Among his nonmaritime publications, Claesz. published Linschoten's great collection of travel narratives (1596) and the first Dutch translation of King James's bestseller *Basilikon Doron* (1603). Schmidt, *Innocence Abroad: The Dutch Imagination* (2001), 161. Verwey calls Claesz. a "central entrepreneur" whose name is "inextricably linked to the rise of Amsterdam as a centre for publishing, bookselling and printing." "Amsterdamse uitgeversbanden van Cornelis Claesz" (1973), 60.

72. Plantin's *Thresoor* (1592) was already being planned when Claesz. produced the new edition of the *Spiegel der Seefahrt* (1589). Schilder, *Early Dutch Maritime Cartography* (2017), 98, 108. Claesz. would also reissue the *Thresoor* (1596).

73. These included Waghenaer's translation of William Bourne's 1574 *Regiment of the Sea*, as *De Const der Zee-vaerdt* (1594); Everaert's translation of Zamorano's *Compendio del arte de navegar* (1581), rendered as *Cort Onderwiis Vande Conste der Seevaert* (1598); and Thomas Hood, *T'ghebruyck van de Zeecaerte* (1602). He also published several rutters (see Schilder, "Cornelis Claesz" [2003], ch. 2) and sold manuscript charts. See his sale catalog, *Const ende caert-register* (1609), described in Van Selm, *Nederlandse boekhandelscatalogi* (1987).

74. Schilder and Van Egmond, "Maritime Cartography in the Low Countries during the Renaissance" (2007), 3.1395, citing *Const ende caert-register* (1609), f. B3. Parts 1 and 2 of the *Spiegel* were also available separately.

75. The work covers the coastline from the Zuider Zee to Cádiz, that is, most of the western coast of Europe. It also shows the North Sea and the Baltic. Waghenaer, *Thresoor der zee-vaert* (1592), 202–4, on merchants.

76. Waghenaer's mnemonic feat was feasible because before the 1609 invention of the telescope only forty-eight constellations and fewer than eleven hundred stars were officially known and named in Ptolemy's second century B.C. *Almagest*.

77. Waghenaer, *Thresoor der zee-vaert* (1592), ch. 4, enumerated nearly thirty criticisms of an early collection of charts by Adriaen Gerritsz. A generation later, it was Blaeu who capitalized on Waghenaer's inaccuracies, declaring his intention "to clense, correct and amende [Waghenaer's charts] of all their faults" in his *Light of Navigation* (1612), "To the Reader."

78. On the differences between Waghenaer's two atlases, especially the larger scale of the *Thresoor*'s charts, see Skelton's introduction to the facsimile, *Thresoor der zee-vaert* (1592; 1965), xiv. The many editions of Ortelius's *Theatrvm orbis terrarvm* (Antwerp, 1570) were also oblong quartos.

79. Waghenaer, *Spieghel* (1584), f. 32 circular diagram; f. 21 volvelle. The latter engraving was sold as a separate broadside with a French title: "Dessein Du Compas Tournant, ou Instrument aux Estoilles" ([Antwerp], 1583). It was reengraved in English for Waghenaer's *Mariners Mirrour*, see note 99 in this chapter.

80. Medina and Coignet, *De Zee-vaert oft Conste van ter Zee te varen* (1589). See prologue, note 33, for editions in other languages. Coignet was involved in publishing numerous editions of Ortelius's famous atlas, tutored private pupils, worked as a wine gauger for the city of Antwerp, and served the Habsburg archduke Albert as a mathematician and engineer. See Meskens, *Practical Mathematics in a Commercial Metropolis* (2013), ch. 2, for Coignet's varied career, and ch. 8, on "The Art of Navigation." On his significant library of 650 books, as well as his instruments, Meskens, 21. See also Verlinden, "De Nederlandse vertaling van het 'Arte de Navegar'" (1987).

81. Medina earned part of his living from making instruments. Coignet's father Gillis was an instrument maker; see Meskens, *Practical Mathematics in a Commercial Metropolis* (2013), 11. Coignet's invention, derived from the "meteoroscope," needed to hang freely, be rotated to align with north and south, and then have the Sun shine through two holes in the alidade. It did evidently see some use; the Casa de la Contratación owned one (Meskens, 151, 154–55). Schilder, *Early Dutch Maritime Cartography* (2017), 243: "The contribution by Coignet increased the value of Medina's book considerably."

82. Coignet, *Nieuwe Onderwijsinghe* (1580) and *Instruction nouvelle* (1581). Each has four volvelles: front and back of Astrolabe (ff. 9, 10; pp. 30, 34); Nocturnal (f. 18; p. 64); age of the Moon (f. 23; p. 89).

83. Publishers stepped into the field in cities such as Franeker, Gouda, Hoorn, Middelburg, and Rotterdam.

84. Among the myriad sobriquets are those of Reyersz., "Liefhebber der Zeevaert" (1637) and Oostwoud, "een Liefhebber van de Konst" (1699). English authors, such as Collins (1659), Gadbury (1659), and Leybourn (1669), preferred "Philomath."

85. Keteltas, *Ghebruyck der Naeld-Wiising* (1609), promoted several his own instruments. Jarichs van der Ley published a series of texts about a new solution to the longitude problem. On the similar strategies of Galileo and other Italian figures, see Marr, *Between Raphael and Galileo* (2011).

86. Veen, *Tractaet Vant Zee-bouck-houden* (1597). On examiners' works, see chapter 4, note 50.

87. Cuningham, *Tractaet des Tijdts* (1605); similar to Cornelis Doedsz.'s *Graedtboeken nae den ouden stijl* (1595) and *Graetboecxken naden nieuwen stijl* (1580s). Cuningham included rhyming questions directed to his former students, "discipulen" who became "clever pilots" and a "sail maker." He cited thirty sources, ranging from the expected — Apianus, Coignet, Medina, Waghenaer, and William Bourne [Boorn] — to the Bible and Hermes Trismegistus.

88. [Académie royale des sciences], *Connoissance des temps* (1679–). [Commissioners of Longitude], *Nautical Almanac and Astronomical Ephemeris* (1766–). The Dutch would begin publishing similar tables in 1788. For an introduction to the errors that frequently distorted these tables, as well as a sense of the mathematical and instrumental facility necessary to compile them, see Roche, "Harriot's 'Regiment of the Sun' " (1981).

89. Davids, "Navigatieonderwijs" (1988), 68–69.

90. Van den Broucke, *Instructie der Zee-Vaert* (1609), 99; (1610), 107. "Note: (Re)read once daily, so that one learns by heart" (*van buyten kan*).

91. Van den Broucke, *Instructie der Zee-Vaert* (1609), "Tot den Leser" and following page. Van Gent, "Het Sterrenlied in Het Hollandse Zeevaartonderwijs" (2005).

92. Van den Broucke, *Instructie der Zee-Vaert* (1610), 40, 68; (1609), f. 21: a very easy method, "sitting in the ship without seeing the sky, without using any cross-staff," involves using the acronymic phrase (on p. 5) that works the same way as the Golden Number: "In Korte Jaren Hebben [De] Hollantsche Ghesellen Gheluckich Ghevaren / Haren Haer Is Connick [= K]" (In few years the Dutch groups have sailed fortunately; their Lord is king). Referring to the first letters of each word in the phrase, the navigator knows that in January he should subtract 9 (I and J are equivalent) degrees, in February 10 (K), and so on. Such phrases could substitute for current almanacs. See other examples in Gietermaker (1659), Van Breen (1662), and Vooght (1695).

93. Van den Broucke, *Instructie der Zee-Vaert* (1609), 56, [62], and subhead below from prefatory poem entitled "'tBoeck tot den Lezer," following the table of contents.

94. Van den Broucke, *Instructie der Zee-Vaert* (1609), 4. In Reyersz.'s popular dialogue, the first thing the experienced Veer offers to teach Jaep is the compass and the zodiac; Reyersz., *Stuurmans-Praetjen Tusschen Jaep en Veer* (1637), 3.

95. See, e.g., manuscripts by Guillaume Le Testu, *Cosmographie Universelle* (Havre de Grâce, 1555), BnF SHD D.1.Z14, after final map; De Vaulx, "Premieres oeuvres" (Havre de Grâce, 1584), f. 8r, BnF Ms. Français 9175; [Le Vasseur?], "Géodrographie" (ca. 1610), PWDRO Ms. 1334, 25 (after Aristotelian elements and diagrams explaining eclipses). Later textbooks *did* include diagrams of compasses (see editions of Gietermaker from 1677 to 1774); on these, as well as on student manuscripts, see chapter 4.

96. *A Discourse of Navigation* (1702), 1. See also Boteler, *Six Dialogues* (1685), 60. The En-

glish authors who did expound upon the weather as it related to navigation (e.g., Leonard Digges, Timothy Gadbury, and Balthazar Gerbier) were astrologers or entrepreneurial educators rather than mariners. In France, Louis XIV's *1681 Ordonnance* (1714), 64–71, stipulated that naval trainees should learn the topic of how "Meteors could predict Storms" within the prescribed curriculum for Hydrography classes but provided no details. French hydrography teacher Pierre Cauvette relied upon Pliny's observations in his *Nouveaux elemens d'hydrographie* (1685), prop. XXII. The learned mayor of Amsterdam, Nicholas Witsen, noted the most extensive list of weather signs in his comprehensive work on shipbuilding. He included stilted renditions of familiar sayings ("The sun red in the evening, results in good weather the next day") on the same page as quotes from Seneca, Pliny, and weather tips of the ancient Egyptians; Witsen, *[Architectura navalis]* (1690), 474.

97. See reference to Robbertsz.'s "Liedeken" in introductory lessons transcribed by Capt. Willem Isbrantsz. Bontekoe, Journal (1622–25), NL-HaNA VOC 5049. Robbertsz.'s verses appeared in every edition of Reyersz.'s *Stuurmans-Praetjen Tusschen Jaep en Veer* between 1614 and 1667. Gietermaker included the "Sterre-liet" in an early version of his landmark work, but he noted that was just to avoid wasting half a sheet of paper; he omitted it from subsequent editions; *Drie boecken . . . Onderwijs der Navigatie* (1666), 28. See Ong, *Orality and Literacy* (1982), on the role of features like rhyme and melodies from familiar songs in early modern memorization and in oral culture generally.

98. Medina, *Arte de Navegar* (1545), f. 78; Medina and Coignet, *De Zeevaert* (1580), ff. 63v–65v.

99. See Waghenaer, *Mariners Mirrour* (1588), f. B2v: "Let the upper part of the booke, which representeth the North, be turned to the South. . . . For so shall these stars stand in order." In the *Thresoor der zee-vaert* (1592), p. xl, the *Wachters* appear only in a table, another indication that working mariners could already visualize them.

100. Van den Broucke, *Instructie der Zee-Vaert* (1609), 14. Contemporary authors who did not include volvelles instead provided large diagrams of the constellation; see Blaeu, *Licht der Zee-vaert* (1608), 35, ch. 19; Reyersz., *Stuurmans-Praetjen Tusschen Jaep en Veer* (1637), 29. Among the few texts that adopted the corporeal language, see BnF Ms. Français 2482, "L'Usage de la navigation" (ca. 1605), f. 17: "Le zenit de lhomme est la poinct au ciel desus sa teste . . ."; Syria, *Arte de la Verdadera Navegacion* (1602), 151.

101. Blaeu, *Licht der Zee-vaert* (1608), is a notable exception; rather than beginning with mnemonics and calculating tools for the Golden Number and the epact, his first chapter is "About the poles of the world" and includes a cosmographical diagram to orient his readers. The Spanish rarely discussed tides; see Medina's theoretical chapter on "pull of the Moon," *Arte de Navegar* (1545), book 7.

102. See chapter 2, note 38, for René Bougard's miscalculation of the Channel tides.

103. For more than a century, Gietermaker included a pair of volvelles (e.g., 1677 ed., pp. 22–23) that help calculate the age of the Moon and the time of the tides. Elly Dekker notes a text published in 1850 that continues to rely upon the epact to calculate the tides; Dekker, "Epact Tables on Instruments" (1993), 323, n. 45. By contrast, virtually no English or French texts include thumb diagrams, although Fournier, *Hydrographie* (1643), 432, XIV "De l'Epacte . . . ," mentions the "racine de pouce," and Newhouse, *Whole Art of Navigation* (1685), illustrates a misshapen thumb. See chapter 3, note 22.

104. E.g., Van Dam, *Nieuwe Hoornse Schatkamer* (1712), "Voorreden": "Comprehension [of subjects like geometry] is the greatest support for a good and vast memory, . . . [while] that which is built without foundation, rests only in concepts, on the testimony of someone else."

2. "BY THE SHORTEST PATH": DEVELOPING MATHEMATICAL RULES —DIEPPE, 1675

1. Denys, *L'Art de naviger, . . . ou Traité des latitudes* (1673), 1, 6. Within the chariot, the four wheels represented the key elements that a navigator must track: "the rhumb of the wind" (compass heading); the "route, and the progress the boat makes towards its goal" (distance); latitude; and longitude.

2. Denys's disciples included Voutremer, who may have been Denys's nephew and taught at Bayonne; Jean-Baptiste Franquelin, who taught in Quebec City; François Gaulette [Cauvette] (possibly related to Pierre Cauvette, the author of *Nouveaux elemens d'hydrographie* [1685]), who taught in Toulon. At least three more of Denys's students published their own textbooks: Le Cordier, who taught at Havre de Grâce, Berthelot at Marseille, and G. Coubert [Coubard] at Brest. Anthiaume, *Denys de Dieppe* (1927), 26–29; Vergé-Franceschi, *Marine et éducation* (1991) 154–155; Vergé-Franceschi, "Entre ciel et mer" (2009), 66.

3. Denys was one of the first authors to champion the idea of a central repository of journals, intended to aid future captains in planning their routes, to say nothing of freeing France from an embarrassing reliance on navigators from neighboring nations. Denys, *L'Art de naviger . . . ou Traicté de la variation* (k1666), 48. Schotte, "Leçons enrégimentés . . . Evolution of the Nautical Logbook in France" (2013), 98–100.

4. As noted by Waters, *Art of Navigation* (1958), 340, the German mathematician Pitiscus coined the term in his *Trigonometria* (Heidelberg, 1595). The "doctrine of triangles" was discussed earlier; see, for instance, Thomas Blundeville's *Exercises* (1594), which included pocket-sized tables and explained trigonometric functions. Stevin, *De Driehouckhandel* [*Trigonometry*], was published in *Wiskonstighe g[h]edachtenissen* (1605–8) but written earlier. See also Taylor, "John Dee and the Nautical Triangle, 1575" (1955).

5. Davis, *Seamans Secrets* (1595), f. C3; Pitiscus, *Trigonometry, or, The Doctrine of Triangles* (1614), subtitle.

6. John Napier dedicated the first English edition of his work, *A Description of the Admirable Table oe* [sic] *Logarithmes* (1616), to the "Worshipfull Company of Merchants of London trading to the East Indies" f. A2; f. [A4].

7. The Oxford mathematician Henry Briggs produced the first tables of common logarithms, *Logarithmorum chilias prima* (1617) covering 1,000 numbers, and *Arithmetica logarithmica* (London, 1624) 1 to 20,000 and 90,001 to 100,000. Adrian Vlacq worked with Ezechiel de Decker to extend these a decade later; see Decker's *Nieuwe Telkonst, inhoudende de Logarithmi voor de Ghetallen beginnende van 1 Tot 100000* . . . (1626–27).

8. See Van Poelje, "Gunter Rules in Navigation" (2004), 11–22; and glossary. Examples include Blondel Saint-Aubin's pioneering work on the sinical quadrant, *Le Véritable art de naviger par le quartier de réduction* (1671), and the multipurpose "universal circle" invented by Havre de Grâce instructor, Boissaye du Bocage II, *Explication et usage d'une partie du Cercle Universel . . .* (Havre de Grâce, 1683). In his *Geometrical Sea-Man* (1652), Henry Phillippes recommended both geometry and paper instruments over books of mathematical tables.

9. In England, according to Collins, *Navigation by the Mariners Plain Scale New Plain'd* (1659) 22, "about 24 of our common English Sea leagues" equal "a degree of latitude under the Meridian." In France, "25 lieues par dégré"; *Dictionnaire de l'Académie française* (1762), vol. 2, s.v. "lieue."

10. Norwood's work was followed quickly by volumes like Skay, *Sea-Mans Alphabet and Primer* (1644), which was intended "for the Better Perfecting of Young Sea-Men in the Art

of Navigation, Viz. in All the Parts of Arithmetique and Geometry, the Doctrine of the Sphaeriques, and of Right-Lined Triangles."

11. Robbertsz., *Onder verbeteringhe* (1600), Dii. was already familiar with spherical trigonometry, noting that "sinus" tables gave more perfect answers than instruments. Lastman, in his *Schat-kamer* (1629), introduced trigonometry as part of his discussion of how to assess the accuracy of a cross-staff.

12. Lastman, *Beschrijvinge van de Kunst der Stuer-luyden* (1657). Basic tide calculations are not treated until proposition 61.

13. See chapter 4, notes 89–90. On the history and applications of spherical trigonometry, see Van Brummelen, *Heavenly Mathematics* (2013), especially ch. 9.

14. Gietermaker, *Driehoex-rekening* (1665). This was a change from his first, more conventional work, *Vermaeck der Stuerlieden* (1659), which opened with the standard calendar questions, and his most popular textbook, *'t Vergulde Licht* (1660), which treated trigonometry only in book II, one hundred pages in. Van Nierop, *Mathematische Calculatie . . . door de Tafelen Sinus Tangents of Logarithmus wiskonstelick uyt te rekenen* (1659). Briggs's dense tables (note 7 in this chapter) continued to be included in numerous books, especially Dutch ones, well into the eighteenth century; see chapter 4, note 99.

15. Spanish and Portuguese mathematicians such as José Zaragoza and António Carvalho da Costa took up the question of trigonometry in the 1670s, but nautical textbooks did not include trigonometric tables until the 1730s; see Navarro-Loidi and Llombart, "Introduction of Logarithms into Spain" (2008), 90.

16. Despite the arguments of Braudel, *Mediterranean and the Mediterranean World* (1972), only select areas of France were oriented toward the water. Le Goff describes the "many diversities" of the country's maritime community; see "Labour Market for Sailors in France" (1997), 293. Breton, Norman, Basque, and Portuguese fishermen rushed to explore the large schools of cod and whales near Labrador by the beginning of the sixteenth century but transmitted their expertise and techniques via nontextual avenues. See Dúo, "L'enseignement de la science nautique en Labourd" (2002). For their part, Mediterranean navigators used few astronomical techniques, a situation that became problematic when trying to redeploy them in the Atlantic, e.g., Milliet de Chales, *L'Art de Naviger* (1677), avant propos: "The regular errors of many navigators [*pilotes*] in the Mediterranean Sea, and their frequent shipwrecks, clearly demonstrate that knowing how to prick a chart and recognize the land [coastline] is not sufficient." An instructor at Marseille, Berthelot, had similar critiques in *Abrégé de la navigation* (1691), Epistre.

17. See Cabantous, *Dix milles marins face à l'ocean* (1991), for the important concept of "maritime cities" and "maritime towns," and Gerard le Bouëdec's expanded definition of "gens de mers," not just those working on boats but in jobs connected to the sea in any manner. Le Bouëdec therefore estimates 350,000 "gens de mers" in the early seventeenth century and 500,000 two centuries later. During that period the population swelled 50 percent to nearly 30 million; thus 17 percent of the French population earned part of its livelihood from the water. Le Bouëdec, *Activités maritimes et societés littorales de l'Europe atlantique* (1997), 254, 257. For a narrower definition, Le Goff, "Labour Market for Sailors in France" (1997), 300: peacetime: 2,000 to 3,000 skilled men in the navy; under Louis XIV, 15,000 at the 1690 peak (but the bulk of these were oarsmen); 43,000 skilled seafarers in France in the late seventeenth century (which grew to 55,000 by 1789).

18. The 1584 Ordonnance proscribed men from carrying out the duties of pilots or masters if they had not publicly proven their abilities in an exam. Cited in Cleirac, *Us et coutumes de la mer* (1647), art. LIV. In 1615, l'Amiral de France, Henry de Montmorency, commissioned Jean

Guérard "master hydrographer and commissioner examiner of pilots" in Dieppe; his manuals on pilotage have been lost. Anthiaume, *Evolution et enseignement* (1920), 1.83–84.

19. Louis XIII's minister Cardinal Richelieu appointed three teachers to instruct sixteen gentlemen "and all the youth who wish to learn." For annual salaries of four hundred livres, these three teachers were directed to hold a series of free public lectures on the subject, three times weekly. Anthiaume, *Evolution et enseignement* (1920), 1.13–14 (code Michau act. 433); Vergé-Franceschi, *Marine et éducation* (1991), 51.

20. Guibert, *Mémoires pour servir à l'histoire de la ville de Dieppe* (1878), 1.348–50. In the Mediterranean sphere, Charles IX founded the first hydrographic school at Marseille in 1571, which became the École nationale supérieure maritime. Oratorians were reportedly teaching hydrography in Nantes in 1625.

21. *1634 Ordonnance du commandeur de la porte*. BnF N.A.Fr. 9490, f. 13, cited in Anthiaume, *Evolution et enseignement* (1920) 1.14. The French term *pilote* was used for those who carried out small *or* large navigation. On medieval French terminology, see Barbier, "Old French *laman, loman* pilot . . ." (1949), 12–22. Vincent Avril, who served as first pilot (*premier pilote*) on the *Lys* sailing to India in 1747–49, noted in his journal (TNA HCA 32/257) the presence on board of a *pilote pratique* and a *pilote costier*, whose purviews were harbor and coastal sailing.

22. This number would expand to 5,254 vessels by 1776. Le Goff, "Labour Market for Sailors in France" (1997), 294.

23. See, e.g., the extensive discussion of Gallic accomplishments in Fournier, *Hydrographie* (1643), book VI, "Mémoires de la Marine de France" (289–389); and on the maritime-themed ballets performed for Louis XIII, see Canova-Green, "Dance and Ritual" (1995), 398.

24. Anthiaume, *Pierre Desceliers, père de l'hydrographie* (1926); Anthiaume, *Denys de Dieppe* (1927), 42; Toulouse, "Les hydrographes normands" (2004); Van Duzer, *The World for a King: Pierre Desceliers' Map of 1550* (2015).

25. Vergé-Franceschi in Charon, Claerr, and Moreau, *Le livre maritime* (2005), 31–32, notes that Dieppe held a "quasi-monopoly for publishing texts related to the sea," especially voyage accounts; he names it "the realm's first maritime crucible, scientific, literary, and geographic in nature." See also Vergé-Franceschi, "Entre ciel et mer" (2009), 66.

From 1618 until the 1640s, Nicolas Acher published French translations of foreign works: Nicolas Le Bon's edition of Manoel de Figueirdo *Hydrographia, Exame de Pilotos* (Lisbon, 1608) and Jean Le Telier [Tellier]'s interpretation of John Davis's voyage to the Indies, among others.

26. Denys, *Traité de la variation* (1666), 48.

27. Denys, *Façon nouvelle de naviger par les nombres* (1648), no copies known; *Table de la déclinaison du soleil . . . et deux tables pour l'estoille du Nord* (1663) and two additional editions); *L'Art de naviger par les nombres* (1668) and two additional editions [hereafter *Nombres*]; *L'Art de naviger perfectionné par la cognoissance de la variation de l'aimant, ou Traicté de la variation* (1666) with three additional editions in both Havre de Grâce and Dieppe [*Variation*]; *L'Art de naviger dans sa plus haute perfection, ou Traité des latitudes* (1673), one edition only [*Latitudes*].

28. "Grand maître" César de Bourbon named Denys to the "office of commissioner-examiner of pilots for all the realm, and of professor of hydrography" in 1664. Colbert expedited Denys's formal commission, which was signed by Louis XIV on November 22, 1665. (Colbert appointed Georges Boissaye du Bocage to the l'École royale d'hydrographie at Havre de Grâce the following year.) Asseline and Guérillon, *Antiquitez et chroniques de la ville de Dieppe* (1874), 2.322; Anthiaume, *Evolution et enseignement* (1920), 1.132, 137, 191; Anthiaume, *Denys de Dieppe* (1927), 22–23.

29. Le Corbeiller, "Denys à Colbert" (1916), 49 #8 (Sept. 9, 1667), Mme. de Sévigné's son paid Denys the compliment of attending a lecture with some friends en route to England.

30. After founding more schools around the country, Colbert collected similar information from other instructors: see series of "Mémoires" in AN-Paris MAR/G86. For a discussion of Colbert's operational style, see Soll, *Information Master* (2009).

31. Le Corbeiller, "Denys à Colbert" (1916), clock: 44, #2 (Oct. 1, 1665); magnets: 53–54, #13 (June 10, 1669); renovations: 52, #11 (Oct. 21, 1668); noise: 54, #13 (June 10, 1669).

32. Le Corbeiller, "Denys à Colbert" (1916), book: #11 (Oct. 21, 1668): "I'm not sure if you received the rest of *L'Art de naviger par les nombres* — which I sent to Mr. Perrau[l]t in Paris for you." Also, Depping, *Correspondances*, 559 (Dec. 1, 1665), Denys to Colbert concerning the preface of *Traicté de la Variation* ("nostre épistre liminaire"), and notes 54, 55 in this chapter. On his students: Le Corbeiller, 46, #5 (Feb. 21, 1666): Virtually all "cappitaines [et] pilottes" in the region had studied with Denys (they were "de sa façon") and neighboring provinces followed his lessons. He gave warm recommendations about the "vivacity and maturity" of the young men; Le Corbeiller, 48, #7 (Nov. 2, 1666).

33. Le Corbeiller, "Denys à Colbert" (1916), 49, #8 (Sept. 9, 1667), noted that he had spent two years running the school without a break. In September 1672 he reported to Colbert that he had 100 students; in 1678 he had 210; Anthiaume, *Denys de Dieppe* (1927), 16, 24. In 1673 he claimed 200 or 300 since the school's founding; Le Corbeiller, 54, #14 (Nov. 24, 1673).

34. Le Corbeiller, "Denys à Colbert" (1916), 52, #12 (Mar. 31, 1669). Anthiaume, *Denys de Dieppe* (1927), 14. Such "auditeurs libres" (who included a few priests) were enthusiasts similar to Dutch *liefhebbers*. Le Corbeiller, 50–51, #10 (Dec. 1667), list of new students (noting students as young as eleven and as old as thirty-four); 54–55, #14 (Nov. 24, 1673), career prospects.

35. Between 1669 and 1795, the French navy was manned by a rotating group of mariners (*système des classes*). Under this "inscription system" (*inscription marine*), all sailors, including fishermen and merchant marines, were enrolled on a list according to their home port and expected to serve in the navy on a rotating basis. Asher, *Resistance to the Maritime Classes* (1960). Colbert reported that those navigators who were up for "the most difficult sailing" "preferred to give themselves to merchants." French sailors often went in search of better wages and working conditions, signing on with the Spanish or Neapolitan merchant marines or foreign navies. McNeill, *Atlantic Empires* (1985) 67; Chapuis, *A la mer comme au ciel* (1999), 136; Anthiaume, *Evolution et enseignement* (1920), 1.117. The English had similar concerns; see Boteler, *Six Dialogues* (1685), 44–45, on the better wages and greater liberties mariners enjoyed in "private Men of War"; and Flamsteed RGO papers [1700?], CUL 1/69/C, f. 99, noting that "highly instructed people . . . shun[ned] o[u]r Navy Royall."

36. *Oxford English Dictionary*, 1 b.: Euclid/tr. H. Billingsley, *Elements Geom.* (1570) i. f. 7v, "Propositions are of two sortes, the one is called a Probleme [which requireth some action, or doing], the other a Theoreme [where something is proved]."

37. Anthiaume, *Denys de Dieppe* (1927), 14. Le Corbeiller, "Denys à Colbert" (1916), 43–46, #2, 3, 5 "me donne à la verité un furieux embarras." Depping, *Correspondances,* 559 (Dec. 15, 1665), "continuer à des pratiques plus curieuses."

38. Doublet, *Journal du corsaire* (1883); wreck, 39–40. Bougard would later publish *Le Petit Flambeau de la mer* (fourteen editions between 1684 and 1817), and also a map of the Channel (Havre de Grâce, 1699).

39. Doublet, *Journal du corsaire* (1883), 53. De Latre [Delastre] was originally trained as a surgeon but learned enough navigation "during his voyages" to be given a commission in 1673 for the *Droite,* a ten-gun frigate; Doublet, 52, n. 1. The Franco-Dutch War (1672–78) was fought

by France, England, and allies against the Dutch Republic. The age of corsairing would essentially end with the Treaty of Utrecht in 1713.

40. Denys subdivided the standard four-hour watch into "quarters." De Latre on his protégé's alacrity, Doublet, *Journal du corsaire* (1883), 57; on skirmish near Dogger bank, a large sandbank in a shallow area of the North Sea about one hundred kilometers off the east coast of England, Doublet, 54.

41. Doublet, *Journal du corsaire* (1883), 58: fifty livres/month room, bed and laundry, as well as books. Doublet was not the only young mariner to board at his house; see Le Corbeiller, "Denys à Colbert" (1916), 43, #1 (May 22, 1665). This was a common arrangement: Sageran ms. (1747), SHD-V MS 511, thanks his teacher and the man he resided with. Moore, *Practical Navigator* (1791), final page, "N.B. Board and Lodging in the House, (if required)."

42. Doublet, *Journal du corsaire* (1883), 59, "cela vous fortifiera à fonds." Anthiaume, *Denys de Dieppe* (1927), 20, "prévôt de salle."

43. Doublet, *Journal du corsaire* (1883), 59–60. After spending some relatively calm years in the merchant marine, in 1696 the English captured him and held him prisoner at Plymouth (ch. 8). On his 1711 Atlantic crossing, his calculated route was only 34 2/3 *lieues* too long (22). He apologized for the lost logbooks ("Au Lecteur," 25), a record-keeping mishap common enough that in 1778 the French marine made it illegal to "throw . . . papers into the sea"; *Ordonnance du commerce* [1800], 4 (2nd group): "Réglement Concernant la navigation des Bâtimens neutres, en temps de guerre (26 juillet 1778)," art. 3.

44. Le Corbeiller, "Denys à Colbert" (1916), 41. Denys stood as godfather numerous times in the 1660s and had the honor of naming offspring of a number of his students and his publisher, Nicolas Dubuc. Anthiaume, *Denys de Dieppe* (1927), 29.

45. On the strong market for travel narratives, see Vergé Franceschi, "Entre ciel et mer" (2009), 56; Holtz, "Hakluyt in France . . . Travel Writing Collections" (2012).

46. Brouscon, Manuel de pilotage, à l'usage des pilotes bretons (1548), BnF Ms. Français 25374. Among the manuscripts likely produced for elite readers, see De Vaulx, "Premieres oeuvres" (1584), BnF Ms. Français 9175; and Le Vasseur de Beauplan, "Traicté de la Géodrographie, ou art de Naviguer" (ca. 1608), BnF Ms. Français 19112.

47. Charon, Claerr, and Moreau, *Le livre maritime* (2005), 135; the authors contend that the late arrival of rutters and other textual navigation aids indicates a strong and persistent oral tradition.

48. Nicolai translated Medina's work as *L'Art de naviguer* (1554), with Jean de Seville making additional revisions ca. 1602; see Charon, Claerr, and Moreau, *Le livre maritime* (2005), 16. The Portuguese humanist, Diogo de Sá, published a critique of Pedro Nuñes's important *Tratado da sphera* (Lisbon, 1537) in Paris in 1549, but his impassioned assessment garnered little response. By contrast, nearly twenty editions of Sacrobosco's *De Sphaera* were published in Paris over the sixteenth century (with the first French translation in 1570).

49. Of Paul Yvounet, the "merchant from La Rochelle" who translated Pieter Goos's *Le grand & nouveau miroir ou flambeau de la mer* (Amsterdam: J. Robijn, 1684) from the Dutch, the Sieur de Viviers cattily declared that the poor quality of the French translation made it all too clear that Yvounet must not understand either French or Dutch and that he was "neither a pilot nor a navigator"; Viviers, "Mémoire" (1681) AN-Paris MAR/G86 [hereafter "Mémoire" (1681)], f. 53. See also Willem and Joan Blaeu's atlas, *Le theatre du monde, ou Nouvel atlas* (Amsterdam: W. and J. Blaeu, 1650), and Claas Gietermaker's textbook *Le Flambeau reluisant* (Amsterdam, 1667). The Dutch also established a pipeline of awkwardly translated guidebooks and atlases to the English-language market; see Schepper, " 'Foreign' Books for English Readers" (2012).

50. See, e.g., Bessard's early printed portolan, *Dialogue de la longitude* (1574). Other works on related sciences or instruments include a work on magnets by Nautonier, *Mécométrie* (1603), and Dounot's critical response, *Confutation de l'invention des longitudes* (1611). Two distinct works on the astrolabe were published in Paris within a year of each other and saw several editions — Jacquinot, *L'Usaige de l'astrolabe, avec un traicté de la sphere* (1545), and Focard, *Paraphrase de l'astrolabe* (1546), but only the latter included any maritime material.

51. Seville, *Compost manuel* (1586). Michel Vergé-Franceschi points to Champlain's two different texts — the *Voyages* and the much more technical *Traitté* — as evidence of the French transition "from narrative of adventure to instruction of rules." Vergé Franceschi, "Entre ciel et mer" (2009), 58.

52. Fournier's *Hydrographie* (1643) saw two later editions, 1667 and 1679. Basic definitions (421; 319 of 1667 ed.) appeared after biblical accounts and stirring paeans to French innovation; there were no practice questions. Fournier, the son of a law professor, taught René Descartes at La Flèche (1629–35) and published several mathematical and scientific works in the 1640s. Anthiaume, *Evolution et enseignement* (1920), 1.106.

53. Viviers, "Mémoire" (1681), ff. 53, 48; Milliet de Chales, *L'Art de naviger* (1677), avant propos.

54. *Traicté de la variation* (1666) was reviewed in the *Journal des sçavans* (1666), 315–16. The reviewer noted that it was to be had in Paris "chez Seb. Mabre Cramoisy." The brief review emphasized Denys's practical audience: "As this book was made particularly for teaching Navigators, this Author endeavored to explain things that are customary; without delaying over those that are more curious than necessary." Le Corbeiller, "Denys à Colbert" (1916), 47, #5 (Feb. 21, 1666), Denys reminds Colbert in a postscript: "Because I have sent you only six copies of my book, you may ask for as many as you wish."

55. In 1665 Denys had been anxious about the *Traicté de la variation,* so a supportive Colbert forwarded it to the Académie française, whose members provided constructive feedback to elevate the writing style; see Le Corbeiller, "Denys à Colbert" (1916), 40 (Chapelain to Colbert, Dec. 20 and 28, 1665).

The Latin mottos and digressions on Ovid seem to originate with Denys rather than the Académie; see discussion linking "sein" (breast) and "Sinus" as the "the most beautiful vital parts" of man and mathematics, *Nombres* (1668), 5. In *Latitudes* (1673), 495, Denys closed with musings on the parallels between a navigator looking at the sky and a Christian looking at God, both gaining guidance for the voyage ahead. In the same text, we see hints of his classroom demeanor when he noted that something was "too hard to balance exactly," and "I'll come back to this other topic on a different morning" (85).

56. Denys, *Latitudes* (1673), 2, 5 (see note 1 in this chapter). These four-wheeled "chariots" were prosaic and multipurpose rather than martial; see *Dictionnaire de l'Académie française* (1694), v. 1, "chariot." Denys, *Nombres* (1668), 33, "triangle rectangle."

57. Fournier, *Hydrographie* (1643), 498.

58. Denys, *Nombres* (1668), 27–28; 6, also 167. For further discussion of the preference for numbers over instruments, see E. Wright, *Errors in Navigation* (1599), praeface.

59. Denys, *Nombres* (1668), 2–4, 6–7. When people "talked about Sines, Tangents and Secants without knowing what they are," he wasn't sure if that was ignorance or stupidity.

60. Such rules were familiar from Renaissance arithmetic books, beginning with the thirteenth-century "libri d'abbaco." Some writers, including Denys, *Nombres* (1668), 38, referred to six components; however, given the possible constructions of a triangle, two of these are functionally duplicates.

61. Edmund Gunter complained that an English edition of Napier's tables was "published

without figures or instructions" (*A Canon of Triangles* [1620], "To the Reader"). Gunter cleverly included a diagram on the title page of his Latin edition that appeared that same year, but (presumably to keep costs down) the English edition of his work was *not* illustrated. His 1623 work on his eponymous sector, by contrast, had many illustrations (figure 2.2). Milliet de Chales emphasized the importance of diagrams. He felt Denys should have included more than two diagrams in his 1673 work, although the Jesuit's own *L'Art de naviger* (1677) was only minimally illustrated; avant-propos (f. e*v*).

62. Denys, *Nombres* (1668), the diagram first appears on p. 5; see also pp. 33, 35, 36.

63. Denys, *Nombres* (1668), 34. See also, e.g., p. 35.

64. See Spink, Navigation workbook" (ca. 1697–1705) NMM NVT/47, for a rare manuscript example of ships sailing upon mathematical diagrams; plate 3.

65. Denys, *Nombres* (1668), 54, plus five pages of tables after p. 236: trigonometric tables, logarithmic tables, then "latitudes croissantes," for measuring distance on Mercator projection maps. The appendix (219–35) offers multiple solutions (e.g., how to "point" one's routes by trigonometry and logarithms; how to find "parallel M[eridian]" by sines as well as the table of meridional parts).

66. Denys, *Nombres* (1668), 217, 160. See also Glos, *Manuel des pilotes* (1678), 2. According to Blondel Saint-Aubin, *Véritable art de naviger* (1679), 78–79, when a navigator compared his estimate to "the latitude found from the altitude [observation], if they are equal, the estimate is very good, but if they are not equal, the estimate is worthless."

67. Denys, *Nombres* (1668), 169–170. The Dutch allegorical author Mathijs Sijverts Lakeman had recognized this conundrum years earlier when assessing the discrepancies between dead reckoning and observed position: "Either the pilot made good estimates, which didn't match the degrees south and north on the plain chart, or the pilot estimated confusedly but these confusions matched each other." *Een Tractaet, seer dienstelijck voor alle Zee-varende Luyden* (1597), Bii. Lastman, *Beschrijvinge van de Kunst der Stuer-luyden* (1657), 196, complained about the "nonsensical" (*ongerijmt*) process of reconciling two conflicting positional estimates.

68. Denys, *Nombres* (1668), 160–218; chapter 10 is the book's longest. On latitude, 160: "The four parts that make up Navigation are so interwoven [*enchaisnées*] one with the others, that if one of these four pieces is lacking, one must say that the rest are similarly lacking." Of these pieces, he continues, "we can only be sure of latitude." Longitude, by comparison, is "very frail, and scarcely certain" (170).

69. Glos, *Manuel des pilotes* (1678), 98.

70. Jarichs van der Ley, *'T Ghesicht des grooten Zeevaerts* (1619), 81, offered alternative rules to follow if one's altitude observation was "not trustworthy." Denys agreed with other seventeenth-century authors that Jarichs deserved the title of "first *Inventeur* of the Corrections" for promising to give "infallible rules for properly correcting the shortcomings and errors which one could have committed in one's navigation." *Nombres* (1668), 164.

71. He believed that rules were essential, especially when it came to longitude calculations: "In such cases, do not guess. Or soon the differences in longitude will fall irregularly." Jarichs van der Ley, *'T Ghesicht des grooten Zeevaerts* (1619), 80.

72. The panel judging the (ultimately unsuccessful) longitude contest in the United Provinces rejected Jarichs van der Ley's method, but he received ongoing royalties for his invention. Davids, *Zeewezen en Wetenschap* (1986), 82. See Metius, *Astronomische ende Geographische Onderwijsinghe* (1632); Snellius, *Tiphys Batavus* (1624).

73. *Belast* in Dutch is "tax" or "load," in German "burden," "strain," or "weight." On the multiplier effect where small adjustments produced risky discrepancies, see Lastman, *De Schat-kamer* (1629), ff. 60–61; Collins, *Navigation by the Mariners Plain Scale New Plain'd*

(1659) (ser. 2) 24: "A small mistake in the estimating of the distance, will cause a considerable falteration in the Rumbe." This geometric asymmetry stems from the fact that latitude was relatively straightforward to measure; see note 68 in this chapter and in chapter 1.

74. In 1624 Snellius maintained that if your observed longitude was in agreement with your estimated position, you need not correct your position at all. In 1632 Metius contended that agreement between your observation and your estimation was no guarantee that you had not made two errors that had simply canceled each other out. He advocated that the navigator check every stage of his course using trigonometry and the rule of three. Davids, *Zeewezen en Wetenschap* (1986), 113–14. A generation later, Denys preferred the simplicity of Snellius's solution for his students but suggested that Metius's method could be made considerably faster if one used a graphical instrument like the "quartier de réduction" rather than "numbers." *Nombres* (1668), 169.

75. Lastman noted that if a navigator did not begin with a "certain" altitude, he might undermine his observation in the process of taking it. Lastman, *Beschrijvinge van de Kunst der Stuer-luyden* (1657), 105.

76. Collins, *Navigation by the Mariners Plain Scale New Plain'd* (1659) (ser. 2) 26, 21–23. The "experience of others" can also help corroborate "the truth of your judgement"; see further at note 114 in this chapter.

77. Denys, *Nombres* (1668), 162–64, 169. Despite the absence of any records, Denys mused that earlier pilots must have had *some* kind of system to "regulate their affairs," just as merchants kept accounts before double-entry bookkeeping. After all, navigators, like businessmen, sought "more method, and more certainty" (162).

78. Denys, *Nombres* (1668), 217.

79. Denys, *Latitudes* (1673), 10.

80. Denys claimed that only the "most ingenious" pilots ("Pilotes les plus spirituelle") included minutes in their route calculations; standard computations, which worked only in terms of full degrees, were adequate in most cases. *Nombres* (1668), 12. The Dutch method required numerous calculations, and Denys found at most a minimal difference in the final positions — he thus declared it "not worth the trouble" (92–93); "good mother" (49).

81. Le Corbeiller, "Denys à Colbert" (1916), 47–48, #6 (Oct. 21, 1666).

82. Though obliged by a 1662 treaty to assist the Dutch Republic in a war with England, Louis XIV delayed the French entry until January 1666, midway through the two-year conflict; Rommelse, *Second Anglo-Dutch War* (2006).

83. Le Corbeiller, "Denys à Colbert" (1916), 47–48, #6 (Oct. 21, 1666).

84. Men who hurried to get a commission in the navy often encountered a very physical limitation before they acclimated to the swells of the sea. Le Corbeiller, "Denys à Colbert" (1916), 54, #14 (Nov. 24, 1673): "In their first voyage they see that they could not get used to the sea, due to being incapacitated by [seasickness] which they continuously suffer there but they were not free to quit."

85. Le Corbeiller, "Denys à Colbert" (1916), 49 #8 (Sept. 9, 1667). Boteler's sixth and final suggestion of a suitable breeding ground for skilled sailors was the Royal Naval fleet during wartime. Although he did not endorse violence, in his view armed conflict could quickly teach many important skills. *Six Dialogues* (1685) 4, 59. However, most administrators recognized the folly of sending raw recruits to sea during the height of political conflict. At the same time, it could be equally difficult to train mariners during peacetime, as funding for naval vessels and expeditions was drastically scaled back. Crisenoy, "Les écoles navales" (1864), 775–79.

86. La Isla de Aves, the Leeward Islands north of Venezuela, in the Caribbean Sea.

87. Méricourt to Colbert, Relation du naufrage de l'île d'Aves (June 1678), AN-Paris MAR/B4/8, 136; Lutun, "Ecoles de marine" (1995), 6.

88. This legislation had a broad mandate: Louis XIV, *1681 Ordonnance* (1714), preface. See also Louis XIV, *1689 Ordonnance* (1847), No. 62.

89. Denys, *Nombres* (1668), 8.

90. Arch. du Min. de la Marine, Ordres du Roy: Colbert to de Seuil (Sept. 1, 1678), cited in Anthiaume, *Evolution et enseignement* (1920), 2.89.

91. By 1681, Colbert, Seignelay, and Louis XIV had established six schools for officers and five public institutions. See Anthiaume, *Evolution et enseignement* (1920), vol. 1, bk. 2, ch. 1, and vol. 2.75; Lutun, "Ecoles de marine" (1995), 5–7. Another half dozen institutions would be established in the eighteenth century; see chapter 5, note 25.

Colbert initially sought out teachers from Holland rather than priests, "who had never been to sea"; cited in Neuville, *Etablissements scientifiques* (1882), 538. However, in 1686 Louis XIV gave Jesuits the exclusive right to teach hydrography to the *gardes de la marine* at "marine seminaries" throughout the country. By 1690, they were also appointed as "royal chairs" in mathematics and hydrography at half a dozen colleges. See Dainville, *Géographie des humanistes* (1940), 435, 439, 219; Audet, "Hydrographes du Roi" (1970), 15; Vergé-Franceschi, *Marine et éducation* (1991), 155. While the king may have been concerned about national security after repealing the Edict of Nantes, the Jesuits did have a tradition (from the early seventeenth century) of teaching the rudiments of hydrography (and the related spherical astronomy, etc.) as an offshoot of geography, which was directly useful to their foreign missions. For more on how the program for Jesuit education, the *ratio studiorum*, was applied to mathematics lessons, see Dainville, *L'Éducation des Jésuites* (1991); Karp and Schubring, *Handbook* (2014), ch. 7, esp. 134–35; Price, "Mathematics and Mission" (2016).

92. This legislation expanded on certain passages from older documents, e.g., 1584 Ordonnance: see Cleirac, *Us et coutumes* (1647), 491–93. Louis XIV, *1681 Ordonnance* (1714), 65–70, "Du professeur d'hydrographie"; 157–162, "Du pilote." Viviers, "Mémoire," f. 48, had weighed in on the importance of choosing a skillful teacher, one who had experience sailing as well as in tailoring the subject matter to the level of his pupils; however, Colbert did not include these criteria.

93. Louis XIV, *1689 Ordonnance* (1847), art. 7, "Des gardes de la marine"; art. 19, "des officiers et gardes de la marine." For instance, a 1682 report from Rochefort (AN-Paris MAR/G86, ff. 25–26) closely foreshadows the language and schedule adopted by the *1689 Ordonnance*. Viviers referred to the Dutch "écoles de pilotes," evidence that several institutions had developed international reputations; "Mémoire" (1681), f. 47. See also anonymous "Mémoire" (1707) MAR/G86, ff. 69, 67: "All the navy [*marinne*] knows that navigation [*pilotage*] is absolutely neglected in France, [and] that the Dutch and English do the opposite."

94. Classrooms: Louis XIV, *1689 Ordonnance* (1847), 67. Examinations: Louis XIV, *1681 Ordonnance* (1714), 157; *1689 Ord.*, 245. Journals: *1681 Ord.*, 68; *1689 Ord.*, 245–46, also notes monthly class meetings (*conférences*) mandatory for *gardes de la marine*. On rewards: after the d'Estrées disaster in 1678, the king offered an extra ten livres a month to "experienced" (*anciens*) *gardes de la marine* who agreed to go to hydrography school; Vergé-Franceschi, *Marine et éducation* (1991), 182. Whereas the penalty for losing a ship had once been death, after 1681 negligent captains were simply banned from future commissions; Anthiaume, *Evolution et enseignement* (1920), 2.81–82.

95. Viviers suggested that these potential teachers be judged on the basis of their new hydrography textbooks; "Mémoire" (1681), ff. 53, 48. Some seem to have indeed written ba-

sic manuals before taking up their posts, e.g., Coubert [Coubard], *Abrégé du pilotage* (1685). Dutch authors did this far more consistently than French ones; see chapter 4, note 50.

96. Milliet de Chales, *L'Art de naviger* (1677), avant propos; his methodical approach led him to include, for example, no fewer than twelve different methods of measuring compass variation (*déclinaison de la boussole*), 99–101.

97. Thoubeau (1653–1728) wished to increase the length of the course of study to a year — instead of a mere two weeks just before the students shipped out. He taught humanities, philosophy, and mathematics at Brest and Toulon before being named librarian of the collège Louis-le-Grand in 1720 (Sommervogel). On his curriculum and teaching methods, see Dainville, "L'Instruction 1692" (1956), 332–36.

98. Guillaume Clément de Viviers (fl. 1668–1701), first marshal of lodgings and galley captain in the French navy, inspected the galleys and the navigation school of Marseille in 1685 and 1690. He warned instructors "not to embarrass them by teaching overly difficult material" but did acknowledge that some among any group might have an aptitude; those could be enlisted to teach their peers; "Mémoire" (1681), f. 52. The intendant in Marseille felt similarly: those officers who were not compelled to attend classes "forget what one has taught them, they do not advance at all, and force the Mathematician to repeat what he's already said"; "Mémoire" (Marseille, 1682) AN-Paris MAR/G86, f. 19v.

99. This meant that he would have a "means of subsisting by studying, and [should] lose no time earning his living"; a quintessential work-study program. Viviers, "Mémoire" (1681), ff. 51v, 48.

100. King's "Réglement" of April 13, 1682, in Receuil des pièces: Hydrographie sous Colbert, BnF N.A.Fr. 9479, f. 199.

101. Dainville, "L'Instruction 1692" (1956), 335–36, 328. Thoubeau's list of subjects to be covered while on ship was of a more applied nature than that of the college. Each lesson was presided over by an expert: a "master" (*maître pilote*) taught hydrography; the master gunner (*maître canonnier*) presided over musket and cannon exercises and theoretical lessons; and a commissioned officer (*capitaine en pied* or *capitaine en second*) supervised naval maneuvers. Le Cordier, *Instruction des pilotes* (1786), 184: Conclusion: "A demonstration makes it much easier to understand things than all the reasoning one can do."

102. (1688–97) War of the Grand Alliance or the War of the League of Augsburg.

103. Tourville, *Exercice en général de toutes les maneuvres* (1693); this saw a Swedish translation in 1698. He presented the abstruse vocabulary of rigging and tackle, as well as detailed descriptions of how to hoist sails and cast anchors.

104. See chapter 3, note 51, for an overly optimistic proposal for a Royal Mathematical School instructor. By contrast, one French instructor bemoaned that without constant practice, commissioned navigators (*pilotes entretenus*) would forget everything: "In order for a man to become a good pilot, he must be perpetually at sea"; "Mémoire" (1707) AN-Paris MAR/G86, f. 69.

105. See note 2 in this chapter.

106. Louis XIV, *1689 Ordonnance* (1847), 243–44; the beginners would learn basic arithmetic and geometric definitions; the middle group would learn the sphere, tides, compass, altitude instruments, and the basics of estimating a course and correcting for magnetic variation. Finally, the most advanced group would be taught to "calculate routes using the 'quartier de réduction,'" while only the strongest of these would learn "geometry, mathematics, and other sciences."

107. Blondel Saint-Aubin, *Trésor de la navigation* (1673), bk. 2, p. 1. Numerous editions of his *Véritable art de naviger par le quartier de réduction* were issued by Gruchet between 1671

and 1763. Cauvette's less popular *Nouveaux elemens d'hydrographie* (1685) also eschewed trigonometry.

108. See, e.g., quote from unidentified textbook transcribed by F.-M. Chautard, first lieutenant on the merchantship *Prince de Conti* in 1756: "9e Cayer," TNA HCA32/257, ch. 32, "Du Pilotage": "Trigonometry is a science absolutely necessary for a navy officer."

109. Anthiaume, *Evolution et enseignement* (1920), 1.151–54, 193; "Correction of the Estimate" formed the subject of a third of the hydrography course at Marseilles, "Mémoire" (Marseille, 1682) AN-Paris MAR/G86, f. 19. French sailors regularly incorporated the Corrections into their day's work; see, e.g., AN-Paris MAR/4JJ/27/5, the logbook of the *Jason* (1724), by Jean Noël Passart/Vieillecourt "premier Pilote": e.g., Monday May 24: "At noon having corrected all my routes since my departure from France until the present— [at] noon I found, following the 3rd Correction, that the Route is 10# south and the corrected Distance is 203#1/2."

110. *Les Principes de la navigation* [1762?], BnF Ms. Français 22046, f. 145; see also Collins, *Navigation by the Mariners Plain Scale New Plain'd* (1659), 26.

111. Glos, *Manuel des pilotes* (1678), 99, "The first is used when the Route was between NNE and NNW; and between SSE and SSW. The second is used between ENE and ESE." Glos, whose book was among the most rudimentary, included a dialogue about the Corrections (99–103) and devoted the final chapter to them (132–43). The numerically inclined Denys provided ranges of degrees; *Nombres* (1668), 184: e.g., "If the rhumb does not exceed 22 deg. 30 min. one uses the 1st Correction."

112. E.g., Le Cordier, *Journal de navigation* (1708), 101, explains how to make the corrections with an instrument.

113. Le Cordier, *Journal de navigation* (1683), 105. He admits (98–99) that the second Correction is still necessary, because even skilled observers had trouble taking accurate latitude observations (being off eight or ten minutes could lead to an error of eighteen to twenty miles over a one-hundred-mile stretch).

114. Collins, *Navigation by the Mariners Plain Scale New Plain'd* (1659), 22–23.

3. HANDS-ON THEORY ALONG THE THAMES —LONDON, 1683

1. John Pepys, clerk, to Navy Commissioners (Oct. 18 and 21, 1670), Admiralty Papers: Masters' certificates (1660–73), TNA SP 46/137/249, 253.

2. Strype, *Survey of London (1720)* [online] (2007), 287. Originating at Deptford as one of England's four seamen's guilds (chartered by Henry VIII in 1514), Trinity House gradually shifted from its religious and charitable duties to become a type of licensing court with authority over the kingdom's lighthouses, in addition to responsibilities for recruiting and regulating Thames river pilots and other sailors. Harris, *Trinity House of Deptford, 1514-1660* (1969); Clarke, "Trinity House and Its Relation to the Royal Navy" (1927).

3. Masters' Qualifications (1660–95), TNA ADM 106/2908. Men could be licensed to sail, for example, "from the Downes westward and southward to & within the Mediterranean as high as Zant"—or Leghorn or "Scanderoone" (Iskenderun, Turkey); others qualified to venture to the West Indies.

4. Cruikshank, *Life of Sir Henry Morgan* (1935), n. 56; Laprise, *Le Diable Volant: l'histoire... des flibustiers... de la Jamaïque...* (2000–2006); Marley, *Pirates of the Americas* (2010), 167.

5. Letter of Oct. 21, 1670, TNA SP 46/137/249.

6. In some cases, piratical experience actually attracted positive attention from the British Admiralty; see Davies, *Gentlemen and Tarpaulins* (1991), 17.

7. "Matters Pertaining to Christ's Hospital," PL 2612 (Mar. 29, 1683), 689–90, "Mr. Colson's Report."

8. Pearce, *Annals of Christ's Hospital* (1901), 100–101. RMS students began at age eight, spending six years learning reading, writing, and Latin at Christ's Hospital. The most adept were then selected to be "King's Boys" at the Mathematical School, where they would spend eighteen months studying navigation and mathematics. Most finished school at age sixteen, at which point they were bound into an apprenticeship with a merchant captain. See Trollope, *A History of the Royal Foundation of Christ's Hospital* (1834); Plumley, "The Royal Mathematical School within Christ's Hospital" (1976); Ellerton and Clements, *Secondary School Mathematics* (2017), 64, 114.

9. PL 2612 (May 1683), 679–80. Transcript of Feb. 28, 1682/83 report on Feb. 23 exam. On apprenticeships, see Pearce, *Annals of Christ's Hospital* (1901); "Christ's Hospital Register of Marine Apprentice Bindings" (1675–1711), LMA CLC/210/F/014/MS12875.

10. Advertisement in Gellibrand, *Epitome of Navigation Containing the Doctrin of Plain & Spherical Triangles* (1674).

11. PL 2612, 689–90. See note 93 in this chapter for Paget's fourteen-point curriculum.

12. Paget was endorsed by Newton, John Flamsteed, Edmond Halley, and John Collins; see PL 2612 513, 529–30, cited in Iliffe, "Mathematical Characters" (1997), 127.

13. PL 2612, 690.

14. See Charles II's 1677 legislation, discussed in Davies, *Gentlemen and Tarpaulins* (1991), 35; Rodger, *Command of the Ocean* (2004), 120–21, noting that Pepys should not be considered the sole instigator of the lieutenants' examinations; see further, Davies, *Kings of the Sea* (2017).

15. George Gale (ca. 1670–1712) and his brother continued their father's successful tobacco import business, eventually moving to Maryland. http://www.findagrave.com/memorial/43407925. Lieutenant Passing Certificate for Mr. Geo[rge] Gale (Oct. 15, 1692), TNA ADM 107/1/15.

16. Gale's Passing Certificate (Oct. 15, 1692), TNA ADM 107/1/15.

17. On the need for officers to be self-reliant, see Boteler, *Six Dialogues* (1685), 3. To help prepare fledgling officers, a subgenre of navigational manuals developed with glossaries and explanations of daily maneuvers: see, e.g., Henry Mainwaring, "Tearms in Navegacon" (ca. 1624, NMM SMP/3) published as *The Sea-Mans Dictionary* (1644); Smith, *An Accidence or The Path-way to Experience Necessary for all Young Sea-men* (1626); Saltonstall, *The Navigator* (1636); Milliet de Chales, *L'Art de naviger* (1677); Tourville, *Exercice en général de toutes les maneuvres* (1693).

18. Early seventeenth-century texts, such as John Smith's *Sea Grammar* (1627), 5, and Mainwaring's *Sea-Man's Dictionary* (1644), mentioned knots and rope work as nautical tools, but not until the following century were these illustrated or explained; e.g., the incomprehensible diagrams in Falconer, *Universal Dictionary of the Marine* (1769). Turner and Van de Griend, *History and Science of Knots* (1996), 139. Knots did not receive systematic treatment in formal naval training until the turn of the nineteenth century; see epilogue, note 10.

19. E.g., Lieutenant Passing Certificates for midshipman Joseph Worlidge (Mar. 17, 1700), TNA ADM 107/1/192; and Robert Prosser (Apr. 20, 1703), ADM 107/1/355.

20. See, e.g., the multiple approaches discussed in Moore, *New Systeme of the Mathematicks* (1681), ch. 6, sec. 5 (Charts), 224–36.

21. See, e.g., Moore, *New Systeme* (1681), 256–57.

22. English nautical texts did not depict thumbs, instead teaching calendrical rhymes; see Perkins, *Seaman's Tutor* (1682), 50 (after tables): "this old Rhime: Thirty days hath September, April, June, and November."

23. See chapter 4, for discussion of an exam published by J. Robyn that focused more on assessing a candidate's comprehension than his calculating ability. Exams in Amsterdam often took place on full-size training ships; see figure 4.6. On model ships, see note 85 in this chapter.

24. For a candidate who failed his exam due to his inability to knot shroud lines, see Lieutenant Passing Certificate for Mr. William Gill (Mar. 25, 1692), TNA ADM 107/1/8. For tide on the Thames, Dickinson, *Educating the Royal Navy* (2007), 11, n. 8, citing Pepys, *Tangier Papers* (1935), 131; Davies, *Gentlemen and Tarpaulins* (1991), 30.

25. E.g., Mr. John Preake (Dec. 23, 1692), TNA ADM 107/1/21; Mr. Cremer, Mr. Rycaut, and Mr. Saunders, whose "Commanders . . . omitted to rate them midshipmen for the time they acted as Midshipmen" (Mar. 28, 1704), ADM 107/2/3; Mr. Jenkinson (July 28, 1712), ADM 107/2/6. For logbook excuses, e.g., Joseph Reeve "produces no Journall but sayes he kept one & that it was spoiled in the latter ship by bad weather" (Feb. 18, 1691), ADM 107/1/5.

26. Flamsteed reported a case of a skilled "Commander [tha]t can neither write nor reed" but "when ye book of Seacharts is opend [he] can call every small Island ever as readily as if he was sayleing by it." He betrayed his illiteracy only "because the wrong end of the book hapen[ed] to be towards him." Flamsteed RGO papers [1700?], CUL 1/69/C, f. 99.

27. E.g., the papers of William Fearne, who upgraded from fifth rate in 1693 to third rate in March 1700, to "any of his Majesty's ships" in November 1701; Nathaniel Browne improved his qualifications from fourth to third rate within nine months (May 1689 to Feb. 1689/90). This accreditation was cumulative: masters would typically carry all subsequent updated documents with them; see Masters' Qualifications (Trinity House), TNA ADM 106/2908.

28. Thomas, "Numeracy in Early Modern England" (1987); Levy-Eichel, " 'Into the Mathematical Ocean' " (2015).

29. Enacted in 1651, renewed in 1660, and modified several times through 1673, the Navigation Acts would remain in force for nearly two hundred years. See Justice, *Sea-Laws* ([1709?]), appendix, 594 pp., for a contemporary response; Israel, *Conflicts of Empires* (1997), 305–18, 349–60; Sawers, "The Navigation Acts revisited" (1992). The volume of British exports rose more than sevenfold from 1720 to 1820, whereas Dutch foreign trade dropped 20 percent. (During this period, French exports rose nearly threefold.) Maddison, *World Economy* (2001).

30. Baugh, "English Navy and Its Administration" (2017), 855. Rodger, *Command of the Ocean* (2004), 636–39, appendix 6, records approximately 15,000 mustered annually in the Royal Navy during peacetime (e.g., 1726, 1775); during the War of Spanish Succession, 35,000 to 45,000 (1702–10) men were mustered, but this was eclipsed by a force of 70,000 to 120,000 during later periods of conflict (e.g., 1779–82, 1793–1802). Estimates of the total number of mariners range widely, e.g., 50,000 in 1700; but anywhere from 150,000 to 300,000 (of a population of 12 million) ca. 1800; see Hope, *New History of British Shipping* (1990), 216, 248; Hope suggests that one family in six may have been directly dependent on the sea.

31. Enthusiasm about the sea typically focused on its commercial potential; sailors were mere necessities in such trade, e.g., A. B., *Gloria Britannica; or, the Boast of the Brittish Seas* (1689); Justice, *Sea-Laws* ([1709?]), preface. See also Rodger, "Queen Elizabeth and the Myth of Sea-Power" (2004). Pepys returned repeatedly to the problem of insufficient navigators: *Naval Minutes* (1926), 38 ("want of able pilots"); also 23 ("encourage . . . our English pilots" as their number was "very low"), 45. On mariners' preference for the merchant marine over the navy, see chapter 2, note 35.

32. On Bourne and the earliest English authors, see Waters, *Art of Navigation* (1958), 127–43. Medina, *Arte of Navigation* (1581).

33. Addison, *Arithmeticall Navigation* (1625), Epistle Dedicatory (f. A2v).

34. Blundeville, *Exercises* (1594), "To the Reader"; Saltonstall, *The Navigator* (1636), 13. Already by 1624 in his *Speculum Nauticum: A Looking Glasse, for Sea-Men,* John Aspley felt that "the projection of the Sphere in plano" qualified as a "Geometrical" rather than a cosmographical concept. Part of this is a terminological difference — English writers viewed the "sphere" as geometric rather than astronomical.

35. E.g., Spink, Navigation workbook (ca. 1697–1731), NMM NVT/47, includes numerous sections on business matters ("barter or exchange"). Sephtonal, "Workbook" (ca. 1713), NMM NVT/6, begins with arithmetic, then the calendar, and a brief section on trigonometry. After the main section on navigation, it concludes with dialing, gauging, and gunnery. Both include the rule of three.

36. Waters, *Art of Navigation* (1958), 185: "The Edwardian Statutes of 1549 had given greater prominence to the importance of mathematics in the university curriculum." Feingold, *Mathematicians' Apprenticeship* (1984).

37. Mixed mathematics was a pseudo-Aristotelian category that included spherical astronomy, geography, surveying, and mathematical instruments; Dear, *Discipline and Experience* (1995), 39. Thomas, "Numeracy in Early Modern England" (1987), noted that arithmetic began as a practical tool (111) but became "an essential part of [an eighteenth-century] gentleman's education" (n. 51). Deborah Harkness draws attention to the burst of mathematical publications—including navigational titles—that appeared in London in the 1590s; Harkness, *Jewel House* (2007). See also Bennett, "Shopping for Instruments in Paris and London" (2002).

38. The book concludes with 384 pages of tables. See Willmoth, *Sir Jonas Moore* (1993), 201, for further analysis of the work's style and content.

39. Perkins, *Seaman's Tutor* (1682), 175. This too included fourteen different tables, but the volume was only a third the length of Moore's.

40. Thomas Hobbes, writing in the 1640s, declared that "the mathematicall sciences [were] the fountains of navigatory and mechanick employments." *Philosophicall Rudiments* (1651), 204. Even for Astronomer Royal Flamsteed, astronomy was not the answer to naval success; see note 46 in this chapter.

41. Pepys served a term as the society's president in 1684–86. His contemporaries recognized him as a "Patron and Encourager . . . of the improvement of Navigation"; see Boteler, *Six Dialogues* (1685), dedication.

42. Pepys, *Naval Minutes* (1926), 314–15; Thrower, "Samuel Pepys FRS (1633–1703) and the Royal Society" (2003).

43. PL 2612, 391 (Dec. 8, 1681), "A Memorial of a Conference had by M Pepys w[i]th ye Trinity House, touching ye present deficiencies of ye Children, & ye importance of its being remedyd." Pepys repeatedly voiced his criticisms of the *training*: see pp. 250–66, for a lengthy list of "Defects in the Institution" and his proposed remedies.

44. PL 2612 preserves the idealistic plans for instructors and curricula; see note 50. On Flamsteed and Newton, see Iliffe, "Mathematical Characters" (1997), 117; Oughtred, *Circles of Proportion* (1632), "Epistle Dedicatorie."

45. Newton and Cotes, *Correspondence* (1850), 282 (May 25, 1694), to N. Hawes, RMS Treasurer; but see also 284, where he claimed that mathematicians but not sailors were "inventive Artists."

46. Flamsteed, "Doctrine and Practice of Navigation" (1697), PL 2184, 1. In a draft letter on the qualifications of naval officers, he noted that "they distinguish themselves very naturally into the Theory Men & the practicall"—but by "theory" those practitioners did not mean "Geometry or Trigonometry" but rather that they knew how "to take an observation[,] to marke a traverse[,] to correct it . . . & to keep a journall." Ultimately, Flamsteed reported, seamen did

not necessarily see the benefit of formal lessons; instead, successful naval careers often came down simply to luck, favoring those willing to "advanture . . . boldly in an hazardous undertaking"; CUL 1/69/C, RGO papers (1700), f. 99.

47. Pepys, *Tangier Papers* (1935), 120, 110, conversation with Phillips; Pepys, *Naval Minutes* (1926), 228, 375. Phillips felt that the prospect of the sailor's extremely unpleasant life scared off many educated men: "The whole trade and knowledge of it has been kept among poor illiterate hands, who for want of a methodical degree of learning have never been able to improve Navigation." Cited in Iliffe, "Mathematical Characters" (1997), 130.

48. PL 2612, 653. Newton and Cotes, *Correspondence* (1850), 294–95. Authors often advised readers to focus on different sections depending on their level; see, e.g., Collins, *Navigation by the Mariners Plain Scale New Plain'd* (1659), preface; also Robertson, *Elements of Navigation* (1780), xxxi.

49. Newhouse, *Whole Art of Navigation* (1685), began his dialogue between a Young scholar and his Tutor with the Golden Number, the sphere, and instruments—the older model. He felt that "more exact Geometrical Practices, [such] as Trigonometry" should be reserved for "those that aim at a greater perfection of the Art."

50. PL 2612, 535–72 and 631–54, contains job application materials from more than a dozen candidates.

51. Pepys called this plan "folly"; *Naval Minutes* (1926), 182. Further on Paget, note 97 in this chapter. On S. Newton, see Iliffe, "Mathematical Characters" (1997), 136–38.

52. The English were well aware of the Spanish system. See, e.g., *The Accomplish'd Sea-Mans Delight* (1686), 44. They also knew of the French Ordonnance of 1681, translated in Justice, *Sea-Laws* ([1709?]). On the Dutch penetration of the English nautical book market, see Seller, *Practical Navigation* (1669), 310, postscript; Rivington, *Pepys and the Booksellers* (1992), 60–61, and Mount's comments in PL 2643, 83; also Schepper, " 'Foreign' Books for English Readers" (2012).

53. In 1664, only a few years after Pepys began his career as Clerk of the Acts, the navy commissioner William Coventry suggested Pepys write a history of the Royal Navy. In 1679 or 80, he began compiling research questions in his "Minute-Book currant for ye Adm[ira]lty & Navy MSS" (PL 2866, published as *Naval Minutes*, 1926). He frequently noted "[What] imperfect judges we of this nation generally are of anything relating to the sea," 379; also 289, 194, 314, 413. On French knowledge of the loadstone, 422; on the oldest English (rather than Spanish or French) maps, 95. After realizing that "our first navigation-book"—the *Rutter*, the oldest-known English book of coastal routes and one of Pepys's most valued volumes—was "a translation from French," he was driven to ask if "Q: our other first books, [were] all translations?" 415. (He kept substantial bibliographic records, see "Bibliotheca Nautica," PL 2643.)

54. Pepys wished to "enquire into and compare the several naval strengths of England, France and Holland in 1688," naming the specific officials who could provide the best answers; *Naval Minutes* (1926), 399. See 359, on Dutch sheathing and borrowed words; 113, cartography; 86, shipboard life.

55. Pepys, *Naval Minutes* (1926), 240, 355–56, 360. Pepys also admired French battle command, fireships, methods of enrolling seamen, and particularly the king's financial support for enlisted sailors aboard merchant ships during peacetime. Pepys traveled with his wife to France and filled his library with a large selection of French-language titles, including "the King of France's new Book of the Ordinance Marine": the *1689 Ordonnance* (PL 693).

56. On Dutch schools, Pepys, *Naval Minutes* (1926), 420, 425, 183. Pepys acquired several published sources about Spanish schools: the PL 2140 volume includes Carlos II, *Foundacion del Semin[a]rio de Niños en la Arte Marit[im]a en Sevilla* (1681), and A. García de Cés-

pedes, *Regimiento de Naveg[acion]* (Madrid: Juan de la Cuesta, 1606). (For all PL holdings, see Latham, *Catalogue of the Pepys Library,* vols. 1 and 5 [1978, 1981]). Pepys was particularly impressed by the series of lectures at the "Contratacion-house," which Richard Hakluyt wished to emulate in London; *Naval Minutes,* 415, 424. He made careful notes about Spanish classroom and shipboard lessons, which blended "art" and "practice"; see Pepys, *Tangier Papers* (1935), 254–57.

57. In Pepys's view not only were masters, pilots, and average tars poorly trained but officers were guilty of "being men either of pleasure, or at least generally men of quite another education than that of the sea"; *Naval Minutes* (1926), 183, 421. On the larger debate, see Davies, *Gentlemen and Tarpaulins* (1991).

58. Pepys excitedly noted various mathematical gadgets in his diary throughout the 1660s and introduced his wife to them (he taught her to use globes and a microscope and also instructed her in arithmetic), e.g., Pepys, *Diary* (1970–83), 4.85, 124, 302, 343–44, 434; 5.6–49. He also made reports to the Royal Society about deep-water diving and watches used for determining longitude. For an episodic account of Pepys's interest in modern instruments, see Nicolson, *Pepys' Diary and the New Science,* chaps. 1, 2 and appendix.

59. His lessons with Cooper, mate of the *Royal Charles,* whom he had met on his expedition to the Netherlands to collect Charles II, extended through the summer of 1662; see Pepys, *Diary* (1970–83), 3.128–49, 255; 4.2, 85, 406 (July 4–30, 1662, and sporadically in 1663), e.g., 3.134, "Up by four o'clock, and at my multiplicacion-table hard, which is all the trouble I meet withal in my arithmetique."

60. Pepys, *Naval Minutes* (1926), 23, 420, 50, also 401 ("imperfectness"); 124 (Norris). Rodger, *Command of the Ocean* (2004), 123: Pepys was "in some ways the English Colbert, trying to do what his French counterpart succeeded in doing; to create an accountant's navy whose prime function was to balance the books and observe the formalities." Davies, *Kings of the Sea* (2017), is more skeptical of his reputation as "saviour of the navy."

61. Pepys initially felt that Trinity House should also make provisions for the "translating of marine books out of foreign languages," but he became skeptical of projects it oversaw. Pepys identified valuable foreign titles and made preliminary efforts to have them updated (Davis's 1595 *Seamans Secrets*) or translated into English (Witsen's 1671 "*Book of Shipwrightry,*" i.e., [*Architectura navalis*]), *Naval Minutes* (1926), 127, 183, 423.

62. Pepys, *Naval Minutes* (1926), 124; "[Norris wished to] have a school erected on purpose for the instructing of those the King means to entertain as *reformadoes* [volunteers] before they enter into the service." Pepys lamented that neither the universities nor Gresham College paid sufficient attention to navigation; *Naval Minutes,* 124, n. 5; 414; 421. On domestic institutions modeled on Christ's Hospital, such as Sir William Boreman's Naval Foundation, 288, 423; low status accorded to navigation in England, 94, 261, 396; enthusiasm for Robert Slyngesbie's *Order of Sea-Knighthood,* 53, 90, 119.

63. Thomas Gresham founded his eponymous college in 1597 to foster scientific discourse. "Orders for the Mathematick-School," PL 2612, 194–95, "so those Youths . . . may be inhibited from Rambling, or being Idle, etc."

64. Flamsteed, "Qualifications for the Mathematical Master" (Nov. 30, 1681), PL 2612, 648.

65. The RMS accounts note various classroom expenses: "Royal Mathematical School Account Book," LMA CLC/210/C/011/MS12874, May 31, 1679; May 27, 1682. Clifton, "London Instrument Makers" (2003), 26, discusses the quadrant gifts. See also Thomas Child's invoice, note 86 in this chapter. Dainville, "L'Instruction 1692" (1956), 333, lists the equipment Thoubeau's French students needed to "work on drawing and on all the things where practical geometry is necessary." Valin, *Nouveau commentaire sur l'Ordonnance de 1681* (1760), reported

that students in French schools must bring their own cross-staff, compass, sinical quadrant, and an atlas of marine charts; see Turner, "Rochefort" (2005), 532. While most navigators had their own backstaffs, etc., large company ships would provide all the necessary equipment down to the blank paper and preruled notebooks for shipboard record keeping; see, e.g., Heren XVII, *Lyste van de Kaerten en Stuurmans gereetschappen* (1673), NL-HaNA VOC 1.04.02 inv. nr. 5017.

66. On the dialogue genre in general, see Burke, "The Renaissance Dialogue" (1989). Early authors of nautical dialogues include Cortés (ca. 1550s), García de Palacio (1587), Bourne (1592, tr. Waghenaer 1594), Linschoten (1596). They continued to be popular through the seventeenth and eighteenth centuries; epilogue, note 23.

67. The most common type of classroom recitation involved regurgitating memorized definitions. In a proposal for an officers' school in Paris, the anonymous author suggested that "every day one takes a convenient hour for everyone to repeat the same lessons, particularly those Naval ones." *L'Académie royale dite de la marine* (1677), 14. See also Dainville, "L'Instruction 1692" (1956), 338; Boissaye de Bocage, "Mémoire" (Havre de Grâce, 1682), AN-Paris MAR/G86, No. 4.

68. See chapter 1, note 90, for Van den Broucke's list. Seller, *Practical Navigation* (1680), 5: sect. 5 "Of Multiplication."

69. Jackson, *An Introduction [to the] Rudiments of Arithmetick* (1661), "Courteous Reader." Martin, *Young Trigonometer's Compleat Guide* (1736), 68–70, a dense three-page list of "12 Precepts" that "must be well imprinted in the memory of all who would be ready and dexterous in this most excellent Art."

70. Dozens of these student workbooks survive, with those produced in Dutch classrooms hewing closest to published works; see chapter 4. Documents often bore internal evidence that they were produced by dictation: notebook pages whose contents do not match the preruled space or preprepared subtitles, or spelling irregularities that suggest oral transmission.

Teachers might also lend their notebooks to students to copy at home. The instructor at the *école d'hydrographie* in Rochefort divided the course material into a set of seven small, ten-page booklets ("cahiers"), which presented the "ordinary practices of navigation" in an "intelligible enough manner" (arithmetic and geometric definitions were followed by the sphere, maps, the calendar and tides, latitude, and how to calculate routes using the "quartier de réduction"). Demuyn, "Mémoire de l'intendant," Rochefort (May 3, 1682), AN-Paris MAR/G86, f. 10, No. 3. Other French teachers also taught officers and children using similar sets of notebooks; see Boissaye de Bocage, "Mémoire" (Havre de Grâce), MAR/G86, f. 20. Such "cayer" were taken to sea: epilogue, notes 22, 27. Further, Blair, "Student Manuscripts and the Textbook" (2008).

71. "Account Book," LMA CLC/210/C/011/MS12874, noting at least three book purchases from Obadiah Blagrave, Perkins's publisher. On September 19, 1696, the school paid "Christopher Hussey for 60 *Ideas of Navigation*, 12 — — ," the textbook written by Samuel Newton, then the current mathematical master. (The following month Samuel Gent received one pound for binding the books.)

72. E.g., those embarking on a military expedition copied out Tourville's "*Mémoires des Manoeuvres*"; Demuyn, "Mémoire" (1682), f. 10, No. 3. See also the published book copied by Guillaume Gloanic; TNA HCA 32/257.

73. Père Thoubeau preferred to acquire three or four hundred copies of Paul Hoste's substantial mathematics textbook rather than have the guards make error-ridden copies that would "steal" the time for "explanation and repetitions"; Dainville, "L'Instruction, 1692" (1956), 334–35. In 1765 the instructors at Rochefort were told to have their students use a printed text-

book for similar reasons; Turner, "Rochefort" (2005), 545. Authors, who admittedly had financial interests in requiring students to purchase published books rather than copying their own, eagerly emphasized the problems inherent in students' dictation practices: "Many do not write correctly, others do not do so legibly, which only causes difficulties for comprehension and greatly slows their progress"; Dulague, *Leçons de navigation* (1768), iv.

74. French teachers emphasized the importance of explaining the dictated lessons. Demuyn, "Mémoire" (1682), ff. 25–26. See also Levot, "Les écoles" (1875), 168–69, for Coubert's similar routine of dictation, explanation and review.

75. Monson, *Naval Tracts [1624]*, (1902), 393.

76. The term "demonstration" was multivalent in the early modern period; see *Oxford English Dictionary*. Oughtred, *Circles of Proportion* (1632), "Epistle Dedicatorie," juxtaposed "Demonstration" and teaching through "Instruments."

77. Coubert felt that working through multiple problems was essential for his audience of officers in training "to thoroughly understand the exercises." *Abrégé du pilotage* (1685), f. 2v. See also "Mémoire" (Marseille, 1682), AN-Paris MAR/G86, f.18v: "He made each take his quill in hand."

78. Le Cordier, *Instruction des pilotes* (1786), 7, "Avertissement"; Martindale, *Country-Survey-Book* (1692), 48, 17. Phillippes, *Geometrical Sea-Man* (1652), f. [A3], emphasized the value of geometric diagrams *and* instruments. Moore, *New Systeme* (1681), 67; he also provided concrete tips, e.g., "For methods sake make always the fore Part of the Book, Slate, or whatever else you draw or work upon, the North Part" (276).

79. "Minute and Memoranda Book," LMA CLC/210/B/006/MS12873/001, Letters Patent (1673), "Books, Globes, Mapps and other Mathematicall Instruments as shall be found necessary for the better Instruc[ti]on of the said children in Arithmatique, and in the Art of Navigation." For a bequest from a benefactor, see LMA CLC/210/B/007/MS12873A, 48 (Mar. 27, 1691), "Mr. Edward Brewster's gift to the Mathematicall schoole of a manuscript called Algebra." Another entry (Apr. 4, 1688) noted "Simon Chapman a Mathematical Instrument maker, [wished to] be permitted to serve this house with such comodities." "Account Book," LMA CLC/210/C/011/MS12874: "Paid 24 Nov. 1680 to John Marke ffor Math'l Instruments ye bill 007 05 04; Paid 22 Aprill 1681 to Wm Palmer for Instrument 004 07 04." Also, "Elizabeth Seller, paid 9 June 1697 8 10; Paid ditto to Jane Hayes for Mathematical Instruments, bill N. 185 2 15."

80. Iliffe, "Mathematical Characters" (1997), 124, citing Dr. Wood's Articles, PL 2612, 370. Because of the lengthy installation delay, the youths' instrumental training through the 1680s was limited to the quadrant and nocturnal.

81. Le Danois, "Mémoire" (Havre de Grâce, May 4, 1682), AN-Paris MAR/G86, f. 12v, No. 4. See Dainville, "L'Instruction 1692" (1956), 333, for Thoubeau's extensive wish list of classroom equipment.

82. "Mémoire" (Rochefort, 1682), AN-Paris MAR/G86, no. 3, ff. 10–11. Thoubeau similarly expected every shipboard lesson to include not only rote repetition (note 67 in this chapter) but also hands-on practice; Dainville, "L'Instruction 1692" (1956), 337. In a private review class, students moved from the simplest "quadrant" to the "much more accurate and universal" sinical quadrant, and — better yet — a "very easy instrument" invented by their teacher; "Mémoire," (Marseille, 1682), AN-Paris MAR/G86, f. 19.

83. Levot, "Les écoles," 167. See also "Mémoire" (Marseille, 1682), AN-Paris MAR/G86, f. 19v. Jean Deshayes, an académicien who taught hydrography at the Jesuit collège in Québec, took his students in small boats on the St. Laurence River to practice estimating distances and drawing charts; see Deshayes, "Riviere de St. Laurens" (1686), QQS Polygraphie II.34.

84. See chapter 5, note 56, for Lieut. Riou's difficulties with various instruments. On the shifting market, see Schotte, "Ships' Instruments in the Age of Print" (forthcoming).

85. RMS had at least two models of full-rigged ships, although Pepys complained that they were "little use[d]"; *Naval Minutes* (1926), 186, 374. "Account Book," LMA CLC/210/C/011/MS12874, "Paid 20 Jan 1681 to John Raven for mending the Shipp in the Math'l School ye bill 001 07 41." For French examples, see M. Le Danois, "Mémoire" (Havre de Grâce, 1682), AN-Paris MAR/G86, ff.12–15: the carpentry master requested a small model to demonstrate the various parts of a ship. In Brest a decade later, Thoubeau described separate models for learning rigging and carpentry, as well as various machines to teach fortification and more. Dainville, "L'Instruction 1692" (1956), 327–28, 334.

86. See inventory, note 96 in this chapter. LMA CLC/210/C/011/MS12874, "paid 19 Sept 1696 to James Moxon for mending globes used in the school by bill No. 23 3 10." Pepys excitedly bought a pair to teach his wife; see note 58 in this chapter. The mayor of London, Sir Francis Child, purchased a set from Samuel Newton for his son's private lessons; "Letter and invoice received by Sir Francis Child and Thomas Child from Mr [Samuel] Newton" (1704), LMA CLC/B/227/MS28949. Thoubeau estimated that globes would last five or six years in a school setting; Dainville, "L'Instruction 1692" (1956), 407. Jean Deshayes had a small wooden one among his accoutrements in Québec; see Post-mortem Inventory of Jean Deshayes, *Greffe* of Florent de la Cetière (Dec. 22, 1706), QQA cote CN301, S146, folder 672.

87. Dekker, *Globes at Greenwich* (1999), 4, 7. Barlow, *Navigators Supply* (1597), f. [K2v], deemed them "The onely good methode of teaching and learning Cosmography." See Wright, *Certaine Errors* (1599), 128, on the downsides to carrying globes on board ship.

88. Martín Cortés's *Arte de Navegar* (1551) also includes his important *Breve compendio de la sphera*, while Willem Blaeu's *Tweevoudigh Onderwiis van de Hemelsche en Aerdsche Globen* (ed. princ. 1634) was rapidly translated into three other languages.

89. Van den Broucke, *Instructie der Zee-Vaert* (1609), 43–45. See also Reyersz., *Stuurmans-Praetjen Tusschen Jaep en Veer* (1637), 21. Such observations must have seemed particularly necessary after the supernova of 1604.

90. Hues, *Globes* (1639), 165. Compared to Milliet de Chales's vote for images as aids to comprehension (chapter 2, note 61), Coubert preferred globes; in his view they required less "imagination" to understand various concepts. For him, "they will serve if one wants to remind oneself, and use them for demonstrations, once one has already understood the [concepts] on the Globe, etc."; *Abrégé du pilotage* (1728), 7. Boissaye de Bocage and Le Cordier were also proponents of using three-dimensional globes to provide a better understanding; Le Danois, "Mémoire" (Havre de Grâce, May 4, 1682) AN-Paris MAR/G86, No. 4 (globes in first lesson on the sphere).

91. See the anonymous Dutch navigation workbook (ca. 1717–26), HSM S.4312 (07), which still contains within its pages a translucent protractor. In Spain, the Real Academia de Guardias Marinas de Cádiz ("Escuela Real de Navegación," f. 1717) had its students produce dozens of nautical charts as part of their schoolwork; preserved at the Library of Congress (Maggs purchase map collection, 1729–1824). For more on the Jesuit's *ratio studiorum*, the pedagogical model influential in France and elsewhere that emphasized repetition but also hands-on learning, see chapter 2, note 91.

92. Newton and Cotes, *Correspondence* (1850), 283. Newton had been asked by the board of Christ Hospital to review Edward Paget's new curriculum in the 1690s shortly after RMS had had a falling out with Pepys.

93. "Improvement and Inlargement of the Mathematicall Maister's Instruccions and Undertakings," PL 2612 (1681), 679–80: 1 geometrical definitions; 2 geometrical ratios; 3 decimals;

4 trigonometry; 5 calendar and tides; 6 latitude/altitude; 7 plane chart; 8 logarithms; 9 Gunter's scale; 10 projection of the sphere, from three-dimensional globes onto two-dimensional charts; 11 arithmetic, beyond the rule of three; 12 drawing; 13 instruments; 14 observations of Sun, Moon, stars. Compared with the curriculum Paget was expected to teach, Robert Wood's "articles" before 1682 did not include items 8, 11, 12, 13, or 14.

94. PL 2612 (Mar. 15 and 16, 1682/83), 684–85. The school clerk, William Parrey, had evidently forwarded an incomplete copy of the new curriculum to Trinity House in advance of Long's and Crockett's exams. Pepys scoldingly rectified this in a letter to Mr. Parrey, reiterating the importance of testing each candidate on all of the instruments: "Without [them] the purpose of that usefull Article would be wholly prevented."

95. PL 2612, 689–90, see Colson's assessment, note 7 in this chapter.

96. PL 2612, 371–78, 1680 Inventory. For the centralized institutions, funding concerns regularly overshadowed those related to curricula; see Grier, "Navigation, Commercial Exchange . . ." (2018), ch. 1. Pepys was eager to determine just how much it cost to educate each blue-coat boy (including "ye Child's entertainment, cloathes & summer to set him out a Prentice being encluded"), and even perhaps economize by replacing the specialized mathematics instructor by "a Common Teacher"; PL 2612, 394.

97. Iliffe, "Mathematical Characters" (1997), 129, Paget to India; 136, resigned in Feb. 1694/95.

98. Such dials, designed by William Oughtred, not only told the time but also demonstrated the motion of the Sun through the day and the year. (They were inscribed with lines of solar declination, the ecliptic and the right ascension of the Sun). *New and Complete Dictionary of Arts and Sciences* (1763), 2.981.

99. In the first twenty-five years, "the regime at the Mathematical School had been spectacularly unsuccessful, having produced only one captain of a Man of War and three or four lieutenants"; Iliffe, "Mathematical Characters" (1997), 142. But on alternate careers in Virginia: LMA CLC/210/B/007/MS12873A (Sept. 7, 1687); Muscovy: CLC/210/B/008/MS12873B (May 27, 1698). See also Cavell, *Midshipmen and Quarterdeck Boys* (2012). William Spink, the young man who lavished attention on his workbook during his navigational lessons, may have been the son of a chandler or become one himself, putting his commercial arithmetic skills to work as much as his nautical ones; see Spink, Navigation workbook (ca. 1697–1731), NMM NVT/47; Spink, Ship chandler's notebook (1718 [or 1735?]-1751), MALSC DE629/2. Ellerton and Clements, *Secondary School Mathematics* (2017), argue persuasively for the long-term pedagogical impact of the institution.

100. Davis, "Sixteenth-Century French Arithmetics" (1960). Thomas, "Numeracy in Early Modern England" (1987), notes that in eighteenth-century England, many people still had to teach themselves math; but again, see Harkness, *Jewel House* (2007), for a far livelier market.

101. Stevin published a landmark pamphlet on decimal fractions, *De Thiende* (1585); Bartjens, *De Cyfferinghe* (1604), was a general arithmetic that started with addition, subtraction, multiplication, and division before moving on to money, measures, and weights. Kool, *Die conste vanden getale* (1999); Davids, "The Bookkeeper's Tale" (2004).

102. For information on other mathematical tutors, see Johnston, "Making Mathematical Practice" (1994); Taylor, *Mathematical Practitioners of Tudor & Stuart England* (1954) and *Hanoverian England* (1966); Bryden, "Evidence from Advertising for Mathematical Instrument Making" (1992).

103. A midshipman would receive an additional twenty pounds per year if he agreed to instruct young gentlemen "not only in the theory but the practical part of Navigation" and to "instruct the Youth in the Art of Seamanship." See Sullivan, "Naval Schoolmaster" (1976), 317,

and Dickinson, *Educating the Royal Navy* (2007), 13–18, for the earliest certified instructors (forty-two in 1713, forty-seven in 1715). These teachers were required to pass exams of their own at Trinity House (the contents of the exam are not known).

104. See chapter 2, note 15.

4. PAPER SAILORS, CLASSROOM LESSONS —THE NETHERLANDS, CA. 1710

1. Van Asson, "Schatkamer" (ca. 1705), OTYA 16275, f. 71v. One particular question required Van Asson to use logarithms, spherical trigonometry, and a calculation method employing the "half sun." He also quickly jotted down the standard geometric diagram, familiar from Pedro de Medina, to analyze the Sun's position relative to the equator and determined the magnetic variation, an object of continued importance to Dutch navigators. Van Asson was most likely from Maassluis, near the port of Delftshaven, in the province of South Holland. See Will of A. van Asson (Aug. 6, 1718), TNA PROB 11/565/44. The questions copied from Gietermaker (e.g., f. 46) have dates in the 1690s, suggesting a turn-of-the century edition. De Graaf cited on f. 58, Van Dam on f. 84.

2. Wijkman, "Ex Samen Der Sturliedn" (1709), TNA HCA 32/176 II. Visscher copy of Gietermaker, *'t Vergulde Licht* (1710), MMR ARCH 4D16. Gietermaker editions, see note 9 in this chapter.

3. Van Dam, *Nieuwe Hoornse Schatkamer* (1712), "Voorreden." On autodidacts, see also Oughtred, *An Addition unto the Circles of Proportion* (1633), 22–23.

4. For census of manuscripts, see Supplemental Bibliography online.

5. Lastman's *Schat-kamer, Des Grooten See-vaerts-kunst* (Treasury of the Art of Great Navigation) (Amsterdam, 1621) was inspired by Waghenaer's *Thresoor der Zeevaert* (Treasure [chest] of Navigation) (1592). The term *schat + kamer*, or treasure chamber, is translated on Waghenaer's frontispiece as "armoire"; cf. the first French translation *Thrésorerie ou cabinet* (Calais, 1601).

6. Winschooten, *Seeman* (1681), "To the Reader."

7. On Claas Gietermaker (1621–67), see Davids, Van der Veen, and De Vries, "Van Lastman naar Gietermaker" (1989).

8. Gietermaker, *'t Vergulde Licht* (1660), thumb on page 4; volvelles on page 22 and 23 (he devotes twenty-three pages to the calendar and tides calculations); star songs, 43, 44–45, 89; cosmographical definitions, 46–48.

9. The VOC's standard equipment list (*lijste*) evolved over time: see Heren XVII, *Lyste van de Kaerten en Stuurmans-gereetschappen* (1673–1747), NL-HaNA VOC 1.04.02, inv. nr. 5017, 5018. The 1747 list gave mariners a choice between "3 Gietermakers [or Lastman's] *Kunst der Stuurlieden* [sic]." These texts were published repeatedly: Lastman saw at least eleven editions, Gietermaker twenty-one, and De Vries fifteen by 1818. See Peters, *Crone Library* (1989).

10. Graaf, *De kleene Schatkamer* (1688), "Voorrede."

11. In order to avoid "superfluities," "the usual tables or numbers that serve this science are noticeably shortened." Graaf, *De kleene Schatkamer* (1688): "We call it the Small Treasure Chest, because in truth it is nothing more: It does include the whole Art, but not everything that one generally adds to it." Later authors carried such parsimony to the extreme; see Pieter Holm's cryptic 1748 textbook.

12. In the English-language translation of Euclid by Milliet de Chales, *Elements of Euclid* (1685), 304, when it came to choosing propositions to teach, he "made [the] choice of some of the plainest and easiest to conceive." His translator, Reeve Williams (a onetime candidate for

the position at the Royal Mathematical School), cited other instances of mathematicians who "reduced the Propositions of these Books to a much lesser Number, and yet have thought them a compleat foundation to all the Sciences Mathematical."

13. Among those who took such classes were Jan Zoet, a "theologian-innkeeper poet," and Isaac Beeckman, the atomist and physician who took a three-month course in arithmetic, geometry, and navigation. Davids, "Ondernemers in kennis" (1991), 46.

14. Dutch society had long recognized the hypocrisy of landlubbers criticizing *stuurlieden* for running into trouble at sea. The earliest citation of the ironic motto, "The best navigators stand on shore," appears in Jacob Cats in 1632, although he may not have originated the phrase. See Cats, *Spiegel van den ouden en nieuwen Tydt* (1712), 254. The motto circulated beyond the Low Countries; see Diderot and d'Alembert, *Encyclopédie* (1765), 12.624, s.v. "Pilote (*Marine*)." This saying applied to those men who "bragged of being experts about pilotage, [but] who were ignoramuses when they got to sea."

15. Van Lottum, "Maritime Work in the Dutch Republic" (2017), 842, 851, and chapter 1, note 9.

16. Gietermaker, *Flambeau reluisant* (1667), "Beloved Reader." See also Van Dam, *Nieuwe Hoornse Schatkamer* (1712), cited in epilogue.

17. Oostwoud, *Vermeerderde Schoole der Stuurluyden* (1712), "Aen den Leser."

18. Van Dam, *Nieuwe Hoornse Schatkamer* (1751), "Voorreden." Some authors, such as the Portuguese Manoel Pimentel, *Arte de Navegar* (1712), "Ao Leitor," provided details about each half of the dichotomous categories they perceived: "on the one side, scientific," fell the rules and instructions for observations, variation, chart use, and mathematical principles. For him, that left familiarity with rutters, coasts, currents, and local geography to "the other [side,] experience."

19. Van Breen, *Stiermans Gemack* (1662), f. 1. "It is necessary to have good experience in the things that man cannot learn well through lessons alone."

20. Crone, *Cornelis Douwes* (1941), 105, 31. After Makreel's tenure (1672–74), the Amsterdam Admiralty let the position lapse until it hired Cornelis Douwes in 1749. The Rotterdam Admiralty began examining its officers in 1702, thirty-five years before that city's VOC chamber made exams mandatory. Bruijn, *Schippers van de VOC* (2008), 135; Davids, *Zeewezen en Wetenschap* (1986), 295–99; Warnsinck, *De Kweekschool* (1935).

21. Davids, "Het navigatieonderwijs" (1988), 65. On Douwes's Zeemanscollege (1748–73): Crone, *Cornelis Douwes* (1941), 122–43. Bruijn, *Commanders* (2011), 184–86 and 306, notes that Dutch naval officers had to make their own way . . . when it came to actual training." This was less "systematic" and more expensive than the training available for French and Swedish officers. However, he does note the resultant openness of the hierarchy could benefit "ship's boys who wanted to get ahead."

22. Haan, "De 'Academie de Marine' te Batavia" (1895); Van Oosten, " 'Hear Instruction and Be Wise': The History of a Naval College on Java (1969); Nieborg, *Indië en de zee* (1989), 28–32; [Van Imhoff], "Project — Tot het opregten van een Corps Cadets de Marine voor Zee-Dienst in Indien" (1743), NL-HaNA Radermacher 1.10.69, inv. nr. 382A. Johannes Siberg operated his academy from 1788 to 1812. Students who passed the annual public examination each December would be posted to the naval service or artillery.

23. Van Berkel, Van Helden, and Palm, *History of Science in the Netherlands* (1999), 34; as "professor extraordinarius of mathematics," Metius taught surveying, navigation, fortification, and astronomy. He published numerous texts in Latin and Dutch and also made instruments.

24. Davids, *Zeewezen en Wetenschap* (1986) 314, 321; Van der Krogt, *Advertenties Voor Kaarten* (1985). See also J. Oostwoud's *Maandelijkse Mathematische Liefhebberij* (1754–69), a teacher's trade journal that ran advertisements and announcements about exams.

25. Mörzer Bruyns, *Schip Recht door Zee* (2003), 79 on Holm and 155–204 on his account book. Ruelle, *Voorlooper des Zee-Quadrants* (1693): "I could teach someone (who has experience) so much in 8 or 14 days that in the art of navigation he would never need another teacher"; cited in Crone, *Cornelis Douwes* (1941), 97.

26. On authors with respectable social status, see, e.g., Nicholas Witsen, who wrote a substantial book on shipbuilding, *[Architectura navalis]* (1671), and then followed in his father's footsteps as mayor of Amsterdam.

27. Iberia was no longer viewed as the center of expertise; in fact, the Spanish were actively seeking out northern texts; see Seixas y Lovera, *Theatre naval hydrographique* (1704), 4 (translated from the 1688 Spanish original). As early as 1619, Christian IV of Denmark appointed the Dutch explorer and cartographer Joris Carolus to teach navigation to sailors in royal duty, and he was also permitted to offer private lessons; Davids, "Ondernemers in kennis" (1991), 39. Samuel Pepys corresponded with Joseph Hill in Rotterdam, hoping to get Dutch nautical books and information about their schools; *Naval Minutes* (1926), 424. Joseph Moxon, printer and hydrographer, spent time in the Netherlands as a youth and learned both his trades there, in part from Joan Blaeu; Sumira, *Globes* (2014), 23.

Tsar Peter I studied with Jan Albertsz van Dam and Nicholas Hartsoeker and apprenticed in the VOC shipyard as well as one in Zaandam. Fontenelle, "Eloge of N. Hartsoeker [1699]" (1825), 150. Hartsoeker owned several navigational textbooks, including *Dictionaire de Marine* (1702), Gietermaker, *Kunst der Zeevaart* (1660), and many titles by Van Nierop, including *Onderwijs der Zeevaart* (1673). Phillips, *The Founding of Russia's Navy* (1995); Driessen-van het Reve, "Wat tsaar Peter de Grote . . . leerde van Holland" (2009).

28. Davids, *Zeewezen en Wetenschap* (1986), 322: Nanne Oenes (d.1649), a schoolmaster in Harlingen, owned a globe and a cross-staff. Jean Deshayes, who taught navigation in Quebec City (1702–6), had a range of instruments, including a wooden globe; see Roy, "Jean Deshayes" (1916), and Schotte, "Hydrographer's Library" (2007).

29. Jan Willem Sleutel, "Konstige oefeningen begrepen in drie boecken," Manuscript schatkamer (Hoorn, 1675–77) MMR H631, 125.

30. Warnsinck, *Kweekschool* (1935): "Seminary for Navigation"; Bruijn, *Commanders* (2011), 186, boys ages twelve to sixteen.

31. Kruik, *Gronden der Navigatie* (1737), 10. See chapter 2, note 61, for Milliet de Chales's comments on the importance of images.

32. Gietermaker, *'t Vergulde Licht* (1710), 150, and Visscher ms. leaf facing p. 151, MMR ARCH 4D16; Van Asson "Schatkamer" (ca. 1705), f. [45v], OTYA 16275.

33. For Medina's early series of circular diagrams demonstrating the observer's position relative to the Sun, see prologue, note 17; Gietermaker had similar images, note 35 in this chapter.

A traverse, also known as a "compound" or "composed" course in English, a "cours composé" in French, and "koppel koers" in Dutch, makes clear how multiple elements are *coupled* together. In the late 1600s, mariners started using columnar "log boards" in place of compass-pointed "traverse boards." These were a hinged pair of boards on which the particulars of a ship's log were noted. Sturmy, *Mariners Magazine* (1669), iv. ii. 14, and see figure E.1. Navigators would transpose details from the log-board into log books or "traverse books" ("ruled and column'd just as the Log-Board is." Chambers, *Cyclopaedia* [1728], s.v. "log").

34. Tapp, *Seamans Kalender* (1602); Aspley, *Speculum Nauticum: A Looking Glasse, for Sea-Men* (1624); on both, see Waters, *The Art of Navigation* (1958), 241, 440; Newhouse, *The Whole Art of Navigation* (1685), 195, Prop. XL, "How to Correct . . . a Composed Course." Dutch examples include De Vries, *Schat-kamer ofte Konst der Stier-lieden* (1702); Van Dam, *Nieuwe Hoornse Schatkamer* (1751).

35. Gietermaker used numerous illustrations elsewhere in his textbook, particularly circular diagrams to help visualize the Sun's declination; basic Euclidean triangles (e.g., for heights of towers); bisected circles to illustrate principles of spherical trigonometry in part III; and at the beginning, a "thumb" image, and two volvelles. Sleutel (1675–77) MMR H631, ff. [86–88]; "In order to place this traverse course in the book, I will use the following little table." This is exactly copied from Gietermaker (p. 79 in 1710 ed.).

36. Anonymous Gietermaker ms. (1760s), MMR H629. A diagram of a six-leg traverse course, again taken from Gietermaker, appears with slight variations in three Dutch manuscripts produced nearly a century apart: Sleutel (1675–77) MMR H631, Grootschoen (1728) HSM S.0712, W. de Graaf (1763) MMR H633.

37. One diagram drawn by Cornelis Boombaar in his manuscript depicts the distance between northern Holland and the Shetland Islands as approximately one hundred miles, a fifth of the actual distance. Boombaar, "Onderwijs Der Zeevaert ofte de konst der stuurluijden" (1727–32), HSM S.1386. On the "rhetorical strength" of naturalistic images, compared to the more common "generic" emphasis of diagrams, see Vanden Broecke, "The Use of Visual Media in Renaissance Cosmography" (2000), 136.

38. Anchors appear in navigation workbooks by Spink (ca. 1697–1705), NMM NVT/47; Sephtonal (1713) NMM NVT/6 (but different distances from Spink anchor); Denoville (1760), BMR Ms gg8. See other playful shapes in Downman (ca. 1685–86), NMM NVT/8; Boombaar (1727–32), HSM S.1386.

39. In some cases, the final image does not seem to be a surprise: young William Spink clearly entitled one page of his workbook "The Traverse of a Casstle," presumably before he even began to draw the image; Navigation workbook (ca. 1697–1705), NMM NVT/47.

40. As Karel Porteman has noted in his study of seventeenth-century Jesuit rhetorical education, the "playful . . . creative use of illustrations" facilitated both acquisition and memorization of new material. Porteman, "The Use of the Visual in Classical Jesuit Teaching and Education" (2000), 191. Not only students but *liefhebbers* appreciated images: Piero Falchetta in Long et al., *Michael of Rhodes* (2009), 203: "This distinct [elite] audience required objects that possessed formal features no longer aimed at simple functionality." Morrice, *The Art of Teaching* (1801), 103, reminded teachers that taking a light-hearted approach could lead to better comprehension.

41. For examples of extended answers, including a single question that took nine pages to solve, see Wijkman's (unpaginated) *schatkamer*, "Ex Samen Der Sturliedn" (1709), TNA HCA 32/176 II.

42. The extant logbooks examined contain only tabular course calculations rather than graphical ones. See, e.g., *Amazoon* log (1791–95), TNA HCA 30/763 II; Logbooks (1758), HCA 32/176 III.

43. Kennerley and Seymour, "Aids to the Teaching of Nautical Astronomy" (2000), 161, discusses such graphical lessons. The exercise of graphing traverse courses probably sufficed to convince practicing navigators of the value of mathematics: by being made to draw out the course, and then painstakingly calculate and add up the distances of each component, only to find that numbers in a table produced the same result, a student received a clear take-home message — numbers could solve complicated problems without taking up as much time or paper.

44. On translating between two types of representation to solve a problem, see Larkin and Simon, "Why a Diagram Is (Sometimes) Worth Ten Thousand Words" (1987). Of course, the decision to make students learn with diagrams or tables could be either an instructor's personal preference or an institutional convention.

45. Davids skeptically notes that naval exams did not become mandatory in the Amster-

dam Admiralty until September 18, 1749 (for *commandeurs* and *luitenants*), and across all the chambers the Netherlands until the 1780s; even after it was universally required, a survey of merchant captains in 1851 revealed that 63 percent had *not* taken the exam. Davids, "Het Zeevaartkundig Onderwijs" (1985), 175.

46. The phrase "to comfortably test the knowledge of trainees" in the text heading is from Makreel, *Lichtende leydt-starre* (1671), table of contents. Makreel wished to test the "Leerlingh," which can be translated either as "student," "trainee," or "ship's boy."

47. W. J. Blaeu was an examiner for six years, until his death in 1638. Lastman first earned 150 florins a year, then 350; he also taught in Amsterdam from 1619 until 1652. Crone, *Cornelis Douwes* (1941), 101. Over the ensuing two centuries, at least sixty individuals are known to have held formal positions across the United Provinces. For a chronological list of examiners, arranged according to the companies and admiralties for which they worked, see Davids, *Zeewezen en Wetenschap* (1986), appendix 2, 398–405.

48. Such familial employment patterns were similar to those in France: see, e.g., the Boissaye de Bocage father and son, and the Le Cordier family.

49. Davids, *Zeewezen en Wetenschap* (1986), 297.

50. Gietermaker, Dirk Makreel, and Abraham de Graaf all published textbooks within a year of their appointments as examiner. Klaas de Vries and Jan Albertsz. van Dam also published textbooks in conjunction with their tenure. (All of these texts contained exams. The significant exception to this trend are Lastman's pioneering textbooks, which predated Gietermaker's and never included any form of exam.)

51. Gietermaker, *'t Vergulde Licht* (1684; also 1710), 148–49, 150–52, "Toetse de Navigatie, ofte Exame der Stuurlieden." This was adapted from his first publication, the *Vermaeck der Stuerlieden* (1659), which included "Questions for my disciples to practice." The *Vermaeck* was a pastiche from other authors' works and was never republished.

52. Candidates were expected to know how to derive an answer from the year's epact; see Gietermaker, *Vermaeck der Stuerlieden* (1659), 84.

53. Dirk Makreel's thirty-one-question exam appeared without solutions in the *Lichtende leydt-starre* (1671), but Makreel himself went through copies and inked in many answers (MMR ARCH 4C21). Abraham de Graaf's twenty-one-question exam, in the *De kleene Schatkamer* (1688, and likely 1680), was the first to provide printed answers.

54. Jonkers, *Earth's Magnetism in the Age of Sail* (2003).

55. This would change in the 1790s; see chapter 5, note. 27.

56. Gietermaker, *'t Vergulde Licht* (1710), 148–150. Gietermaker included the exchange in his tract on trigonometry (*Driehoex-rekening*, 1665), and it was translated into French, in *Flambeau reluisant* (1667), despite its limited relevance to French navigators (the written exam was omitted).

57. E.g., Vooght, *Zeemans Wegh-Wyser* (1695), 234. Some texts included directions for negotiating the Vliestroom, the tricky Friesland passage to the open ocean beyond Texel (see plate 1). English treatises included similar details about the Channel; see John Hamilton Moore (chapter 5 in this book). The VOC issued a pamphlet containing details of the route to Java in the winter and summer, which was soon included in textbooks; e.g., Makreel (1671), 189–90 (including details about the English Channel, 191); De Vries (1702), 323–32; Gietermaker (see, e.g., 1677, 1774). These instructions included only a few lines about the nearby coastal regions tested on the exam.

58. *Nieuwe uytgereckende Taafelen* ([1701]), 8–16, "D'Examinateur, der Stuerlieden," bound with *'t Nieuw Stuermans Graed-boek* (1697).

59. Certain navigators were expected to be adept at mental math; for instance, late-

eighteenth-century French candidates were examined in front of public audiences, presumably orally, on trigonometry and other math. Anthiaume, *Evolution et enseignement* (1920), 2.293.

60. He grew more critical of the textbook as time went on, a sign of his misplaced confidence rather than actual errors in the printed text; see Van Asson, "Schatkamer" (ca. 1705), ff. 92, 103, 105, OTYA 16275.

61. Van Asson, "Schatkamer" (ca. 1705), f. 118, OTYA 16275.

62. Wijkman "Ex Samen Der Sturliedn" (1709), unpaginated. Question 19: "Een stuurman, zaat twij maatroosen, tot haar onder wys." The hefty, vellum-bound folio ended up on a WIC vessel in the Caribbean that was seized by the British during the Seven Years' War (see TNA HCA 32/176 II); it is unlikely that Wijkman's career stretched into the 1750s, although there were other Swedes on board the *Concordia*, so perhaps he had bequeathed it to them, or abandoned it, voluntarily or not, at some point in his career.

63. Compare Gietermaker, *'t Vergulde Licht* (1677), 150–52, with Makreel, *Lichtende leydtstarre* (1671), 192–94.

64. Vos (1748–53) MMR H632; West (ca. 1760) HSM S.0187; and Kok (1763) HSM S.3003 were all based on De Vries. Other extant manuscripts include Boombaar (1727–32), HSM S.1386, based on Gietermaker; and Grootschoen (1728), HSM S.0712, based on Van Dam.

65. Warius, *Examen* (ca. 1731); Van Kinckel, "Luytenants Examen" (Middelburg, 1767), NL-HaNA Adm. coll. 1.01.47.11, inv. nr. 13. In both, "Gety/Tye Rekening" remained the first question.

66. Gietermaker, *'t Vergulde Licht* (1710), 91–120.

67. West (ca. 1760) HSM S.0187, 170, "We will skip numbers 5 to 9 due to the easiness of the topics"; these are the basic observations of the Sun's altitude.

68. Metius, *Astronomische ende Geographische Onderwijsinghe* (1632), 180; cited in Crone, "How Did the Navigator Determine the Speed," *Revista* (1969), 8.

69. Denys, *L'Art de naviger par les nombres* (1668), 93. See also Coubert, *Abrégé du pilotage* (1728), 89–90, 93; Blondel Saint-Aubin, *Le Véritable art de naviger* (1679), 78; Perkins, *Seaman's Tutor* (1682), 242.

70. On Snellius, Van Berkel et al., *History of Science in the Netherlands* (1999), 33; Crone, "How Did the Navigator Determine the Speed," *Revista* (1969), 5. By convention the knots were spaced 7 fathoms (42 feet / 7.71 meters) apart. The French measure was "7 or 10 *brasses*"; Fournier, *Hydrographie* (1643).

71. Perkins, *Seaman's Tutor* (1682), 243. Waters, "Development of the English and the Dutchman's log" (1956), 85, underscores the ongoing need for "a certain amount of judgment and skill."

72. Lastman, *Schat-kamer* (1629), 54.

73. Van Breen, *Stiermans Gemack* (1662), ch. 8. He recommended the same strategy Cornelis Lastman had thirty-five years earlier: pay careful attention to how long it takes the ship to sail between two known visible landmarks. Out of sight of land, "some use the chip and knotted line."

74. Van Breen, *Stiermans Gemack* (1662), 113.

75. Van Breen calculated an entire day's count to be 155,520 (one 24-hour day comprises 86,400 seconds). Used in combination with Snellius's mile, that gave a figure of 409 for every 60 feet. Van Breen gave further instructions on where to stand to make the observation: choose a place far enough back to avoid spray coming over the bow, but where the forward mark was still visible; and at night the distance should be halved to compensate for reduced visibility. Finally the navigator might find it helpful to place a metal hook sticking out near the waterline, to make it easier to see when the white foam passed the marks. Van Breen, *Stiermans Gemack* (1662), 113–14.

76. Van Breen, *Stiermans Gemack* (1662), "gewoonelijcke tel." Dutch speakers needed to eliminate the inequalities that one-syllable digits would cause, by beginning the count not with "one, two, three" but with "one and twenty, two and twenty" up to "two and ninety" for a total of seventy-two numbers. See also Holm, *Stuurmans Zeemeeter* ([1748]), 172, a count up to seventy-four; Crone, "Pieter Holm and His Tobacco Box" (1953).

77. Vooght, *Zeemans Wegh-Wyser* (1695), 147.

78. Lastman, *Beschrijvinge* (1657), Voorstel 37. The sandglass could run faster one day than the next, and even altitude measurements could be less certain than one would like, so it was wise to arithmetically compute the average.

79. Land-based readers were reminded that water appears to move more rapidly the closer one is to it, so being low to the water rather than on a high deck would give a different impression. Similarly, water immediately in the lee of the vessel could be artificially calm, as it was "carried along by the boat." Van Breen, *Stiermans Gemack* (1662), 114.

80. Lastman, after carefully describing the strategies for training the eye and the mind, ultimately recommended using the log and line, because it was substantially faster to learn, and — he contended — more certain; *Schat-kamer* (1629), 54–55. Bion, *Traité de la construction* (1723), 275, is a rare later example of a French text that explains how to track a ship's speed with a half-minute glass.

81. Crone, "Pieter Holm and His Tobacco Box" (1953). Approximately twenty such boxes survive; see example held by National Museum of Ireland (Dublin), DM:1971.365.

82. Holm, *Stuurmans Zeemeeter* ([1748]), 149.

83. Bruijn, *Commanders* (2011), 184, translates the school's name as "Ship Dead on Course."

84. Bruijn, *Commanders* (2011), 185. Holm's records cover the period 1737–74. In December 1758, VOC commander Willem Hoogland noted that three of his crew had taken lessons from Holm. Similarly, three of Commander Jan Spek's officers did so in August 1759.

85. Holm, *Stuurmans Zeemeeter* ([1748]).

86. The textbook included instructions on how to find Holm's school. Crone, "Het gissen van vaart en verheid" (1928–29), 154–55, declared it "worthless as a nautical text." See also Mörzer Bruyns, *Schip Recht door Zee* (2003), 68.

87. Holm, *Stuurmans Zeemeeter* ([1748]), facing p. 1, "Tot de Liefhebbers."

88. Gietermaker's work was first published by Doncker (1660–90), with an edition by Pieter Goos (1666), and nine additional editions from Van Keulen (1697–1749). The VOC used the Blaeu enterprise as their publisher (see chapter 1, note 62), followed by Paulus Matthysz., Adriaen Wor, and others in the eighteenth century; NL-HaNA VOC 1.04.02, inv. nr. 5165.

89. Veur, *Zeemans schatkamer* (1755), "Voorreden Aan den Leezer." See Davids, "Van Anthonisz tot Lastman. Navigatieboeken" (1984).

90. Veur, *Zeemans schatkamer* (1755). Veur was adamant: "This order and approach [*manier*] are better for the Master as for the Student." Veur's view of Gietermaker as straightforward, despite the pioneer's emphasis on challenging questions, attests to how ubiquitous Gietermaker's template had become by the mid-eighteenth century. Veur directed his readers to other Van Keulen publications, both as a way to keep his own work brief and to promote others available from his publisher.

91. Steenstra, *Grond-beginzels der Stuurmans-kunst* (1779). Davids, *Zeewezen en Wetenschap* (1986), 399.

92. Steenstra, *Grond-beginzels* (1779), xiii, xvi, xix, 2. Trigonometry would facilitate both dead reckoning and tracking compass variation. On *denkbeelden*" ("ideas," lit. "thought-pictures"), e.g., "In these schools they were supposed to learn the relationship between Heaven and Earth but they didn't get the least image." See Van Dam, *Nieuwe Hoornse Schatkamer*

(1712), "Voorreden," for a similar discussion of the importance of comprehending foundations — in Van Dam's view, geometry was the key, without which the *leerling* "otherwise would not only learn with more difficulty but would also find it harder to remember."

93. See prologue. French instructors offered different courses of study for men who were or were not already literate; Des Nos, "Mémoire sur l'escole du canon de Brest" (1682), AN-Paris MAR/G86, f. 5v.

94. Kruik, *Gronden der Navigatie* (1737), "Voorreden." This progression was similar to the Royal Mathematics School, which scheduled apprenticeships after one and a half years spent studying navigation and mathematics, see chapter 3, note 8.

95. Davids, *Zeewezen en Wetenschap* (1986), 323–325, finds evidence of navigational lessons in sixty Dutch towns and cities in the later eighteenth century.

96. Mörzer Bruyns, *Schip Recht door Zee* (2003). The Lijsten reflected this shift: in 1747 every ship was equipped with three octants; by 1788, two octants and one sextant. Bruijn, *Commanders* (2011), 287.

97. Douwes's method was first published in the *Actes de l'Academie de Haarlem* (1754). Bowditch included it in his *Improved Practical Navigator* until 1826. Silverberg, "Fathers of American Geometry" (2012), 11.

98. Denoville's 1760 manuscript included a staggering 83 pages (of a 150-page manuscript) of exercises about working with the "quartier de réduction." Hébert, *Le traité de navigation* (2008), 91.

99. See Peters, *Crone Library* (1989), 615, 642, for a list of nautical works that included Briggs's and Vlacq's tables.

100. Van Lottum, "Maritime Work in the Dutch Republic" (2017), 848.

101. Van Kinckel, "Luytenants Examen" (1767), NL-HaNA Adm. coll. 1.01.47.11, inv. nr. 11A, 13. Baron van Kinckel, from Heilbronn, Germany, had elitist views about who should become a midshipman ("no one whose parents keep a shop or pursue a trade or some lowly office") and similar disdain for merchant mariners; Bruijn, *Commanders* (2011), 60.

102. Gietermaker, *'t Vergulde Licht* (1774). Material in book 2 related to the "reflection of the oval-surface of the Earth" was not for practitioners but "only to introduce a short method for enthusiasts to carry out rhumb-line sailing" on Mercator projections.

103. For a recent analysis of the disparate programs by proponents of various "mixed mathematics," see Cormack et al., *Mathematical Practitioners* (2017), esp. Schuster, ch. 3.

5. LIEUTENANT RIOU IS PUT TO THE TEST — THE SOUTHERN INDIAN OCEAN, 1789

1. Nash, *Last Voyage of the Guardian* (1989), 72, n. 21: this incident took place 1,300 miles (2,080 km) southeast of the Cape of Good Hope. The *Guardian* was 140 feet long (Nash, xxi). Riou judged the first iceberg "to be about 1½ mile in circuit and about 30 or 40 feet high from the surface of the sea." Nash, 36, citing "Remarks at Sea" from NMM RUSI/NM/235/ER/7/2. See also Clements, *Guardian. A Journal of the Proceedings on Board the above Ship* (1790), 6.

Because Nash did not have access to Riou's logbook from the voyage (now at SLNSW, MLMSS 5711/1 [Safe 1/233b]), she relied upon the transcription by Ludovic Kennedy, "Log of the Guardian" (1952); hereafter Nash, "Journal" (where no year is cited, the events occurred in fall 1789 and the first two months of 1790).

2. Clements, *Guardian* (1790), "two-thirds allowance of every kind." Some of these men had been convicted for stealing (one was a horse thief); their sentences were either seven or fourteen years, or life (Convict Transportation Registers, 1787–1870, Australian Joint Copying

Project. Microfilm Roll 87, Class and Piece Number HO11/1, pp. 23–24). There were seven "superintendents" to guard the convicts. One of these men, Philip Schaffer, was accompanied by his daughter Elizabeth, the only woman on board. Nash, *Last Voyage*, 231. McHugh, *Drovers* (2010), ch. 1, errs in claiming three hundred passengers.

3. Clements, *Guardian* (1790), 11.

4. Nash, *Last Voyage*, 64–65, n. 9: two chain pumps, and two smaller Cole's suction pumps.

5. Clements, *Guardian* (1790), 15.

6. Clements, *Guardian* (1790), 10, 16. Falconer, *Universal Dictionary of the Marine* (1769), "fothering: a peculiar method of endeavouring to stop a leak in the bottom of a ship while she is afloat." The sail would be lined with oakum (rope yarn), or other loose material, with the goal of partially filling the leak. See Nash, *Last Voyage*, xxviii, n. 31.

7. Nash, "Journal," 71, Dec. 25. Approximately thirty men left the ship in those small vessels. The jolly boat, a smaller open boat used as a working vessel, filled with water almost immediately and sank. An additional twenty or more men seem to have fallen overboard and drowned in the desperate aftermath of the crash; see note 12 in this chapter.

8. Clements, *Guardian* (1790), 20. He described the desperate efforts to gather adequate food and equipment. One boat traded a quadrant with another vessel "and received a cheese in lieu" (25), while the ill-fated jolly boat "brought with her neither provision, water, compass, or quadrant" (26).

9. Nash, "Journal," 73–74, Dec. 25–26; this was Cook's account of his second expedition, *Voyage towards the South Pole* (London, 1777). Throughout the logbook, Riou carefully differentiated between "observed" and "Dead Reckoning" positions; see his daily records of "Co[urse]," "Dist[ance]." and "Obs. Lat." compared to "D.R. Long." or "Long. adduced by D.R."). (Nash, perhaps following Kennedy, reports this as "Guessed" longitude and "Guessed" course, but Riou did not use that term.)

10. The larger of the two is Marion Island (46°46′ S 37°51′ E). Before 1789, they had been visited by European ships on only three occasions, including by Cook in 1776; this was when Riou likely learned of them. Nash, *Last Voyage*, 72, n. 21. Riou was not optimistic about locating—or surviving on—these small desert islands, even though they were significantly closer than Africa's southern coast, which Clements (*Guardian* [1790], 27) deemed to be a "vast distance of 411 leagues in a boisterous ocean."

11. Clements, *Guardian* (1790), 20–21.

12. Clements, *Guardian* (1790), 23–28. Of the original 120 souls on board, only 60 remained with Riou, including 21 of the 25 convicts.

13. Clements, *Guardian* (1790), 30. On December 27, they purportedly boiled one small goose in an iron pot, divided it into fifteen parts, and blindfolded one man so he could serve out the morsels equitably.

14. Clements, *Guardian* (1790), 37–38. Point Natal is just north of Durban, South Africa.

15. Clements, *Guardian* (1790), 39. *Narrative of the Distresses and Miraculous Escape of His Majesty's Ship the Guardian* (1790), 43. See the *General Evening Post* (London), no. 8820 (Apr. 22–24, 1790), for a typical news report.

16. Nash, *Last Voyage*, xxxii; *The Times* (Apr. 29, 1790).

17. In addition to Clements's *Guardian* (1790) and *Narrative of the Distresses* (1790), a slightly different account credited to "one of the Officers present" would appear in the *Naval Chronicle* 5 (1801), 482–92.

18. *The World* (London), no. 1411 (Saturday, July 9, 1791): advertisement for "a great variety of Entertainments, . . . A favourite historical representation, in two parts, in which is given, a living Picture of THE GUARDIAN FRIGATE, Commanded by Lieut. RIOU. . . ." Ballads in-

cluded: "The Unfortunate voyage of the Guardian man of war" (1790) SLNSW Dixson Library Drawer item 339; and Shapter and Moulds, "The Forecastle Sailor or the Guardian Frigate" (ca. 1790), with the verses reissued by W. Collard (ca. 1860s), SLNSW Dixson Library Drawer item 304.

19. Most of Riou's papers are held at the Caird Library, NMM (RUSI/ documents); however, see SLSNW (Sydney) for Riou's narrative of the wreck of the *Guardian*, in vol. 2 (ca. 1790), MLMSS 5711/2, and the original logbook of the voyage, MLMSS 5711/1, hereafter "Log Book."

20. Nash, *Last Voyage*, 24, Nov. 4, 1789: "Pray did I leave behind me an old Log Book of the *Mediator* & the *Salisbury*..." This volume is now among Riou's papers at the Caird Library: NMM RUSI/NM/235/ER/2/2.

21. Englishman John Hadley (1682–1744) and American Thomas Godfrey (1704–49) are credited with inventing the octant in the 1730s. John Bird then modified their device in the 1750s to create the larger and more versatile sextant. Mörzer Bruyns, *Schip Recht door Zee* (2003).

22. The idea that longitude could be determined by measuring the precise distance between the Moon and the fixed stars was first suggested by Germans Johann Werner (1514) and Peter Apianus (1524, 1533). It took more than two hundred years to develop instruments and astronomical tables accurate enough to carry out this method. Dunn and Higgitt, *Ships, Clocks, and Stars* (2014), 51–53. On Mayer's method and tables, H. Derek Howse, "Lunar-Distance Method," in Andrewes, *Quest for Longitude* (1996), 154.

23. According to Dunn and Higgitt, *Ships, Clocks, and Stars* (2014), 96–99, the lunar distance method was accurate to the order of 0.5°, compared to 10° with dead reckoning. The navigator would use the tables in the *Nautical Almanac,* which predicted the positions of the Sun and Moon several years in advance, to convert the observed angle to a geocentric lunar distance. The almanac gave data for Greenwich time in three-hour intervals; it was necessary to interpolate between those values to get the actual time aboard ship. On the problems of atmospheric refraction and lunar parallax, Howse, "Lunar-Distance Method" (1996), 151. Howse (156) estimates that the position would take at least thirty minutes to compute, whereas without Maskelyne's tables it could take four hours, but see Jane Wess's skepticism about Howse's claims and the general use of lunars: "Navigation and Mathematics" in Dunn and Higgitt, *Navigational Enterprises* (2015), 206.

24. Vergé-Franceschi, *La marine française au XVIIIe siècle* (1996); Gaastra, *Dutch East India Company* (2003).

25. Garralón, "La formación de los pilotos" (2009), 160, 165; Rodger, *Command of the Ocean* (2004), 265; Dickinson, *Educating the Royal Navy* (2007), 33; Warnsinck, *De Kweekschool voor de Zeevaart* (1935); and on the short-lived VOC schools in Amsterdam and Batavia, see chapter 4, notes 21, 22. On the French marine in the eighteenth century, see Vergé-Franceschi, *La marine française au XVIIIe siècle* (1996); Chapuis, *A la mer comme au ciel: Beautemps-Beaupré* (1999).

26. The VOC produced new guides for both examiners and test-takers to ensure consistency in the updated certification process. The company published model written and oral exams to assist "all those who wish to give themselves in service to the VOC" with their preparation. In addition to sample questions, these pamphlets provided a bibliography of recommended authors and footnoted page references to explicate tricky problems. The authorities clearly wished their candidates to pass. VOC Heren XVII, *Examen voor de Capitain en Capitain-lieutenants* (1794), HSM S.4793(342). On the "tougher" content of these updated exams, Bruijn, *Commanders* (2011), 293.

27. Heren XVII, "Instructie voor de Examinateurs der Stuurlieden" (1793), NL-HaNA VOC

1.04.02 inv. nr. 5026, 4. Extensively revised from the earlier eighteenth century (see chapter 4), the "Mond-Examen" now relegated the entire contents of the traditional oral exam into the penultimate of the twenty-three questions; Heren XVII, *Examen* (1794), HSM S.4793(342), 14. Before addressing "the route from and to the Indies . . . &c.," the candidate fielded twenty-one questions where he was expected to demonstrate the use of traditional and new instruments.

28. Heren XVII, *Examen* (1794), 12–14; Heren XVII, "Instructie" (1793), 7, 9, 11; see further Schotte, "Expert Records" (2013), 300.

29. Anthiaume, *Evolution* (1920), 2.99, 127; Fauque, "Les écoles d'hydrographie en Bretagne (2000), 392–94.

30. Aman, *Les officiers bleus* (1973), 162. Physics was a new addition, no longer just for loading ships but for dealing with storm forces.

31. See Rodger, *Wooden World* (1986), 265, for debates over education and the amount of theory required of officers.

32. Moore, *Practical Navigator* (1791), 272–95. Moore borrowed heavily from one of his competitors, the navigation teacher John Adams. In his own popular textbook, *The Young Sea-Officer's Assistant, both in his Examination and Voyage* (1773), Adams credited his former students for something "never before made public," the "Questions and Answers relative to the managing of a Ship" (vii–viii).

33. Moore, *Practical Navigator* (1791), 282, 287. Navy men who read the *Practical Navigator* disputed the contents of this section, suggesting that certain shipboard tasks were difficult to routinize (295). Toward the end of the questions (e.g., 283), Moore touched on military strategy, gesturing to what the French found of paramount importance.

34. Moore, *Practical Navigator* (1791), 273, 276, 285: Q: "How do you splice your cables?"

35. Moore, *Practical Navigator* (1791), 273–77, for sines, etc.

36. The Royal Navy exam appeared in the early American editions of Moore's work, published by Edmund Blunt (1799). The chapter title was altered to reflect the absence of an exam in the United States — "The Substance of Information every Candidate for the American Navy ought to be acquainted with previous to his being appointed" — but while the book was updated with American information, the "exam" maintained its exclusive focus on British waters. (When Nathaniel Bowditch issued his famed revisions of Moore, under the title *The New American Practical Navigator* [1802], he did not include the exam section.)

37. See, e.g., R. N. Lieutenants' Passing Certificates (1809), TNA ADM 107/40. On the changing social status of navy officers, Rodger, "Honour and Duty at Sea" (2002).

38. E.g., Morrice, *Young Midshipman's Instructor* (1801), and see chapter 4. Rodger, *Wooden World* (1986), 115, and ch. 7. Vergé-Franceschi, *Marine et éducation* (1991), 181, in the French navy the officers became more "scientific" but lost "color and autonomy." See further, Taillemite's profiles of French naval officers (1982, 1999).

39. Riou was born November 20, 1762. In 1774 he served on the *Barfleur* at Portsmouth, and then the *Romney*, based at Newfoundland. Robson, "Edward Riou (1762–1801)" (2011), 47; Nash, *Last Voyage*, xvi.

40. Edward Riou, "Mathematical Lessons for a Young Sailor . . . under Instruction of James Paterson and Thomas Adams" (1774), NMM RUSI/NM/235/ER/3/1.

41. Robertson, *Elements of Navigation* (1754); the textbook saw seven editions in fifty years.

42. Robertson, *Elements of Navigation* (1754), preface. He treated Astronomy in bk. 8 (433–60); Day's Work, bk. 9 (504–603, including "various methods of correcting the account of longitude," 529–35).

43. Robertson, *Elements of Navigation* (1754); as the subtitle noted, Robertson intended it "*For the Use of the Royal Mathematical School at Christ's Hospital, and the Gentlemen of the*

Navy." He addressed the different level of learners in the preliminary "Advertisement" (unpag.). In his preface, Robertson claimed to be committed to "brevity and perspicuity," tailoring the "elementary parts" for "beginners," but felt that later in the six-hundred-page book "the practical parts" justified being "more fully treated on, and intended to include every useful particular, worthy of the mariner's notice."

44. Robertson, *Elements of Navigation* (1754), 107; the text is the same, but the questions on this page are given different values in the manuscript.

45. Riou's lessons included military tactics alongside trigonometry; see the loose sheet inserted after p. 75 of "Mathematical Lessons" (1774), NMM RUSI/NM/235/ER/3/1: "The new method used in the French artillery for their 12 lb field pieces — drawn by 15 men." Second workbook: NMM RUSI/NM/235/ER/3/3. Of the forty Astronomical definitions that appear in section III of Robertson, *Elements of Navigation* (1754), Riou transcribed thirty-five.

46. *Narrative of the Distresses* (1790), vii; Nash, *Last Voyage*, xvi; Skelton, "Captain James Cook as a Hydrographer," *Mariner's Mirror* (1954). During a stop in New Zealand, Riou adopted a native dog that then bit some of the crew; the "cannibal dog" was given a mock trial and killed. See Salmond, *Trial of the Cannibal Dog* (2003).

47. Kindred, "James Cook: Cartographer" (2009). See Dunn and Higgitt, *Ships, Clocks, and Stars* (2014), 141, on the "longer-term legacy" of Cook's voyages "for embedding astronomical and timekeeper methods within naval practice." William Bligh also became a top-notch navigator during the third voyage.

48. See note 10 in this chapter. During Cook's second voyage (Dec. 1773), icebergs were encountered in latitude 50°57′ S, longitude 20°45′ E. See Cook, *Voyage towards the South Pole* (1777), 1. ch. 6; Nash, *Last Voyage*, xxvii, n. 30. Riou well knew Cook's efforts to "survey . . . Ice"; see his "Ship's Note, Dec. 22" (erroneously noting 52°52′ S), in Nash, 36–37.

49. Riou passed his lieutenant's exam on October 28, 1780. He spent some time on half pay and visited France; see Nash, *Last Voyage*, xxii.

50. NMM RUSI/NM/235/ER/2/2. On the penultimate leaf of the journal-*cum*-sketchbook from the voyage of the *Mediator*, Riou's complaints are almost poetic: "Trifling Misfortunes 19 Nov 1786 / No sun / lose at backgammon / A Guinea going every day / Salt beef / No mutton / and less cheese / A foul wind, / a lame foot, / And damn the drop of port wine."

51. Nash, *Last Voyage*, xxi, 228–33. The voyage from Cape Town to Sydney was 13,000 miles (21,000 km). The crew hailed primarily from the British Isles, but included two Danes, two Americans, one German, and one mariner from Minorca. Muster Book for HMS *Guardian* (TNA ADM 36/11005), ages listed ranged from eighteen to forty-four (but ages of midshipmen — usually young teens — were not specified).

52. The supplies were valued at 70,000 English pounds. Nash, *Last Voyage*, 29, 31. Banks sailed with Cook on the *Endeavour* (1768–71) and served as president of the Royal Society (1778–1820). The "garden" of plants and seedlings was chosen for being "useful in food or physic." For Banks's plans for the 174-foot-square "Coach," see Nash, 6–7. Two of the convicts' superintendents were gardeners; one, James Smith, sent itemized reports of the ninety-three original specimens back to Banks; Nash, 26–30.

53. Robertson, *Elements of Navigation* (1754), 263, 270: "This method of finding the latitude from the meridian altitude of a celestial object is the most easy, when it can be put in practice: but as this is not always the case at sea . . . ; Take one, two, or more altitudes of the Sun in the forenoon, noting the time of each by a watch. . . . Let the morning altitudes . . . be taken at equal intervals of time, if possible."

54. The "horizon" Riou mentions may have been similar to the modification devised by John Elton in 1732, where two bubble levels were used to keep backstaffs and other instru-

ments properly oriented; Dunn and Higgitt, *Ships, Clocks, and Stars* (2014), 95. The double-altitude method is used when a "noon sight" is not available and the latitude is uncertain; Silverberg, "Fathers of American Geometry" (2012). Nash, "Journal," 105, Feb. 20, "looking sharp out for double altitudes," but Riou fortunately "got a good meridian observation" instead.

55. Riou, *Guardian* "Computation book" (1789–90), NMM RUSI/NM/235/ER/3/8, front fly leaf. Riou determined that a daily loss of 2′38″6 would result in the watch slowing by one minute and six seconds every quarter of an hour. Riou, "Log Book," Sept. 16, 1789, e.g., "The result of which distances when corrected of Refraction and Parallax gave the Longitude: 8°26′45″.

56. Riou used compasses by the prominent instrument makers George Adams and Kenneth McCulloch. He noted other instruments crafted by [John] Troughton and Jesse Ramsden — most likely a sextant from the former and perhaps an octant or a variation compass from the latter. *Guardian* "Computation book" (1789–90), Sept. 24, notes a frustrating error with the Ramsden. Cotter, *History of Nautical Astronomy*, 89.

57. Royal Society, "Directions for Observations and Experiments to Be Made by Masters of Ships, Pilots, and Other Fit Persons in Their Sea-Voyage" (1667).

58. Riou, "Log Book," Sept. 13 (gun powder and biscuit); Sept. 23 (fourteen azimuth and amplitude observations); temperature records *passim*. He made similar observations in earlier voyages, see, e.g., Journal of the *Salisbury* (NMM RUSI/NM/235/ER/2/2), Aug. 18, 1786, when he noted various thermometer readings.

59. Nash, *Last Voyage*, 10, Letter from Riou to Banks, Portsmouth, July 15, 1789. John Harrison perfected his famous marine timekeeper in 1760 but did not receive his reward until 1773; see Dunn and Higgitt, *Ships, Clocks, and Stars* (2014); Howse, "Lunar-Distance Method," in Andrewes, *Quest for Longitude* (1996), 159, on three timekeepers.

60. *Guardian* "Computation Book," NMM RUSI/NM/235/ER/3/8, e.g., Sept. 11, "The watch I suppose to be losing daily oh 3′0″ pr diem"; Sept. 16, "Turn over for Correction of the Watch . . ."; cf. Riou, "Miscellaneous Computations" (1782–87), NMM RUSI/NM/235/ER/3/4 (Nov. 29, 1782).

61. Riou, *Guardian* "Computation Book," Sept. 24: he noted perplexedly, "These distances are well taken by the Master. — gave 4′ error by the 2 suns I cannot make it above 1¼ or ½ by the Horizon. I don't know what can make the difference tho' I've worn myself out in that [trouble]." (The "2 suns" may refer to a double altitude reading; see note 54 in this chapter.) On September 22, Riou admitted to "some mistakes in reading the sextant."

62. E.g., *Guardian* "Computation book," Sept. 28. Nash, *Last Voyage*, 223: TNA ADM 1/2395, letter of June 22, 1791. The father of John Gore (1774–1853) sailed with Cook as his first lieutenant on the third voyage and named his friend Joseph Banks as his son's guardian. See further Madge Darby, "John Gore's 'Young One' " (1998), 1550; Thornton, "John Gore's 'Young One' — An Update" (2007), 34.

63. Nash, "Journal," 66, 72, Dec. 24–25.

64. Nash, "Journal," 74, 77–78, 87–90: Dec. 26, 28, 29, Jan. 20–24. Cook, *Captain Cook's Journal* (1893), 276–77.

65. E.g., Nash, "Journal," 70, 73, Dec. 25 ("my watches, silver buckles, pistols and sword gone"); 97, Feb. 5; 104, Feb. 19 ("people were negligent, sleepy and had stolen wine").

66. The provisions evidently included ostrich eggs, which the crew consumed in late December. Nash, "Journal," 79 (on the wealth of provisions), 96 (talk of rations and the infirmary, Feb. 4), also 36 and xxiv (kraut and scurvy).

67. Nash, *Last Voyage*, 222; "Journal," 75, 77, Dec. 26–28. Prompted by that memory, Riou consulted John Hawkesworth's *Account of the Voyages . . . for Making Discoveries in the Southern Hemisphere* (1773), which included a brief account of the *Tamar*, a sloop that had also lost her rudder.

68. Nash, "Journal," 80–81, Jan. 1; 93, n. 38, Jan. 30. Ross was a thirty-five-year-old seaman from London. The *Guardian*'s carpenter, Murray Sampson, corroborated his idea. See Pakenham, *Invention of a Substitute for a Lost Rudder* (1793).

69. Nash, "Journal," 95, Feb. 3 (fish); 105, Feb. 20; he estimated Cape Lagullus [Aguhlas] to be "five leagues distant" (170 kilometers southeast of Cape Town).

70. Nash, "Journal," 88, Jan. 21.

71. *Guardian* "Computation book," dates skip from Dec. 7, 1789, to Feb. 4 and 5, then to Feb. 20, 1790. The lack of readings in December suggests that some of the uncertainty throughout January stemmed from the weather more than the conditions after the crash. Riou did have a second rough book that he filled with various jottings during that stage of the voyage (NMM RUSI/NM/235/ER/3/9), which might contain some of the absent observations.

72. Nash, "Journal," 98, Feb. 7. By the eighteenth century, navigators were very familiar with the regular patterns of the intercontinental trade winds and took advantage of their predictable steadiness to cross the oceans.

73. Nash, "Journal," 88, Jan. 21. On February 7, Riou noted that he "turned [his] book over" to the next page in the chart book. Nash, 99, n. 45, suggests Riou used J.-B.-N.-D. d'Après de Mannevillette, *Neptune Oriental* (1775).

74. Nash, "Journal," 88, Jan. 21 (men doubled Cape); 103, Feb. 18 (danger); 99, Feb. 8; 101, Feb 12 (anxious); and 103, Feb. 18 (mast).

75. Riou injured his hand several times; Nash, "Journal," 67, Dec. 25; 106, Feb. 21. Parts of the logbook page are neatly written, but other numbers and comments are scrawled across the columns. Riou may have gone back to fill in details at a more composed moment. He made no note of the nadir of the voyage in his prepared narrative, SLNSW MLMSS 5711/2.

76. Nash, "Journal," 104–5, Feb. 19.

77. Nash, "Journal," 105, Sat. Feb. 20. Gannets are coastal birds. See note 53 in this chapter about meridian observations.

78. Nash, "Journal," 105–7, Feb. 21. He was more exuberant in the polished account: "I saw it directly, and thank God, it was the only land I knew on the coast, it was the Cape of Good Hope! The ship was running right for it."

79. Whale boats: small, open boats. Nash, "Journal," 107–9; *Last Voyage*, 114, Letter #34. Upon reaching land on February 22, Riou hastily penned a brief note to his mother and sister (Nash, *Last Voyage*, #33): "Dearest, God has been merciful. I hope you will hear no fatal accounts of the *Guardian*—I am safe, I am well, notwithstanding you may hear otherwise."

80. Nash, *Last Voyage*, xxx, xxxiii and n. 42. Twenty convicts survived (one died at Cape Town), but six evidently died en route to Australia, in the notoriously grim conditions of the transport ships. A small portion of the original cargo (including a number of hardy sheep) was transferred to the *Lady Juliana*, a boat carrying female convicts to Botany Bay. *The Times* (London), May 16, 1791, cited in Nash, *Last Voyage*, 221, #70.

81. *Whitehall Evening Post* (London), no. 6673 (May 19–21, 1791), reported on the court martial. Back in London, Riou wrote a preface to an account of another ill-fated ship, the *Grosvenor,* which was published as Jacob van Reenan, *Journal of a Journey from the Cape of Good Hope* (London: G. Nicol, 1792).

82. Nash, *Last Voyage*, xxxix: Danish guns "virtually cut him in two." According to his will, Riou's small estate was valued at approximately one thousand pounds and was to be divided among his mother (who predeceased him), his sister, and a friend.

83. *Narrative of the Distresses* (1790), 45, 37, vi, vii.

84. Clements, *Guardian* (1790), 46–47, "[The face of] Lieutenant Riou . . . indicates all the heroism he has displayed. He is irritable in his temper; but his passion, easily inflamed,

as soon subsides, and leaves no pique, sulkiness, or animosity behind it." Nash, *Last Voyage*, xxii.

85. Riou's notebooks are full of examples of his skill and judgment — for instance, his ability to estimate a coastal landmark to be "five leagues distant"; see note 69 in this chapter.

86. E.g., Wise, *Values of Precision* (1995); Daston and Lunbeck, *Histories of Scientific Observation* (2011).

87. Dulague, *Leçons de navigation* (1768), 268, para. 726: A "Pilot destitute of the aid of Longitude" could fail to correctly determine the vessel's position for any number of reasons, including inattentive carpenters, unusual foreign design elements, the sudden swerves to which a boat is subject, or the disproportionate relationship between a short length of knotted logline and the miles covered by the ship.

88. Dulague, *Leçons de navigation* (1768), 268, para. 726. He felt that compasses were too small for absolute certainty.

89. Dulague, *Leçons de navigation* (1768), para. 727, 728. Dulague based his proposal on the "new corrections" included by the astronomer N. L. de Lacaille in his edition of Bouguer, *Nouveau Traité de Navigation* (1760).

90. Dulague, *Leçons de navigation* (1768), 202, para. 571. Cornelis Lastman had taken a similar approach: *Beschrijvinge van de Kunst der Stuer-luyden* (1657), ch. 62: "What natural mistakes are implied in the common method of calculating tides, and how one corrects them oneself."

91. Dulague, *Leçons de navigation* (1768), 268, para. 726; 291–92, para. 775.

92. Moore, *Practical Navigator* (1791), 287–88, 295.

93. Robertson, *Elements of Navigation* (1754), advertisement.

EPILOGUE. SAILING BY THE BOOK, CA. 1800

1. Vergé-Franceschi, "Un enseignement éclairé au XVIIIe siècle" (1986), 53.

2. Willaumez, *Projet pour former des élèves marins à Paris même* (1800). See also the proposal in the 1786 Ordonnance for third-class "élèves" to begin their training on an armed naval corvette. Crisenoy, "Les écoles navales" (1864), 98. This had an extremely hands-on focus: "They will stand watch like the sailors, be expected to climb up onto the mast-platforms, to the crow's nest and onto the yardarms; they will go in the ship's boat when one anchors the boat fore and aft or raises the anchors, and take part in all the maneuvers with the sailors." On the challenges of running such training vessels in times of war *and* peace, see chapter 2, note 85.

3. "Reflexions sur les ecoles de marine" [ca. 1761], AN-Paris MAR/G86, No. 16 "Ecoles flotantes."

4. On the "young gentlemen" who entered naval colleges scarcely able to read, see Vergé-Franceschi, "Un enseignement éclairé au XVIIIe siècle" (1986), 50.

5. David Morrice, *Art of Teaching* (1801), 97, who at one point taught aboard a naval frigate, admitted that a schoolmaster might be better equipped to "teach them good manners and behaviour" rather than "mere navigation, which, after all, is more frequently better taught them practically by the officers, than theoretically by himself."

On the Royal Navy program for shipboard teaching, see chapter 3, note 103, and Sullivan, "The Naval Schoolmaster" (1976), who estimates five hundred to six hundred teachers between 1712 and 1824. In the decentralized United Provinces, no evidence survives of such formal vetting of instructors. However, by the 1760s Dutch mathematics teachers had to pass a test to secure employment; see the practice exams in Van Steyn, *Liefhebbery der Reekenkonst* (1768).

6. Professor Pierre Lévêque, dissatisfied with his working conditions at Nantes, wrote impassioned statements about the important public role fulfilled by instructors like himself: Arch. dép. Loire-Atlantique, C403, 2, see also 10–14, cited in Fauque, "Écoles d'hydrographie" (2000), 388.

7. See Davids, "Maritime Labour" (1997), 50–54; chapter 4, note 100. VOC Heren XVII, *Instructie voor de Examinateurs der Stuurlieden* ([1780]); VOC Heren XVII, *Examen voor de Capitain* ([1794]), HSM S.4793(342), and see chapter 5.

8. Dickinson, *Educating the Royal Navy* (2007), 33; Rodger, *Command of the Ocean* (2004), 265, on the Portsmouth Academy's poor reputation.

9. See Pritchard, *Anatomy of a Naval Disaster* (1995); Vergé-Franceschi, *Marine française au XVIIIe siècle* (1996). Jonathan Dull, *Age of the Ship of the Line* (2009), treats the crucial role of the navies in the seven conflicts between France and England over the course of the eighteenth century. On the reforms under the Maréchal de Castries, see Crisenoy, "Les écoles navales" (1864); Vergé-Franceschi (1996).

10. "Idées d'un marin sur les connaissances théoriques et pratiques à exiger des officiers de la Marine royale" (ca. 1802) SHD-Vincennes MS 516, Art. 9 and 11.

11. See, e.g., Van Dam, *Nieuwe Hoornse Schatkamer* (1712), "Voorreden": "Experience and Art [theory] must go together in navigation, and never be separated from the other, for man to be able to bring a Ship across the sea." He specified that "in small navigation experience is more relevant than Art, but in large [navigation] both of these must be known."

On the continued oscillations of the pendulum in the nineteenth century, see Charon, Claerr, and Moureau, *Le livre maritime* (2007), 126, and *Histoire de l'École navale* (1889), 161: Napoleon launched two training ships ("écoles flotantes"), but soon thereafter Louis XVIII's naval minister De Bouchage declared that men being trained for the navy could not have "practical and theoretical education" simultaneously.

12. See chapter 3 on Paget's lack of experience and the minimum time at sea for Royal Navy officers. The French period of nine months is noted in both Louis XIV, *1681 Ordonnance* (1714) and the 1785 reforms; see Fauque, "Écoles d'hydrographie" (2000), 393. Comprehensive school programs lasted one or more years, while entrepreneurial teachers like John Hamilton Moore promised to teach navigation "in a few Weeks"; see *Practical Navigator* (1791), final page. On average, however, classes lasted for one season — either the summer, when weather was fine for naval training expeditions, or the winter months, when men had time in between voyages; see details on Denys, Holm, and others discussed earlier.

13. Stavorinus and Wilcocke, *Voyages to the East-Indies* (1798), 3.465–67.

14. Fauque, "Écoles d'hydrographie" (2000), 388.

15. Bezout, *Cours de mathématiques à l'usage des gardes* (1764), iii, 'Préface.'

16. Boer, *Zeemans Oeffening* (1769), 22–23.

17. See chapter 5, notes 26–28.

18. On the introduction of marine timekeepers in the 1780s, see chapter 5, note 59. On vernier scales, see Mörzer Bruyns, *Schip recht door Zee* (2003), 38. Compared to the simple tools that a navigator took to sea in the sixteenth century (see chapter 1), the list had ballooned; see Le Cordier, *Instruction des pilotes* (1786), 186.

19. *Principes de la navigation [Le petit manuel du pilote]* (n.p., n.d.), 2, BnF Ms. Français 22046.

20. See, e.g., a prefatory essay by Edmund Wilson in Robertson, *Elements of Navigation* (1764), xx; he aligned himself with "the best writers on navigation" who "always complained" of sailors' "obstinacy in adhering to inveterate mistakes." Morrice was equally critical of the "sea youth" he taught; *Young Midshipman's Instructor* (1801), 110. Vergé-Franceschi calls this

late eighteenth-century shift in curriculum and attitude an "increased 'scholarization,'" noting that the most junior naval officers were reclassified as students rather than guards. "Un enseignement éclairé au XVIIIe siècle" (1986), 30.

21. On Blaeu, Van Netten, *Koopman in Kennis* (2014). Denys's Dieppe publisher Nicolas Dubuc shared editions with Jacques Gruchet in nearby Havre de Grâce. (See also the extended publishing history of Samson Le Cordier's works.) Adams, *Non-cartographical Maritime Works Published by Mount and Page* (1985), catalogues two centuries of that firm's "mathematical and sea-books."

22. F.-M. Chautard, "9e Cayer" (ca. 1750s) TNA HCA 32/257, ch. 32, "Du Pilotage" (quote from unidentified textbook).

23. See Stokoe, "Elements of Navigation" (1790), LMA CLC/521/MS01226, for an example of a late eighteenth-century RMS student manuscript. The substantial volume opens with arithmetic before moving on — like so many business handbooks — to "Money Weights & Measures." Dialogues continued to be popular; e.g., late revisions of Le Cordier's work, *Instruction des pilotes* (e.g., 1734, 1786), were written as a 180-page series of questions and answers. See also Pietersz., *Handleiding, tot het Practicale of Werkdadige Gedeelte van de Stuurmanskunst* (1790); Van Cleeff, *Catechismus der Zeevaartkunde* (1840). Moore, *Practical Navigator* (1791), 272, 295, believed that didactic dialogues "refresh[ed] Memories," and "fix[ed] attention and facilitate[d] the speedy improvement of young practitioners in seamanship."

24. See chapter 3, note 79, and the general RMS emphasis on instruments, and note 90, on Boissaye de Bocage's curriculum. In Robertson, *Elements of Navigation* (1764), Wilson, a medical doctor turned maritime historian, underscored anew the importance of globes for sailors (xx). Morrice also emphasized hands-on lessons: "In telling a young pupil that a circle is round, the teachers should at the same time demonstrate it to him, and convince his *eyes* by putting into his hand a round piece of money, wood, or any other substance." He suggested the same strategy for globes. *Young Midshipman's Instructor* (1801), 105; also 108.

25. Philips Jacobsz., *Zeemans Onderwijzer in de Tekenkunst* (1786). Driver and Martins, "J. S. Roe and the Art of Navigation" (2002). For a different type of pedagogical benefit from working through textbooks, see Eddy, "The Shape of Knowledge: Children and the Visual Technologies of Paper Tools" (2013).

26. Van Oosten, "Hear Instruction" (1969), 260, gives an example of a pupil from Johannes Siberg's academy in Semarang, Java, who was able to safely guide a ship into harbor after his captain died en route to Ceylon in 1788.

27. E.g., Jean Bezie, a gunner, enthusiastically identified the rules he thought were "bonne regles" in his "Cahier Mathématiques" (ca. 1732–56), TNA HCA 32/257.

28. Bouguer, *Nouveau traité de navigation* (1753), viii, It was not enough just to "utter the rules and practices"; that would make the navigator's work "merely a matter of memory and, if one can say that, of routine." He suggested "linking together" (*enchaîner*) topics to help retain them in one's mind (*l'esprit*), which in turn made it easier to understand related subjects and made one "an Inventor."

29. E.g., the classic tide volvelles in the anonymous manuscript schatkamer (ca. 1760); MMR H629. V. J. Warnar worked through a copy of Oostwoud, *Vermeerderde Schoole der Stuurluyden* (1712), interleaving his answers in the volume in 1806–7; MMR ARCH 8D21.

30. See Gietermaker, *'t Vergulde Licht* (1774), Voorrede. However, throughout much of the eighteenth century, the French used the tables from the British Nautical Almanac, not computing their own until 1790. See Andrewes, *Quest for Longitude* (1996), 156.

31. Schotte, "Sailors, States, and the Co-creation of Nautical Knowledge" (forthcoming).

Bibliography

MANUSCRIPTS

For a Census of Nautical Manuscripts and Workbooks (*Schatkamers*) to 1800, see Supplemental Bibliography online at http://hdl.handle.net/10315/35728.

AUSTRALIA

Dixson Library, State Library of NSW (Sydney)
Drawer item 339 "The Unfortunate voyage of the Guardian man of war." Ballad verses. London: n.p., 1790.
Drawer item 304 [Shapter, Mr.] "The Forecastle Sailor or the Guardian Frigate." Ballad verses. Bristol: W. Collard, ca. 1860s.

Mitchell Library, State Library of NSW (Sydney)
5711/1 (Safe 1/233b) Riou, Edward. "Log Book of a Voyage to Port Jackson in New South Wales, performed by Lieutenant Edward Riou Commanding his Majesty's Ship Guardian. . . ." 1789–90.
5711/2 (Safe 1/233b), vol. 2 Riou, Edward. Narrative of the wreck of the *Guardian*. [Manuscript account] ca. 1790.

National Library of Australia (Canberra)
Australian Joint Copying Project. Microfilm Roll 87, Class and Piece Number HO 11/1. Convict Transportation Registers, 1787–1870.

AUSTRIA

Albertina Museum, Vienna
Inv. 13372 Vinckboons, David. "[Nautische Gesellschaft/Nautical Company]." ca. 1610.

CANADA

York University Libraries, Clara Thomas Archives and Special Collections (Toronto, ON)

16276 Asson, Assuerus van. "Schatkamer Ofte Konst Der Stuurlieden." Netherlands [Maassluis?], ca. 1705.

Bibliothèque et Archives nationales du Québec (Québec City, QC)
cote CN301, S146, folder 672. *Greffe* of Florent de la Cetière. Post-mortem Inventory of Jean Deshayes. Dec. 22, 1706.

Musée de la civilisation, Bibliothèque du Séminaire de Québec, Fonds ancien (Québec City, QC)
Polygraphie II.34. Deshayes, Jean. "Riviere de St. Laurens." 1686.

FRANCE

Archives nationales de France (Paris)
MAR/4JJ/27/5 Passart, Jean Noël/Vieillecourt, "premier Pilote." Logbook of the *Jason*. 1724.
MAR/B4/8 Méricourt to Colbert. Relation du naufrage de l'*île d'Aves*. June 1678.
MAR/G86 Mémoires et projets. Rapports sur l'enseignement des écoles établies dans les ports, hydrographie, construction, canonnage. 1682–1700.

Bibliothèque municipale de Rouen (Rouen)
Ms gg8 Denoville, Jean-Baptiste. "Le traité de navigation." 1760.

Bibliothèque nationale de France, Richelieu (Paris)
Ms. Français 2482 "L'Usage de la navigation." [Mélange] ca. 1605.
Ms. Français 9175 Vaulx, Jacques de. "Les premieres oeuvres de Jacques de Vaulx pilote pour le Roy en la marine contenantz plusieurs reigles praticques segrez et enseignementz très necessaires pour bien et seurement naviguer par le monde. . . . [Havre] de Grâce, 1584.
Ms. Français 19112 Le Vasseur de Beauplan, Guillaume. "Traicté de la Géodrographie, ou art de Naviguer." ca. 1608.
Ms. Français 25374 Brouscon, G. Manuel de pilotage, à l'usage des pilotes bretons. 1548.
Ms. N.A.Fr. 9479 Recueil des pièces: Hydrographie sous Colbert. 1671.
Ms. N.A.Fr. 9490 Ordonnance du commandeur de la porte. 1634.
SHD D.1.Z14 Le Testu, Guillaume. "Cosmographie universelle, selon les navigateurs, tant anciens que moderne. . . ." [Havre] de Grâce, 1555.

Service historique de la défense (Lorient)
Marine C6 Sous-série 2P 37-III.5. Rôle du *Prince de Conti*. 1754–56.

Service historique de la défense (Vincennes)
MS 511 Sageran, Lacoste. "Recueil de cours de navigation et de pilotage, de journaux de navigation et de comptes." 1747–49.
MS 516 "Idées d'un marin sur les connaissances théoriques et pratiques à exiger des officiers de la Marine royale et sur les avantages que présenteraient un collège de Marine établi à Port-Louis pour l'instruction des élèves." ca. 1802.

NETHERLANDS

Nationaal Archief (Den Haag)
Admiraliteitscolleges XVIII / Verzameling H. A. baron van Kinckel, 1.01.47.11
inv. nr. 7 Aanteekeningen en berekeningen van zeevaartkundigen aard. "Eenige Exempelen van Navigatie getrokken uyt Klaas de Vries" n.d.
inv. nr. 11A Generale rol en schutrol van 's Lands schip *St. Maartensdijk*. [Includes Van Kinckel's completed lieutenant exam. 1767.]
inv. nr. 13. Zeevaartkundige vraagstukken, examen voor luitenant. May 1767.

Radermacher 1.10.69, inv. nr. 382A Gedrukte ordners, reglementen, plakkaten enz. van de Hoge Regering in Indië, 1647–1743: Imhoff, G. W. van. *Project — Tot het opregten van een Corps Cadets de Marine voor den Zee-Dienst in Indien, en tot het aanqueken van jonge Zee-Officiers, midsgaders het stigten van een Academie voor de selve*. N.p., n.p. 1743.

VOC 1.04.02

inv. nr. 5017 Heren XVII, *Lyste van de Kaerten en Stuurmans-gereetschappen*. 1673.

inv. nr. 5018 Heren XVII, *Lyste[n] van de Boeken, Kaarten en Stuurmans-gereetschappen*. 1675, 1747.

inv. nr. 5026 Heren XVII, *Instructie voor de Examinateurs der Stuurlieden by de edele Oost-Indische Compagnie*. 1793.

inv. nr. 5049 Bontekoe, Willem Isbrantsz. Journal. 1622–25.

inv. nr. 5165 *Instructie ende Ordre Op 't houden van de Scheeps-soldie en Guarnisoen-boecken*. Amsterdam: J. Blaeu, 1656. Middelburgh: Pieter van Goetthem, [n.d.]. Amsterdam: Adriaan Wor, 1751.

Maritiem Museum Rotterdam (Rotterdam)

H629 Anonymous. Navigation workbook based on Gietermaker. 1760.

H631 Sleutel, Jan Willem. "Konstige oefeningen begrepen in drie boecken." Hoorn, 1675–77.

H632 Vos, Hendrik de. "Schatkamer ofte Const der Stuurliede int ligt gegeven door Claas de Vries door mijn Hendrik de vos 1748." 1748–53.

H633 Graaf, Willem de. "Schat kaamer ofte konst der stuurlieden." 1763.

H891 Hoffman, H. M. "De Stuurmans Kunst, Uitgewerkt door H.M. Hoffm bij R[asmus?] Dekker te Amsterdam." 1818.

Het Scheepvaartmuseum (Amsterdam)

(Former call numbers noted in parentheses.)

A.4300 (Hs 0988) Komis, Pieter Rutsz. "Het derde deel van Klaes Heyndricksz Gietermakers Schatkamer ofte Konst der stuerlieden." 1708–13.

S.0187 (Hs 0968) West, Dirk de. "Schatkamer ofte konst der stierlieden. Door Klaas de Vries uitgewerkt door Dirk de West." ca. 1760.

S.0712 (Hs 0969) Grootschoen, Maarten Janszoon. Navigation workbook. 1728.

S.1386 (Hs 0989) Boombaar, Cornelis Jansz. "Onderwijs Der Zeevaert ofte de konst der stuurluijden." 1727–32.

S.3003 (Hs 0985) Kok, J. "Schatkamer ofte konst der stuurlieden . . . , Uijtgesyfert door mijn in Campen." Kampen, 1763.

S.4312 (07) (Hs 0991) Anonymous. Navigation workbook. ca. 1717–26.

UNITED KINGDOM

Caird Library, National Maritime Museum (Greenwich, London)

NVT/6 Sephtonal, Daniell. Navigation workbook. ca. 1713.

NVT/8 Downman, William. Navigation workbook. ca. 1685–86.

NVT/47 (formerly MS75/081) Spink, William. Navigation [catalogued as Commonplace] workbook. ca. 1697–1705.

RUSI/NM/235/ER Papers of Capt Edward Riou (1762–1801).

ER/2/2 Journal of *Mediator* and *Salisbury*. 1782–83, 1786–88.

ER/3/1 "Mathematical Lessons for a Young Sailor . . . under Instruction of James Paterson and Thomas Adams." 1774.

ER/3/3 Trigonometry workbook. n.d. (after 1774).

ER/3/4 "Miscellaneous Computations." 1782–87.
ER/3/8 *Guardian* "Computation book." 1789–90.
ER/7/2 "Remarks at Sea." ca. 1789–90.
SMP/3 Mainwaring, Henry. "Tearms in Navegacon." ca. 1624.

Cambridge University Library (Cambridge)
RGO 1/69/C Flamsteed RGO papers, n.d. [1700?].

London Metropolitan Archives (London)
CLC/210/B/006/MS12873/001, 002 Christ's Hospital, Minute and Memoranda Books. 1673–91.
CLC/210/B/007/MS12873A Christ's Hospital, School Committee Minute Book. 1681–88.
CLC/210/B/008/MS12873B Christ's Hospital, School Committee Rough Minute Book. 1697–99.
CLC/210/C/011/MS12874 Christ's Hospital, Royal Mathematical School Account Book. 1681 [recte 1679]-1702.
CLC/210/F/014/MS12875 Christ's Hospital, Register of Marine Apprentice Bindings. 1675–1711.
CLC/521/MS01226 Stokoe, T. W. "Elements of Navigation." 1790.
CLC/B/227/MS28949 Letter and invoice received by Sir Francis Child and Thomas Child from Mr [Samuel] Newton. 1704.

Medway Archives (Strood, Rochester)
DE629/2 Spink, William. Ship chandler's notebook. 1718 [or 1735?]–51.

National Archives of the UK (Kew, Richmond, Surrey)
ADM 1/2395 Admiralty Secretary's In-Letters, Captains "R." June 22, 1791.
ADM 36/11005 Muster Book for HMS *Guardian*. Apr.–Dec. 1789.
ADM 106/2908 Masters' Qualifications. 1660–95.
ADM 107/1 Lieutenants' Passing Certificates. 1691–1703.
ADM 107/2 Lieutenants' Passing Certificates. 1703–12.
ADM 107/40 Lieutenants' Passing Certificates. 1809.
HCA 30/763 II Captured ship papers: Logbook of the *Amazoon*. 1791–95.
HCA 32/176 II Prize court papers: A. Wijkman, "Ex Samen Der Sturliedn." 1709.
HCA 32/176 III Prize court papers: Logbooks. 1758.
HCA 32/257 Prize court papers: Miscellaneous papers, ca. 1732–56, some seized from *Prince de Conti*.
PROB 11/565/44 Will of Assuerus Van Asson, Maassluis, Holland, Aug. 6, 1718.
SP 46/137 pts I–V Admiralty Papers: Masters' certificates. 1660–73.

Pepys Library, Magdalene College, Cambridge University (Cambridge)
PL 2184 Flamsteed, John. "Doctrine and Practice of Navigation." 1697.
PL 2612 Collection of Matters Pertaining to Christ's Hospital. 1673–84.
PL 2643 "Bibliotheca Nautica." 1694–98.
PL 2866 "My Minute-Book currant for ye Adm[ira]lty & Navy MSS." 1680–96.

Plymouth and West Devon Record Office (Plymouth)
Ms. 1334 [Le Vasseur de Beauplan, Guillaume?] "Géodrographie." ca. 1610.

ANNOTATED PRINTED BOOKS

Bibliothèque nationale de France, Richelieu (Paris)
Ms. Français 22046. *Les Principes de la navigation théorique, par demandes & réponses*

[manuscript title: *Le petit manuel du pilote*]. N.p., n.d. Annotated by J. J. de La Lande? [1762?]

Maritiem Museum Rotterdam (Rotterdam)
ARCH 4C21 Makreel, D. *Lichtende Leydt-Starre*. Amsterdam: Hendrick Doncker, 1671. Exercises interleaved by the author.
ARCH 4D16 Gietermaker, C. H. *'t Vergulde Licht*. Ed. Frans vander Huips. Amsterdam: J. van Keulen, 1710. Annotated by C. Visscher in 1710.
ARCH 8D21 Oostwoud, G. M. *Vermeerderde Schoole der Stuurluyden*. Hoorn: Stoffel Jansz. Kortingh, 1712. Exercises interleaved by V. J. Warnar in 1806–7.

Het Scheepvaartmuseum (Amsterdam)
S.4793(342) (Mg 0298). VOC Heren XVII. *Examen voor de Capitain en Capitain-lieutenants by de Oost-indische Compagnie*. [Amsterdam?], [1794].

PRINTED PRIMARY SOURCES

[Académie française.] *Le Dictionnaire de l'Académie française*. Paris: 1694; 4th ed. 1762.
[Académie royale des sciences / Bureau de longitudes.] *La Connoissance des temps, ou calendrier et éphémérides*. . . . Paris: J. B. Coignard, 1679–[1803].
L'Académie royale dite de la marine, établie à Paris l'an 1677, par les soins de MM. les chevaliers associés de l'ordre de N.-D. du Mont-Carmel et de S.-Lazare. Paris: S. Cramoisy, 1677.
The Accomplish'd Sea-Mans Delight Containing: 1. The Great Mistery of Nature Demonstrated by Art . . . 2. The Closset of Magnetical Miracles Unlocked . . . 3. Directions for Sea-Men in Distress of Weather. . . . London: Printed for Benjamin Harris, 1686.
Adams, John. *The Young Sea-Officer's Assistant, both in his Examination and Voyage. In four parts. I. The Substance of that Examination, which every Candidate for a Commission in the East-India Service, or the Navy, must necessarily pass, previous to his Appointment*. . . . London: Printed for Lockyer Davis, 1773.
Addison, Thomas. *Arithmeticall Navigation: Or, an Order thereof: Compiled and published for the advancement of Navigation More particularly, for the benefit of English Mariners, or Seafaring men that delight therein*. London: [G. Purslowe] for Nathaniel Gosse, 1625.
Anthonisz. [Antoniszoon], Cornelis. *Onderwijsinge vander zee, om Stuermanschap te leeren*. . . . Ed. princ. 1544. Amsterdam: [Jan Ewoutszoon], 1558.
———. *The safegard of Sailers, or great Rutter Containing the Courses, Distances, Depthes, Soundings, Floudes and Ebbes . . . with other necessarie Rules of common Navigation*. Tr. Robert Norman. London: John Windet and Thomas Judson for Richard Ballard, 1584.
Apian, Peter. *Cosmographicus liber*. Landshut: J. Weyssenburger, 1524.
———. *Cosmographie, ofte Beschrijvinge de gheheelder werelt*. Amsterdam: Cornelis Claesz., 1598.
Après de Mannevillette, J.-B.-N.-D. d'. *Neptune Oriental*. Paris: Demonville, and Brest: Malassis, 1775.
Aspley, John. *Speculum Nauticum: A Looking Glasse, for Sea-Men: Wherein they may behold a small instrument called the Plaine Scale: whereby all questions Nauticall, and propositions Astronomicall are very easily and demonstratiuely wrought*. . . . London: Thomas Harper, 1624.
B., A. *Gloria Britannica; or, the Boast of the Brittish Seas. Containing a True and Full Account of the Royal Navy of England*. . . . London: Printed for Thomas Howkins, 1689.
Barlow, William. *The Navigators Supply: Conteining Many Things of Principall Importance Belonging to Navigation: With the Description and Use of Diverse Instruments Framed Chiefly for That Purpose, but Serving Also for Sundry Other of Cosmography in Generall*. London: G. Bishop, R. Newbery, and R. Barker, 1597.

Bartjens, Willem. *De Cyfferinghe.* Amsterdam: Cornelis Claesz., 1604.

Berthelot, F. *Abrégé de la navigation, contenant ce qui est nécessaire de sçavoir de la théorie de cet art pour conduire les bâtiments sur la Méditerranée, avec des explications touchant les erreurs attachées à la routine des navigateurs.* Marseille: Henry Brebion, 1691.

Bessard, Toussaint de. *Dialogue de la longitude: est-ouest.* . . . Rouen: Martin le Mesgissier, 1574.

Bezout, Étienne. *Cours de mathématiques à l'usage des gardes du pavillon et de la marine.* 3 vols. Paris: J. B. G. Musier, 1764.

Bion, Nicolas. *Traité de la construction et des principaux usages des instrumens de mathematique. Avec les figures nécessaires pour l'intelligence de ce traité.* La Haye: P. Husson, 1723.

Blaeu, Willem Janszoon. *Het Licht der Zee-vaert.* Amsterdam: Willem Janszoon [Blaeu], 1608.

———. *The Light of Navigation.* Amsterdam: William Johnson [Willem Janzs. Blaeu], 1612.

———. *[Eerste Deel der] Zeespiegel, Inhoudende een korte Onderwysinghe inde Konste der Zeevaert.* . . . Amsterdam: W. J. Blaeu, 1623.

———. *Tweevoudigh Onderwiis van de Hemelsche en Aerdsche Globen* . . . . Amsterdam: W. J. Blaeu, 1634.

———. *Institution astronomique de l'usage des globes et sphères célestes & terrestres, comprise en deux parties.* . . . Amsterdam: J. & C. Blaeu, 1642.

———. *A Tutor to Astronomy and Geography; Or, An Easie and Speedy way to Understand the Use of both the Globes, Celestial and Terrestrial. Laid down in so plain a manner that a mean Capacity may at the first Reading understand it and with a little Practise, grow expert in those Divine Sciences. Translated from the first Part of Gulielmus Blaeu, Institutio Astronomica.* Ed. Joseph Moxon. London: Joseph Moxon, 1654.

Blagrave, John. *The Mathematical Jewel Shewing the Making, and Most Excellent use of a Singuler Instrument so Called . . . The use of Which Jewel . . . Leadeth Any Man Practising Thereon . . . through the Whole Artes of Astronomy, Cosmography, . . . and Briefely of Whatsoever Concerneth the Globe or Sphere.* . . . London: Walter Venge, 1585.

Blome, Richard. *The Gentlemans Recreation.* London: S. Roycroft for Richard Blome, 1686.

Blondel Saint-Aubin, Guillaume. *Le Trésor de la navigation, divisé en deux parties, la premiere contient la théorie & pratique des triangles sphériques, enrichie de plusieurs problèmes astronomiques & géographiques, très utiles aux navigateurs; la seconde enseigne l'art de naviger par la suputation & démonstration des triangles rectelignes & sphériques; tant par les sinus que par les logarithmes.* Havre de Grâce: Jacques Gruchet, 1673.

———. *Le Véritable art de naviger par le quartier de réduction, avec lequel on peut réduire des courses des vaisseaux en mer, & enrichy de plusieurs raretez qui n'ont point encor esté découvertes.* Havre de Grâce: Jacques Gruchet, 1671, 1763, etc.

———. *L'Art de naviger par le compas de proportion.* Havre de Grâce: Jacques Gruchet, 1680.

Blundeville, Thomas. *M. Blundevile His Exercises Containing Sixe Treatises, . . . in Cosmographie, Astronomie, and Geographie, as Also in the Arte of Navigation . . . To the Furtherance of Which Arte of Navigation, the Said M. Blundevile Speciallie Wrote the Said Treatises.* . . . London: John Windet, 1594.

Boer, Jan de. *Zeemans Oeffening over de groote Zeevaart.* Amsterdam: Joh. van Keulen en Zoonen, 1769.

———. *Extract uit de Zeemans Oeffening over de groote Zeevaart.* . . . *Berigt van het Dryfanker.* N.p., n.p., [1773].

Boissaye du Bocage, Georges [II]. *Explication et usage d'une partie du cercle universel de ses tables et echelles. Necessaire à tous pilotes & utile à toutes sortes de personnes.* Havre de Grâce:

Jacques Gruchet; se vend chez Nicolas l'Anglois à Paris, 1683. Havre de Grâce: Jacques Hubault, 1695.

———. *Le Petit Neptune François Or, the French Coasting Pilot. Being a Particular Description of the . . . Coast of France. . . . Translated from the Petit Flambeau de La Mer of Du Bocage. . . .* London: Printed for T. Jefferys, 1761.

Boteler, Nathaniel. *Six Dialogues about Sea-Services between an High-Admiral and a Captain at Sea.* London: Printed for Moses Pitt, 1685.

———. *Boteler's Dialogues.* Ed. William G. Perrin. London: Navy Records Society, 1929.

Bougard, René [Raulin?]. *Le Petit Flambeau de la mer, ou le Véritable Guide des pilotes côtiers.* Havre de Grâce: J. Gruchet, 1684.

Bouguer, Pierre. "Sur les opérations nommées par les pilotes *corrections.*" *Histoire de l'Académie des Sciences* (1752): 125–30.

———. *Nouveau traité de navigation: contenant la théorie et la pratique du pilotage.* Paris: H.-L. Guérin & L.-F. Delatour, 1753.

———. *Nouveau traité de navigation: contenant la théorie et la pratique du pilotage.* Ed. N. L. de Lacaille. Paris: H.-L. Guérin & L.-F. Delatour, 1760.

Bourne, William. *An Almanacke and Prognostication for Three Yeares That Is to Saye for the Peace of Oure Lord. 1571. and 1572. & 1573. Now Newlye Added Vnto My Late Rulles of Navigation. . . .* London: Thomas Purfoote, 1567.

———. *A Regiment for the Sea Conteyning Most Profitable Rules, Mathematical Experiences, and Perfect Knowledge of Navigation, for All Coastes and Countreys: Most Needefull and Necessarie for All Seafaring Men and Travellers, as Pilotes, Mariners, Marchants.* London: [Henry Bynneman for] Thomas Hacket, 1574.

———. *De Const der Zee-vaerdt begrypende seer nootwendighe saecken voor allerhande Zeevaerders als Schippers, Stuerlieden, Piloten, Bootsvolck ende Coop-lieden.* Tr. L. J. Waghenaer. Amsterdam: Cornelis Claesz., 1594.

Bowditch, Nathaniel. *The New American Practical Navigator.* Newburyport, MA: Edmund Blunt, 1802.

———. *The Improved Practical Navigator.* Ed. Thomas Kirby. London: Printed for James and John Hardy . . . and David Steel, 1802.

Breen, Joost van. *Stiermans Gemack, ofte een korte Beschrijvinge vande Konst der Stierlieden . . . tot groote verlichtingen voor den Stierman inde dagelijckse practijcke, ende oock de aenkomende Leerlingen inde Rekenkonst niet wel ervaren synde. . . .* 's Gravenhage: J. van Breen, J. van Tongerloo, 1662.

Briggs, Henry. *Logarithmorum chilias prima.* London: n.p., 1617.

———. *Arithmetica logarithmica.* London: William Jones, 1624.

Broucke, Jan van den. *Instructie der Zee-Vaert.* Rotterdam: Abraham Migoen, 1609, 1610.

Carlos II. *Copia de las Cedulas Reales, Que Su Magestad El Rey N. Señor Don Carlos Segundo deste Nombre, Mandó Expedir lara la Fundacion del Colegio, y Seminario, que Mandò Hazer para la Educacion de Niños en la Ciudad de Sevilla, para la Enseñanza, y Erudicion de Ellos en la Arte Maritima, Artilleria, y Reglas de Marineria, y Dotacion, y Privilegios para este fin.* Seville: Juan Cabezas, 1681.

Cats, Jacob. *Spiegel van den ouden en nieuwen Tydt.* Amsterdam: Wed. van Gysbert de Groot, 1712.

Cauvette, Pierre. *Nouveaux élémens d'hydrographie. Ou par une méthode courte & aisée l'on peut apprendre de soy même & en très peu de tems tout ce qui est nécessaire pour entreprendre & achever une heureuse navigation.* Paris; Dieppe: Nicolas Dubuc, 1685.

Chambers, Ephraim. *Cyclopaedia, or, An Universal Dictionary of Arts and Sciences.* London: W. Innys et al., 1728.

Champlain, Samuel de. *Traitté de la marine et du devoir d'un bon marinier.* In *Les Voyages de la Nouvelle France Occidentale, dicte Canada.* . . . Paris: Louis Sevestre, 1632.

[Claesz., Cornelis.] *Const ende Caert-Register, In welcke gheteyckent staen, alderhande soorten vant Caerten ende Wappen des gantschen Aertbodems, groot ende cleyn . . . alle by Cornelis Claesz te becomen.* . . . Amsterdam: Cornelis Claesz., 1609.

Cleeff, J. van. *Catechismus der Zeevaartkunde, of Vragen en Oefeningen over het Beschouwende en Werkdadige der Stuurmanskunst.* Groningen: J. Oomkens, 1840.

Cleirac, Estienne. *Us et coutumes de la mer, divisées en trois parties: I. De la navigation. II. Du commerce naval & contracts maritimes. III. De la jurisdiction de la marine.* . . . Bordeaux: G. Millanges, 1647.

Clements, Thomas. *Guardian. A Journal of the Proceedings on Board the above Ship: Lieutenant Riou, Commander: As Delivered into the Admiralty Board, by Mr. Clements.* . . . London: Printed for Charles Stalker, 1790.

Coignet, Michiel. *Nieuwe Onderwijsinghe, op de principaelste Puncten der Zee-vaert.* Issued with Pedro de Medina, *De Zeevaert, Oft Conste van ter Zee te varen.* . . . Antwerp: Hendrick Hendricksen, 1580.

———. *Instruction nouvelle des poincts plus excellents & nécessaires, touchant l'art de naviguer. Contenant plusieurs reigles, pratiques, enseignemens, & instrumens tresidoines à tous pilotes, maistres de navire, & autres qui journellement hantent la mer.* Anvers: Henry Hendrix, 1581.

———. *Nieuwe onderwysinghe, op de principaelste puncten der Zee-vaert.* Amsterdam: Cornelis Claesz., 1589, 1593, 1598.

Collins, John. *Navigation by the Mariners Plain Scale New Plain'd: Or, A Treatise of Geometrical and Arithmetical Navigation . . . 2. New Rules for Estimating the Ships Way through Currents, and for Correcting the Dead Reckoning. 3. The Refutation of Divers Errors, and of the Plain Chart, and How to Remove the Error Committed Thereby.* London: Tho. Johnson for Francis Cossinet, 1659.

[Commissioners of Longitude.] *The Nautical Almanac and Astronomical Ephemeris.* London: Printed by W. Richardson and S. Clark, sold by J. Nourse and Mount and Page, 1766–.

Considerations on the Nurseries for British Seamen; the present state of the Levant and carriage-trade in the Mediterranean; and the comparative, military, naval, and commercial powers of the Barbary States. [London?]: n.p., 1766.

Cook, James. *A Voyage towards the South Pole, and Round the World. Performed in His Majesty's Ships the Resolution and Adventure, in the Years 1772, 1773, 1774, and 1775.* 2 vols. London: for W. Strahan and T. Cadell, 1777.

———. *Captain Cook's Journal during His First Voyage Round the World Made in H.M. Bark "Endeavour," 1768-71.* Ed. William J. L. Wharton. London: E. Stock, 1893.

Cortés, Martín. *Breve compendio de la sphera y de la arte de navegar.* Seville: Anton Alvarez, 1551.

———. *Arte of Navigation. Conteynyng a Compendious Description of the Sphere, with the Makyng of Certen Instrumentes and Rules for Navigations.* . . . *Wrytten in the Spanyshe tongue by Martin Cortes.* . . . Tr. Richard Eden. London: Richard Jugge, 1561. Facsimile, ed. David W. Waters. Delmar, NY: Scholars' Facsimiles, 1992.

Coubert [Coubard, G.]. *Abrégé du pilotage, pour servir aux conférences d'hydrographie, que le Roi fait tenir pour ses off[f]iciers de marine.* Brest: Malassis, 1685; Guillaume Camarec, 1728.

Cuningham, Robert. *Tractaet des Tijdts. Almanach, Kalendier, Journael ofte Tijdt-Boeck eeu-*

wich gheduyrende, seer ghedienstich voor alle Menschen, maer byzonder voor alle Navigateurs ende Piloten. Franeker: Dirck Jansz. Prins en Rombertum Doyma, 1605.

Dam, Jan Albertsz. van. *De Nieuwe Hoornse Schatkamer ofte de Konst der Zeevaart.* Bound with: *De Tafelen der Sinuum, Tangentium en Secantium . . . Als ook de Logarithmus Numeri van 1 tot 1000, . . . en . . . de Tafel der Kromstreeken.* Amsterdam: Joannes Loots, 1712. Amsterdam: J. van Keulen, 1751.

Davis, John. *Seamans Secrets.* London: Thomas Dawson, 1595.

———. *Le Voyage de maistre Jean David . . . reduit par luy en tables.* Tr. Jean Le Telier. Dieppe: Nicolas Acher, 1631.

Decker, Ezechiel de. *Nieuwe Telkonst inhoudende de Logarithmi voor de Ghetallen beginnende van 1 tot 10000, ghemaeckt van Henrico Briggio. . . .* Gouda: Pieter Rammazeyn, 1626.

———. *Tweede Deel vande Nieuwe Tel-konst ofte Wonderliicke Konstighe Tafel, Inhoudende de Logarithmi, voor de Getallen van 1 af tot 100000 toe.* Gouda: Pieter Rammazeyn, 1627.

———. *Practyck vande Groote Zee-Vaert.* Gouda: Pieter Rammazeyn, 1631. Rotterdam: Bastiaen Wagens, 1659.

Dee, John. *Elements of Geometrie.* London: John Daye, 1570.

———. *General and Rare Memorials Pertayning to the Perfect Arte of Navigation Annexed to the Paradoxal Cumpas.* London: John Daye, 1577.

Denys, Guillaume. *Façon nouvelle de naviger par les nombres, c'est-a-dire par sinus, destinée a servir seulement dans les navigations de brief cours.* Dieppe: [Nicolas] du Buc, 1648.

———. *Table de la déclinaison du soleil, et des principales et plus remarquables estoilles du firmament et deux tables pour l'estoille du Nord.* Dieppe: Nicolas du Buc [Dubuc], 1663.

———. *L'Art de naviger perfectionné par la cognoissance de la variation de l'aimant, ou Traicté de la variation de l'aiguille aimantée.* Dieppe: Nicolas du Buc, 1666. Havre de Grâce: Jacques Gruchet, 1673.

———. *L'Art de naviger par les nombres.* Dieppe: Nicolas du Buc, 1668.

———. *L'Art de naviger dans sa plus haute perfection, ou Traité des latitudes.* Dieppe: Dubuc, 1673.

Diderot, Denis, and Jean d'Alembert. *Encyclopédie, ou dictionnaire raisonné des sciences, des arts et des métiers, etc.* Paris, 1751–72.

Digges, Leonard. *A Prognostication everlasting of right good effect.* London: Thomas Marshe, 1567.

A Discourse of Navigation; Containing several Rules and Instructions to be Observed in the Practice thereof: And Shewing the use of two Instruments, lately invented, for resolving all Triangles and all Problems, in Sailing, without any Operation by Arithmetick. London: J. Nutt, 1702.

Doublet, Jean. *Journal du corsaire Jean Doublet, de Honfleur, lieutenant de frégate sous Louis XIV, publié d'après le manuscrit autographe.* Ed. Charles Bréard. Paris: Charavay Frères Editeurs, 1883.

Dounot, Didier. *Confutation de l'invention des longitudes ou de la mécométrie de l'eymant. Cy devant mise en lumiere souz de nom de Guillaume le nautonnier.* Paris: François Huby, 1611.

Dulague, Vincent François Noël. *Leçons de navigation.* Rouen: Chez la veuve Besongne et J. J. Besongne, 1768, 1793.

———. *Principes de navigation, ou abrégé de la théorie et de la pratique du pilotage.* Bordeaux: Paul Pallandre, 1788.

Dunton, John. *The Young-Students-Library Containing Extracts and Abridgments of the Most Valuable Books Printed in England, and in the Forreign Journals.* London: for John Dunton, 1692.

Falconer, William. *An Universal Dictionary of the Marine . . . To which is annexed, A Translation of the French Sea-Terms and Phrases.* . . . London: Printed for T. Cadell, 1769.

Figueiredo, Manoel de. *Hydrographia, Exame de Pilotos.* Lisbon: Vicente Álvarez, 1608.

———. *Hydrographie ou examen auquel sont contenuës les regles que tous pilotes doivent garder en leurs navigations, . . . augmenté . . . de la parfaite cognoissance de la variation de l'aiguille pour trouver les longitudes.* Tr. Nicolas Le Bon. Dieppe: Nicolas Acher, 1618, 1640.

Focard, Jacques. *Paraphrase de l'astrolabe, contenant les principes de geometrie, la sphere.* . . . Lyon: par Jean de Tournes, 1546.

Fontenelle, Bernard de. "Eloge of N. Hartsoeker [1699]." In *Oeuvres de Fontenelle. Éloges.* Paris: Salmon, 1825.

Fournier, Georges, S.J. *Hydrographie, contenant la theorie et la pratique de toutes les parties; de la navigation.* Ed. princ. Paris: M. Soly, 1643; 2nd ed., Paris: Jean du Puis, 1667.

Gadbury, Timothy. *The Young Sea-Man's Guide: Or, the Mariners Almanack. Containing an Ephemeris, with the Use Therof. . . . Also the Names and Nature of All the Thirty Two Windes.* London: Printed for Francis Cossinet, 1659.

García de Céspedes, Andrés. *Regimiento de navegacion que mando hazer el Rei.* Madrid: en casa de Juan de la Cuesta, 1606.

García de Palacio, Diego. *Instrución Náutica para el buen Uso, y regimiento de las Naos, su traça, y govierno conforme à la altura de Mexico.* Mexico: P. Ocharte, 1587.

———. *Nautical Instruction: A.D. 1587.* Tr. J. Bankston. Bisbee, AZ: Terrenate Associates, 1988.

Garcie, Pierre. *Le Routier de la mer.* [Rouen]: Jacques Le Forestier, [ca. 1502].

———. *The Rutter of the See.* [London: Thomas Petyt, 1536].

Gellibrand, Henry. *An Epitome of Navigation Containing the Doctrin of Plain & Spherical Triangles, and Their Use and Application in Plain Sailing, Mercator's Sailing, and Great Circle Sailing . . . : Together with Tables of the Sun and Stars . . . of the Latitude and Longitude of Places, of Meridional Parts: Likewise a Traverse-Table . . . and the Logarithm Sines and Tangents.* London: Andrew Clark for William Fisher, 1674.

Gerbier, Balthazar. *The First Lecture Touching Navigation Read Publiquely at Sr. Balthazar Gerbiers Accademy.* London: for Robert Ibbitson, 1649.

Gietermaker, Claas Hendriksz. *Vermaeck der Stuerlieden. Inhoudende De voornaemste stucken der Zeevaert, zijnde alles seer nut ende vermakelick voor alle Schippers ende Stuerlieden: Midtsgaders voor mijn Discipelen ende Zeevarende Persoonen; Toevoeghsel Op het Vermaeck der Stuerlieden; Vervolgh op 't Vermaeck der Stuer-lieden. Inhoudende Practica der Groote Zee-vaert; dienende voor alle Liefhebberen der Navigatie.* Amsterdam: Gietermaker, 1659.

———. *'t Vergulde Licht der Zeevaert, ofte Konst der Stuurluyden. Zijnde een volkoomen en klaare onderwijsinge der Navigatie, bestaande in 't geen een Stuurman hoognodig behoorde te weten.* Amsterdam: Doncker, 1660.

———. *Driehoex-rekening, Bestaende in de verklaringe en ontbindinge der Platte Driehoecken . . . Met een discours tusschen een Schipper en Stuurman, aengaende de Zeevaert.* Amsterdam: gedruckt voor den Autheur, 1665.

———. *De drie Boecken van Claes Hendricksz. Gietermakers Onderwijs der Navigatie, ofte Konst der Stuurlieden, Bestaende in de volmaecktheyt der Zeevaert.* Amsterdam: Gedruckt voor den Autheur, en Pieter Goos, in Compagnie, 1666.

———. *Le Flambeau reluisant, ou proprement thresor de la navigation.* Tr. J. Viret. Amsterdam: Henri Doncker, 1667.

———. *'t Vergulde Licht der Zeevaert. . . . Vernieuwt en verbetert door Claas Hendriksz. Gietermaker, in zijn leven Examinateur van de Geoctroyeerde Oost- en West-Indische Compagnie*

———. *'t vierde Boeck vermeerdert met de Ontbindingh van verscheyde konstige t'saam-gevoeghde Questien.* Ed. Frans vander Huips. Amsterdam: Hendrik Donkker, 1677, 1684.

———. *'t Vergulde Licht der Zeevaert.* Ed. Frans vander Huips. Amsterdam: Jo[h]annes van Keulen, 1710.

———. *'t Vergulde Licht der Zee-Vaard.* Ed. Frans vander Huips. Amsterdam: Jo[h]annes van Keulen en Zoonen, [1774].

Gilbert, William. *De Magnete.* London: Petrus Short, 1600.

Glos, Guillaume de. *Le Manuel des pilotes, ou Introduction à la navigation.* Rouen, Honfleur: D. Dumoulin, 1678; Havre de Grâce: Hubault, 1689, 1694.

Graaf, Abraham de. *De kleene Schatkamer. Het I Boek Van de Konst der Stierlieden.* Ed. princ. 1680. 2nd ed., Amsterdam: voor den Autheur By d'Erfg. van Paulus Matthysz, 1688.

Gunter, Edmund. *A Canon of Triangles, or a Table of Artificiall Sines, Tangents, and Secants, Drawne from the Logarithmes of the Lord of Merchistone. Whereby all Sphericall Triangles are most easily resolved by Addition and Substraction onely.* London: Printed by W. Stansby [for] John Tappe, 1620.

———. *De Sectore & Radio. The description and use of the Sector in three bookes. The description and use of the Cross-Staffe in other three bookes. For such as are studious of Mathematicall practise.* London: William Jones, 1623.

Haeyen [Haeijen], Albert Hendricksz. *Amstelredamsche Zee-Caerten.* Leiden: C. Plantin, 1585.

———. *Een corte onderrichtinge, belanghende die kunst vander Zee-vaert.* Amsterdam: H. de Buck, 1600.

Hakluyt, Richard. *Principal Navigations Voyages Traffiques and Discoveries of the English Nation.* London: George Bishop, Ralph Newberie and Robert Barker, 1598–1600.

———. *The Principal Navigations, Voyages, Traffiques, and Discoveries of The English Nation, Vol. XII., America, Part I.* Ed. Edmund Goldsmid. Edinburgh: E. & G. Goldsmid, 1889.

Hawkesworth, John. *Account of the Voyages . . . for Making Discoveries in the Southern Hemisphere.* 3 vols. London: for W. Strahan; and T. Cadell, 1773.

Hellingwerf, Adriaan Claasz. *Vermaack der Zee-Luy ofte Stuurmans-Regel, sijnde Een Regelstelling op de Navigatie ende de Clootse Reeckening, als meede Een Verklaring, en 't Gebruyk op de Hemelse en Aerdse Globus, en veel meer andere Wercken; aldus by een gestelt.* Hoorn: Elbert Beukelman, 1703.

Histoire de l'École navale et des institutions qui l'ont précédé. Paris: Maison Quantin, 1889.

Hobbes, Thomas. *Philosophicall Rudiments Concerning Government and Society [De Cive].* London: Printed by J.G. for R. Royston, 1651.

Hollesloot, Govert Willemsz. van. *De Caerte vander Zee, om oost ende west te seylen.* Amsterdam: Cornelis Claesz., 1587.

Holm, Pieter. *Stuurmans Zeemeeter.* Amsterdam: voor den Autheur, Anno 5844 [1748].

———. *Mond-Exame Voor de Stierlieden van Het Schip Recht Door Zee.* Amsterdam: [for the author], 5855 [1759].

Hood, Thomas. *T'ghebruyck van de Zeecaerte.* Amsterdam: Cornelis Claesz., 1602.

Hues, Robert. *A Learned Treatise of Globes, Both Coelestiall and Terrestriall with their severall uses. Written first in Latine.* London: by the assigne of T[homas] P[urfoot] for P. Stephens and C. Meredith, 1639.

Instructie ende Ordonnantie voor Commissarissen vande Zeevaert. Amsterdam: Nicolaes van Ravesteyn, 1641.

Instruction pour les jeunes gens désignés pour le service de la marine, et qui doivent être élevés dans les collèges aux frais du roi. Paris: l'Imprimerie Royale, 1786.

Jackson, William. *An Introduction of the First Grounds or Rudiments of Arithmetick Plainly explaining the five Common parts of that most Useful and Necessary Art . . . By way of Question and Answer; for the ease of the Teacher, and benefit of the Learner.* London: Printed for R.I. for F. Smith, 1661.

Jacquinot, Dominique. *L'Usaige de l'astrolabe, avec un traicté de la sphere.* Paris: Jehan Barbé, 1545.

Jarichs van der Ley, Jan Hendrik. *Het Gulden Zeeghel Des grooten Zeevaerts, daerinne beschreven wordt de waerachtige grondt vande Zeylstreken en platte Pas-caerten. . . .* Leeuwarden: Abraham vanden Rade, 1615.

———. *'T Ghesicht des grooten Zeevaerts.* Franeker: Jan Lamrinck, 1619.

———. *Voyage Vant Experiment vanden Generalen Regul des Gesichts vande Groote Zeevaert.* 's Gravenhage: Hillebrant Jacobssz., 1620.

Justice, Alexander. *A General Treatise of the Dominion of the Sea: And a Compleat Body of the Sea-Laws. . . .* 2nd ed. London: D. Leach, for J. Nicholson, and J. and B. Sprint; and R. Smith, [1709?].

Keteltas, Barent Evertsz. *Het Ghebruyck der Naeld-Wiising Tot Dienste der zee-vaert beschreven.* Amsterdam: Barent Otiz. Boeckdrucker, 1609.

Kruik, Dirk. *Gronden der Navigatie ofte Stuurmans Konst, Behelsende een klare Fondaméntéle Instructie in de Konst der Groote Zeevaart. . . .* Rotterdam: Johannes Kuyper, 1737.

Lakeman, Mathijs Sijverts [Siverts]. *Een Tractaet, seer dienstelijck voor alle Zee-varende Luyden, door het t'samen spreken van twee Piloten: Veel dinghen dienstelijck den Piloten gheopenbaert, bysonder die seer begheerlijcke conste om te Schieten Oost ende West, ende d'opservatie van dien.* Amsterdam: voor Laurens Jacobsz. [Ghedruckt t'Haerlem: by Gillis Rooman], 1597.

Lastman, Cornelis Jansz. *De Schat-kamer, Des Grooten See-vaerts-kunst.* Ed. princ. 1621. Amsterdam: C. J. Lastman, 1629.

———. *Beschrijvinge van de Kunst der Stuer-luyden, Daer in Door seeckere Gront-regelen getoont wordt Hoe die selve Kunst na 't behooren ghebruyckt: En langs soo meer magh gebetert worden: Ende Tot dien eynde sijn hier by ghestelt de Tafelen vande Hoeckmaten, Raecklijnen ende Snijlijnen. Met een grondige Instructie des drie-hoecks reeckeningh.* Ed. princ. 1642. Amsterdam: Symon Cornelisz. Lastman, 1657.

———. *Konst der Stuurluiden. In Welkke Door zeekere Grondt-regelen getoondt werdt, hoe die zelve Konst na 't behooren gebruikt, en langhs zoo meer magh gebeterd werden. Tot welkken einde hier by gesteldt zijn de Tafels van de Hoekmaten, Raak- en Snylijnen: nevens het maakzel, gebruik en inhoudt der zelver Tafelen; met een grondige onderwijzinge des Driehoeks reekeningh.* Ed. Frans vander Huips. Amsterdam: By Jacob en Casparus Loots-man, in Compagnie met Hendrik Donkker, 1675.

Le Corbeiller, Edouard. "Quelques lettres de Guillaume Denys à Colbert." *Revue catholique de Normandie* 25.1 (1916): 40–55.

Le Cordier, Samson. *Journal de navigation.* Havre de Grâce: Jacques Gruchet, 1683; Veuve de Jacques Hubault, 1708.

———. *Traité des pratiques journalières des pilotes: divisé en deux parties, dans lequel est plainement enseigné, & clairement démontré l'art & la science des navigateurs.* 2 vols. Havre de Grâce: Jacques Gruchet, 1683.

———. *Instruction des pilotes.* Havre de Grâce: Jacques Gruchet, 1683.

———. *Instruction des pilotes: premiere partie contenant les principes nécessaires pour trouver l'heure de la marée dans tous les ports . . . avec un dialogue ou examen des pilotes en faveur de ceux qui veulent se rendre experts dans la pratique de la navigation, & se mettre au fait des manoeuvres d'un vaisseau.* . . . Ed. Jacques [Nicolas] Le Cordier. Havre de Grâce: P.J.D.G. Faure, 1734.

———. *Instruction des pilotes: premiere partie contenant un abrégé de la sphere, les principes nécessaires pour trouver l'heure de la pleine mer dans les ports . . . avec une instruction générale sur le pilotage, en faveur de ceux qui veulent se rendre experts dans cette science, & se mettre aussi au fait des manoeuvres d'un vaisseau.* Ed. M. Fouray. Havre de Grâce: P.J.D.G. Faure, 1786.

Le Telier [Le Tellier], Jean. *Le Voyage de maistre Jean David.* Dieppe: Nicolas Acher, 1631.

———. *Le Vray moyen de trouver la variation de l'aymant.* Dieppe: Nicolas Acher, 1631.

———. *Voyage faict aux Indes Orientalles.* Dieppe: Nicolas Acher, 1631, 1640, 1654.

Leybourn, William. *Nine Geometricall Exercises, for Young Sea-men and Others That Are Studious in Mathematicall Practices.* London: James Flesher, for George Sawbridge, 1669.

Linschoten, Jan Huyghen van. *Itinerario, voyage ofte schipvaert naer Oost ofte Portugaels Indien.* . . . Amsterdam: Cornelis Claesz, 1596.

———. *His Discours of Voyages into Ye Easte & West Indies Deuided into Foure Bookes.* London: [John Windet for] John Wolfe, 1598.

———. *Histoire de la navigation aux Indes orientales.* . . . Amsterdam: H. Laurent, 1610.

Louis XIV. *Ordonnance de Louis XIV, donnée à Fontainebleau au mois d'Aoust 1681, touchant la Marine.* Paris: Denys Thierry, 1681; Paris: Charles Osmont, 1714.

———. *Ordonnance de Louis XIV. pour les armées navales et Arcenaux de la Marine.* Paris: Estienne Michallet, 1689.

———. *Ordonnance de Louis XIV. pour les armées navales et Arcenaux de la Marine* [1689]. In *Annales maritimes et coloniales* no. 62 (1847): 91–284.

Makreel, Dirck. *Lichtende Leydt-Starre der Groote Zee-vaert. Zijnde een klare, en fondamentele verhandelinge, van alle de voornaemste stucken der Navigatie.* Amsterdam: Hendrick Doncker, 1671.

Mainwaring [Manwayring], Henry. *The Sea-Mans Dictionary: Or, An Exposition and Demonstration of All the Parts and Things Belonging to a Shippe: Together with an Explanation of All the Termes and Phrases Used in the Practique of Navigation.* London: G. M. for John Bellamy, 1644.

———. *The Seaman's Dictionary,* in *The Life and Works of Sir Henry Mainwaring.* Ed. G. E. Manwaring and W. G. Perrin. Vol. 2. London: Printed for the Navy Records Society, 1920–22.

Martin, Benjamin. *Young Trigonometer's Compleat Guide.* London: for J. Noon, 1736.

Martindale, Adam. *The Country-Survey-Book: Or Land-Meters Vade-Mecum.* . . . London: Printed for R. Clavel . . . and T. Sawbridge, 1692.

Maskelyne, Nevil. *The British Mariner's Guide.* London: for the author, 1763.

Medina, Pedro de. *A Navigator's Universe; the Libro de Cosmographia of 1538.* Tr. Ursula Lamb. Chicago: for the Newberry Library by the University of Chicago Press, 1972.

———. *Arte de navegar.* Valladolid: en casa de Francisco Fernández de Córdoba, 1545.

———. *Regimiento de navegación.* Sevilla: Juan Canalla, 1552.

———. *L'Art de naviguer. Contenant toutes les reigles, secrets, & enseignemens necessaires à la bonne navigation.* Tr. Nicolas de Nicolai. Lyon: Guillaume Rouillé, 1554. Rev. ed. Jean de Seville. Rouen: Théodore Reinsart, 1602.

———. *The Arte of Navigation Wherein Is Contained All the Rules, Declarations, Secretes, & Aduises, Which for Good Navigation Are Necessarie & Ought to Be Knowen and Practised.* Tr. John Frampton. London: Thomas Dawson, 1581.

Medina, Pedro de, and Michiel Coignet. *De Zeevaert, Oft Conste van ter Zee te varen, vanden Excellenten Pilote M[ees]ter Peeter de Medina Spaignaert. Inde welcke niet alleene de Regels . . . der selver Consten begrepē sijn: Maer ooc de clare . . . fondementen der Astronomijen . . . verclaert* worden . . . Uit den Spaensche ende Fransoysche in onse Nederduytsche tale ouergheset, ende met Annotatië verciert, by Mr. Merten Everaert . . . Met noch een ander nieuvve Ondervvijsinghe, op de principaelste puncten der Navigatien, van Michiel Coignet. Tr. Martin Everaert. Antwerp: Hendrick Hendricksen, 1580.

———. *De Zeevaert oft conste van ter Zee te varen, vanden . . . Pilote Meester Peeter de Medina . . . Inde welcke niet alleene de Regels, Secreten, Practijcke[n], en[de] constighe Instrumenten der seluer Consten begrepen zijn: maer oock de clare ende oprechte fondamenten der Astronomijen . . . verclaert worden . . .* Amsterdam: C. Claesz, 1589, 1593, 1598.

Metius, Adriaan. *Astronomische ende Geographische Onderwysinghe.* Amsterdam and Franeker: Hendrick Lauwerentz, ghedruckt by Ulderick Balck, 1632.

Milliet de Chales, Claude-François. *L'Art de naviger, demontré par principes; & confirmé par plusieurs observations tirées de l'experience.* Paris: E. Michallet, 1677.

———. *The Elements of Euclid, Explained and Demonstrated in a New and Most Easie Method . . . Done out of French.* Tr. Reeve Williams. London: Printed for Philip Lea, 1685.

Monson, Sir William. *The Naval Tracts of Sir William Monson: In Six Books [1624].* Ed. Michael Oppenheim. London: Navy Records Society, 1902.

Moore, John Hamilton. *The Practical Navigator, and Seaman's New Daily Assistant: Being a Complete System of Practical Navigation.* London: Printed by W. and J. Richardson, 1772; 9th ed., 1791.

———. *New Practical Navigator.* Newburyport, MA: Edmund Blunt, 1799.

Moore, Jonas. *A New Systeme of the Mathematicks.* Ed. Peter Perkins. London: A. Godbid and J. Playford, for Robert Scott, 1681.

Morrice, David. *The Art of Teaching, Or, Communicating Instruction Examined, Methodized, and Facilitated, as well as Applied to all the Branches of Scholastic Education.* London: Lackington, Allen, and Co., 1801.

———. *The Young Midshipman's Instructor (designed as a Companion to Hamilton Moore's Navigation) With Useful Hints to Parents of Sea Youth, and to Captains and Schoolmasters in the Royal Navy.* London: Knight and Compton . . . for the author, 1801.

Moxon, Joseph. *A Tutor to Astronomie and Geographie, Or, An Easie and Speedy Way to Know the Use of Both the Globes, Coelestial and Terrestrial in Six Books.* London: Joseph Moxon, 1659.

Napier, John. *Mirifici logarithmorum canonis descriptio.* Edinburgh: Andreae Hart, 1614.

———. *A Description of the Admirable Table oe [sic] Logarithmes.* London: Nicholas Okes, 1616.

Narrative of the Distresses and Miraculous Escape of His Majesty's Ship the Guardian. London: Sold by J. Forbes, Tavistock-Row, Covent Garden and to be had of all the booksellers in town and country, 1790.

Nautonier, Guillaume de. *Mécométrie de l'eymant: cest a dire la maniere de mesvrer les longitudes par le moyen de l'eymant. . . .* Vénès: chez l'auteur, 1603.

New and Complete Dictionary of Arts and Sciences. London: W. Owen, 1763.

Newhouse, Daniel. *The Whole Art of Navigation in Five Books.* London: for the author, 1685.

Newton, Isaac, and Roger Cotes. *Correspondence of Sir Isaac Newton and Professor Cotes: Including Letters of Other Eminent Men.* Ed. J. Edleston. London: J. W. Parker, 1850.

Newton, Samuel. *An Idea of Geography and Navigation Containing Easie Rules for Finding the Latitude and Difference of Longitude of Places by Observation of the Sun, Moon and Stars . . . Also, Tables of the Sun's Declinaision and Right Ascension for Every Navigation. . . .* London: Printed for Christopher Hussey . . . sold likewise by M. Marsh [etc.], 1695.

Nierop, Dirck Rembrantsz. van. *Onderwys der Zee-Vaert*. . . . Ed. princ. 1661. Amsterdam: Hendrick Doncker, 1670.

———. *Mathematische Calculatie . . . door de Tafelen Sinus Tangents of Logarithmus wiskonstelick uyt te rekenen*. Amsterdam: Gerrit van Goedesberg, 1659.

———. *'t Nieuw Stuermans Graed-boek, Of Kleenen, Schat-kamer, Inhoudende de Tafelen der Son en Sterren Declinatie, uitgereekent op nieuws . . . en meer andere nodige Werken*. Amsterdam: Joannis Loots, 1697. Bound with: *Nieuwe uytgereckende Taafelen . . . Als meede een Almanach*. Amsterdam: Jacobus Robyn, n.d. [1701] And: *Eeuwig Duerende Almanach. Beginnende A:1681*. Amsterdam: Iacobus Robyn, n.d. [1681].

Norwood, Richard. *Trigonometrie. Or, The Doctrine of Triangles*. . . . London: William Jones, 1631.

———. *Sea-Mans Practice*. London: G. Hurlock, 1637.

Oostwoud, Govert Maartensz. *Een T'Samenkouting Tusschen een Stuurman En een Matroos: Gehouden over de Nieu-gepractiseerde Oeffening der Stuurluyden*. . . . Alkmaar: Jan Blom, 1699.

———. *Vermeerderde Schoole der Stuurluyden: waer in kort en klaer wort aengewesen: 't gene een Stuurman behoort te weten, om sijn saken in de groote Zee sijnde wel aen te leggen: door ervarentheyt en ondervindinge ten dienst van de Leerlingen*. Hoorn: Stoffel Jansz. Kortingh, 1712.

Oostwoud, Jacob, ed. *Maandelijkse Mathematische Liefhebberij*. Purmerende, 1754–69.

Ordonnance du commerce, du mois de mars 1673, et ordonnance de la marine, du mois d'août 1681. Bordeaux: Audibert et Burkel, an VIII [1800].

Oughtred, William. *The Circles of Proportion and the Horizontall Instrument*. . . . Tr. William Forster. London: Augustine Mathewes for Elias Allen, 1632. Oxford: W. Hall for R. Davis, 1660.

———. *An Addition Unto the use of the Instrument Called the Circles of Proportion, for the Working of Nauticall Questions Together with Certaine Necessary Considerations and Advertisements Touching Navigation*. London: Augustine Mathewes, 1633.

Pakenham, Edward. *Captain Pakenham's Invention of a Substitute for a Lost Rudder, and to Prevent Its Being Lost. Also a Method of Restoring the Masts of Ships When Wounded or Otherwise Injured*. London: T. Spilsbury, 1793.

Pepys, Samuel. *Samuel Pepys's Naval Minutes*. Ed. J. R. Tanner. London: Navy Records Society, 1926.

———. *The Tangier Papers of Samuel Pepys*. Ed. E. Chappell. London: Navy Records Society, 1935.

———. *The Diary of Samuel Pepys: A New and Complete Transcription*. Ed. Robert Latham and William Matthews. 11 vols. London: G. Bell & Sons, 1970–83.

Perkins, Peter. *The Seaman's Tutor: Explaining Geometry, Cosmography, and Trigonometry. With Requisite Tables of Longitude and Latitude of Sea-Ports, Traverse Tables, Tables of Easting and Westing, Meridian Miles, Declinations, Amplitudes, Refractions, Use of the Compass, Kalendar, Measure of the Earth's Globe, Use of Instruments and Charts, Difference of Sailing, Estimation of a Ship's Way by the Log, and Log-line, Currents, &c.: Compiled for the Use of the Mathematical School in Christ's Hospital-London, His Majesty Charles II, His Royal Foundation*. London: Obadiah Blagrave, 1682.

Philips, Caspar (Jacobsz.). *Zeemans Onderwijzer in de Tekenkunst; of Handleiding, om door Geometrische, Doorzichtkundige, en Perspectivische Regelen, alle Landverkenningen, Kusten, Baaien . . . en wat een' Zeeman meer voorkomt, op het Papier Wiskundig Aftetekenen*. Amsterdam: Elwe en Langefeld, 1786.

Phillippes, Henry. *The Geometrical Sea-Man: Or, the Art of Navigation Performed by Geometry.* London: Robert and William Leybourn, for George Hurlock, 1652.

Pietersz., Cornelis. *Handleiding, tot het Practicale of Werkdadige Gedeelte van de Stuurmanskunst. Zeer nuttig voor de Zeevaarende in het algemeen, en in 't byzonder voor 's E.O.I. Comp's Zee-Officieren, Of die geene, welke zich daar toe zoeken te bekwaamen.* Amsterdam: Gerard Hulst van Keulen, 1790.

Pimentel, Manoel. *Arte de Navegar, em que se ensinam as regras praticas.* . . . Lisboa: Na Officina Real, Deslandesiana, 1712.

Pitiscus, Bartholomaeus. *Trigonometry, or, The Doctrine of Triangles First Written in Latine . . . Wherunto Is Added (for the Marriners use) Certaine Nauticall Questions, Together with the Finding of the Variation of the Compasse.* Tr. Ralph Handson. London: Edward Allde for John Tap[p], 1614.

Reenen, Jacob van. *A Journal of a Journey from the Cape of Good Hope, Undertaken in 1790 and 1791 . . . in Search of the Wreck of . . . the East India Company's Ship the "Grosvenor."* . . . Edward Riou. London: G. Nicol, 1792.

Review of G. Denys, *Traité de la Variation* (Dieppe, 1666). *Journal des sçavans* (Paris, 1666), 315–16.

Reyersz., Heyndrick. *Stuurmans-Praetjen Tusschen Jaep en Veer, Nieuwelijcx vermeerdert door een Liefhebber der Zeevaert.* Ed. princ. 1614(?). Amsterdam: Jacob Colom, 1637.

———. *De vaste Grondt der Loflycker Zee-vaert.* Amsterdam: Dirc Pietersz., 1622.

Robbertsz., Robbert. *"Le Canu." Onder verbeteringhe een waerschouwinghe der zeeluyden, voor Albert Haijen dolinghe in zijn boecxken. Ghenaemt een corte onderrichtinghe belanghende de conste der zee-vaert.* . . . N.p., n.p., 1600.

Robertson, John. *The Elements of Navigation.* London: J. Nourse, 1754, 1764, 1780.

Royal Society. "Directions for Observations and Experiments to Be Made by Masters of Ships, Pilots, and Other Fit Persons in Their Sea-Voyages." *Philosophical Transactions* 2, no. 24 (1666–67): 433–48.

Ruelle, Pieter. *Voorlooper des Zee-Quadrants, ofte Ruyt-kaert, midtsgaders wat tot het gebruyck der Zee-vaert van nooden is. Int Licht gebracht door Johannes van Loon.* Amsterdam: Hendrick Doncker, 1693.

Saltonstall, Charles. *The Navigator Shewing and Explaining All the Chiefe Principles and Parts Both Theoricke and Practicke, That Are Contayned in the Famous Art of Navigation.* . . . London: Printed [by Bernard Alsop and Thomas Fawcet] for G. Herlock, 1636.

Seixas y Lovera, Francisco de. *Theatro Naval Hydrographico, de los Fluxos, y Refluxos, y delas corrientes de los mares.* . . . Madrid: Antonio de Zafra, 1688.

———. *Theatre naval hydrographique, des flux et réflux, des courans des mers.* . . . Paris: Pierre Gissey, 1704.

Seller, John. *Practical Navigation.* London: J. Darby, 1669, 1680.

———. *An Epitome of the Art of Navigation.* London: n.p., 1681.

———. "A Catalogue of Mathematical Books, Maritime Charts, Draughts, Prints and Pictures, Made and Sold by John Seller." [London: J. Seller, ca. 1700].

Seville, Jean de. *Le Compost manuel, calendier, et almanach perpetuel. Recueilli & reformé selon le retranchement des dix jours, avec la déclinaison du soleil reformée, un abbregé de la sphère, & autres choses appartenantes à la navigation, principalement pour la longitude de l'est & ouest.* Rouen: Robert Mallard, 1586.

Shapter, Mr. "The Forecastle Sailor or the Guardian Frigate, Written by Mr. Shapter, and Sung by Mr. Darley at Vauxhall Gardens, Etc." Sheet music, composed by John Moulds. London: G. Goulding, [ca. 1790].

Skay, John. *A Friend to Navigation, plainely expressing to the capacity of the simpler sort the whole misery or foundation of the same art.* . . . London: T. Cotes, 1628.

———. *Sea-Mans Alphabet and Primer.* [London], n.p., 1644.

Smith, John. *An Accidence or The Path-way to Experience Necessary for all Young Sea-men, or those that are desirous to goe to Sea, briefly shewing the Phrases, Offices, and Words of Command, Belonging to the Building, Ridging, and Sayling, a Man of Warre; And how to manage a Fight at Sea.* London: [Nicholas Okes] for Jonas Man, and Benjamin Fisher, 1626.

———. *A Sea Grammar, With the Plaine Exposition of Smiths Accidence for young Sea-men, enlarged . . . Written by Captaine John Smith, sometimes Governour of Virginia, and Admirall of New-England.* London: John Haviland, 1627.

Snellius, Willebrord. *Tiphys Batavus, sive histiodromice, de navium cursibus et re navali.* Leiden: Ex Officina Elzeviriana, 1624.

Stavorinus, Johan Splinter, and Samuel Hull Wilcocke. *Voyages to the East-Indies.* London: G. G. and J. Robinson, 1798.

Steenstra, Pybo. *Grond-beginzels der Stuurmans-kunst.* Amsterdam: G. Hulst Van Keulen, 1779.

———. *Uitgewerkt examen der Stuurlieden, en verscheiden wyzen, om de Breedte en Lengte op Zee te verbeteren.* Middelburg: P. Gillissen en Zoon, 1781.

Stevin, Simon. *De Thiende.* Leiden: Plantijn, 1585.

———. *De Havenvinding.* Leiden: Plantijn, By Christoffel van Raphelinghien, 1599.

———. *Wisconstighe g[h]edachtenissen.* Leiden: Jan Bouwensz., 1605–8.

Steyn, Gerard van. *Liefhebbery der Reekenkonst: Zynde eene Verzaameling van Examens, over de Reekenkonst op verscheiden Vacante Schoolmeesters Plaatsen voorgevallen.* . . . Amsterdam: P. Huart, 1768.

Strype, John. *Survey of London (1720), [online].* Ed. J. F. Merritt. Humanities Research Institute Online, Sheffield, 2007. http://www.hriOnline.ac.uk/strype/.

Sturmy, Samuel. *The Mariners Magazine, Or, Sturmy's Mathematical and Practical Arts Containing the Description and Use of the Scale of Scales . . . and Other Most Useful Instruments for All Artists and Navigators: The Art of Navigation, Resolved Geometrically, Instrumentally, and by Calculation, and by That Late Excellent Invention of Logarithms, in the Three Principal Kinds of Sailing: With New Tables of the Longitude and Latitude of the Most Eminent Places . . . Together with a Discourse of the Practick Part of Navigation.* . . . London: E. Cotes for G. Hurlock, [etc.], 1669.

Stuurman, Cornelis. *Klaare beschryvingh over het gebruyk der quartier de reduction.* Amsterdam: Johannes van Keulen, 1734.

Stuurman, Jan Pietersz. *Sardammer Schatkamer Of 't Geopende Slot der Groote Zee Vaert. Synde een volkomen en klaare onderwysing der Navigatie . . . Als mede, De Examen der Stuurlieden . . . De Declinatie Tafel verlengt tot 1731 en achter aan de Logarithmus Tafelen van Sinus, Tangens en Secans, &c. en met een vermakelijk questie toegepast.* Amsterdam: Hendrik Donker, Boekverkooper en Graadboogmaaker, 1699.

Syria, Pedro. *Arte de la Verdadera Navegacion.* Valencia: J. C. Garriz, 1602.

Tapp, John. *The Seamans Kalender, or An Ephemerides of the Sun, Moone, and Certaine of the Most Notable Fixed Starres Together with Many Most Needfull and Necessary Matters, to the Behoofe and Furtherance Principally of Mariners and Seaman: But Generally Profitable to All Travailers, or such as Delight in the Mathematicall Studies. The Tables Being for the Most Part Calculated from the Yeare 1601. to the Yeare 1624.* London: E. Allde for John Tapp, 1602.

Tollen, G. van der. *Het Stuurmans en Lootsmans Hand-Boek, Zynde een Beschryvinge van de Diepten, daar op eenige Landen of Kusten gesien mag worden . . . Na de beste ondervinding van veel ervaren Stuurlieden en Lootsen.* Ed. Christoffel Middagten. Amsterdam: Joannes Loots, 1723.

Tourville, Anne-Hilarion de Cotentin, Chevalier de. *Exercice en général de toutes les maneuvres qui se font à la mer.* Havre de Grâce: J. Hubault, 1693.

———. *General Exercitier till Siös, Med alla de brukelige Handgrep af Skepps Arbete som uthi alla Occasioner kunna förfalla.* Stockholm: Tryckt Olao Enæo, 1698.

Valin, René-Josué. *Nouveau commentaire sur l'Ordonnance de la Marine, du mois d'août 1681.* 2 vols. La Rochelle: J. Legier, 1760.

Veen, Adriaen. *Tractaet Vant Zee-bouck-houden op de Ronde gebulte Pas-kaert.* Amsterdam: Barendt Adriaenszoon, 1597.

Veur, Adriaan Teunisz. *Zeemans Schatkamer, daar in de Stuurmans-Konst, Niet alleen als voor deezen beschreeven; maar ook verscheide zaken verbetert en bygevoegt zyn.* Amsterdam: J. van Keulen, 1755.

VOC Heren XVII. "Instructie voor de Examinateurs der Stuurlieden by de edele Oost-Indische Compagnie" [Amsterdam: n.p., 1780].

Vooght, Claas Jansz. *De Zeemans Wegh-Wyser, Waar in klaar en volkoomen beschreven wort al 't geene tot 't Onderwijs des Stuurmans konst noodig en nut is. Verthoonende 't Ty-reekenen, hooghte nemen, Kompas-peylen, en 't zelve te vergoeden, 't drieheoks rekenen, platte Paskaarts rekenen, Stroom-kavelen, Kromstreeks reekenen, 't vultige drieheoks reekenen en Starkonstige Werkstukken. Als ook Waar in alle de verloopene Taaffelen vernieuwd . . . Tweede Druck. . . .* Amsterdam: Johannes van Keulen, 1695.

Vries, Klaas de. *Schat-kamer ofte Konst der Stier-lieden, Zynde een klaar Onderwysing der Navigatie van al het geen een Stierman, aangaande de Konst, behoorde te weeten. Hier by gevoegt de Tafelen van Sinus, Tangens, en Secans; en de Logarithmus, Sinus &c. en de Nommer Logarithmus, mitsgaders de Streek-tafel.* Amsterdam: Joannes Loots, 1702.

De Vroolyke Oost-Indies-Vaarder, of Klinkende en Drinkende Matroos. . . . Amsterdam: Erven de Weduwe Jacobus van Egmont, n.d. [ca. 1780].

Waghenaer, Lucas Janszoon. *Spieghel der Zeevaerdt.* Ed. princ. Leiden: Christophe Plantin, 1584.

———. *Speculum Nauticum super navigatione maris occidentalis confectum.* Leiden: Franciscus Raphelengius for Luca Johannes Aurigarius, 1586.

———. *The Mariners Mirrour Wherin May Playnly Be Seen the Courses, Heights, Distances, Depths, Soundings . . . Together W[i]th the Rules and Instrume[n]ts of Navigation. First Made & Set Fourth in Divers Exact Sea-Charts, by That Famous Navigator Luke Wagenar of Enchuisen and Now Fitted with Necessarie Additions for the Use of Englishmen.* Tr. Anthony Ashley. London: John Charlewood, 1588.

———. *Spiegel der Seefahrt.* Amsterdam: Cornelis Claesz., 1589.

———. *Thresoor der zee-vaert. Inhoudende de geheele Navigatie ende Schip-vaert vande Oostersche, Noordtsche, Westersche ende Middellantsche zee, met alle de Zee-caerten daer toe dienende.* Leiden: Plantin, 1592. Facsimile, ed. R. A. Skelton. Amsterdam: Theatrum Orbis Terrarum, 1965.

———. *Thresoor der zee-vaert.* Amsterdam: Cornelis Claesz., 1596.

———. *Thrésorerie ou cabinet de la routte marinesque.* Calais: Bonaventure d'Aseville, 1601.

Warius, P. *Examen der Stierlieden.* Hoorn: Adriaan Brouwer, 1735.

Westerman, Adam. *Groote Christelijke Zee-Vaert, In XXVI. Predicatien, In maniere van een Zee-Postille. In welcken een Schipper, Koopman, Oorloghs-volk, Zee-ende Reysende-man, geleert wort.* . . . Amsterdam: Abraham en Jan de Wees, 1664.

Willaumez, C. Pt. Jean-Baptiste-Philibert. *Projet pour former des élèves marins à Paris même.* [Paris], 1800.

Winschooten, Wigardus. *Seeman: Behelsende Een grondige uitlegging van de Neederlandse Konst, en Spreekwoorden . . . die uit de Seevaart sijn ontleend. . . .* Leiden: Johannes de Vivié, 1681.

Witsen, Nicolaas. *[Architectura navalis et regimen nauticum ofte] Aeloude en Hedendaagsche Scheeps-Bouw en Bestier.* Ed. princ. 1671. Amsterdam: Pieter en Joan Blaeu, 1690.

Wright, Edward. *Errors in Navigation 1 Error of Two, or Three Whole Points of the Compas, and More Somtimes, by Reason of Making the Sea-Chart after the Accustomed Maner . . . 2 Error of One Whole Point, and More Many Times, by Neglecting the Variation of the Compasse. 3 Error of a Degree and More Sometimes, in the use of the Crosse Staffe . . . 4 Error of 11. or 12. Minures [sic] in the Declination of the Sunne, as It Is Set Foorth in the Regiments Most Commonly Used among Mariners. . . .* London: [Valentine Simmes and W. White] for Ed. Agas, 1599.

———. *Certain Errors in Navigation . . . with Many Additions That Were Not in the Former Editions.* London: J. Moxon, 1657. Facsimile ed. Amsterdam: Theatrum Orbis Terrarum, 1974.

Zamorano, Rodrigo de. *Compendio del arte de navegar.* Sevilla: Alonso de la Barrera, 1581. Joan de Leon, 1588.

———. *Cort Onderwiis Vande Conste der Seevaert. . . .* Tr. Martin Everaert. Amsterdam: Cornelis Claesz., 1598.

SECONDARY SOURCES

Adams, Thomas Randolph. *The Non-cartographical Maritime Works Published by Mount and Page: A Preliminary Hand-List.* London: Bibliographical Society, 1985.

———. "The Beginnings of Maritime Publishing in England, 1528–1640." *The Library* 6th ser., 14, no. 3 (1992): 207–19.

———. *English Maritime Books Printed before 1801: Relating to Ships, Their Construction and Their Operation at Sea.* Greenwich, UK: National Maritime Museum, 1995.

Akveld, L. M., S. Hart, and W. J. van Hoboken, eds. *Maritieme Geschiedenis der Nederlanden: Zeventiende Eeuw, van 1585 tot ca 1680.* Bussum: De Boer Maritiem, 1977.

Albuquerque, Luís de. "Portuguese Books on Nautical Science from Pedro Nuñes to 1650." *Revista da Universidade de Coimbra* 32 (1986): 259–78.

Aman, Jacques. *Les officiers bleus de la marine française au XVIIIe siècle.* Paris: Librairie Droz, 1973.

Andrewes, William J. H., ed. *The Quest for Longitude: The Proceedings of the Longitude Symposium, Harvard University, Cambridge, Massachusetts, Nov. 4-6, 1993.* Cambridge, MA: Collection of Historical Scientific Instruments, Harvard University, 1996.

Anthiaume, Albert, Abbé. *Evolution et enseignement de la science nautique.* 2 vols. Paris: Ernest Dumont, 1920.

———. *Pierre Desceliers, père de l'hydrographie et de la cartographie françaises.* Rouen: Impr. J. Lecerf, Société "les Amys du vieux Dieppe," 1926.

———. *L'abbé Guillaume Denys de Dieppe (1624-1689), Premier Professeur Royal d'hydrographie en France.* Paris: E. Dumont, 1927.

Arroyo, Ricardo. "Las ensenanzas de nautica en el siglo xviii." *Revista de Historia Naval [Spain]* 12, no. 46 (1994): 7–30.

Ash, Eric H. *Power, Knowledge, and Expertise in Elizabethan England.* Baltimore: Johns Hopkins University Press, 2004.

———. "Expertise and the Early Modern State." *Osiris* 25 (2010): 1–24.

Asher, Eugene L. *The Resistance to the Maritime Classes: The Survival of Feudalism in the France of Colbert.* Berkeley: University of California Press, 1960.

Ashley, Clifford W. *The Ashley Book of Knots*. Garden City, NY: Doubleday, 1944.

Asseline, David, and A. Guérillon. *Les antiquitez et chroniques de la ville de Dieppe*. Vol. 2. Dieppe: Maisonneuve, 1874.

Audet, Louis-Philippe. "Hydrographes du Roi et cours d'hydrographie au collège de Québec, 1671–1759." *Cahiers de Dix* 35 (1970): 13–37.

Baigrie, Brian S., ed. *Picturing Knowledge: Historical and Philosophical Problems concerning the Use of Art in Science*. Toronto: University of Toronto Press, 1996.

Baldasso, Renzo. "The Role of Visual Representation in the Scientific Revolution: A Historiographic Inquiry." *Centaurus* 48, no. 2 (2006): 69–88.

Barbier, Paul. "Old French *laman, loman* Pilot, Sea-Pilot, Coastal- or Harbour-Pilot." In *Studies in French Language, Literature, and History: Presented to R. L. Græme Ritchie*, ed. Fraser Mackenzie, R. C. Knight and J. M. Milner, 19–23. Cambridge: Cambridge University Press, 1949.

Baugh, Daniel. "The Professionalisation of the English Navy and Its Administration, 1660–1750." In *The Sea in History: The Early Modern World / La mer dans l'histoire: La période moderne*, ed. Christian Bouchet and Gérard Le Bouëdec, 3.852–866. Woodbridge: Boydell Press, 2017.

Bell, Rudolph M. *How to Do It: Guides to Good Living for Renaissance Italians*. Chicago: University of Chicago Press, 1999.

Bennett, J. A. "The Mechanic's Philosophy and the Mechanical Philosophy." *History of Science* 24 (1986): 1–28.

———. *The Divided Circle: A History of Instruments for Astronomy, Navigation and Surveying*. Oxford: Phaidon Christie's, 1987.

———. "The Challenge of Practical Mathematics." In *Science, Culture, and Popular Belief in Renaissance Europe*, ed. Stephen Pumfrey, Paolo L. Rossi, and Maurice Slawinski, 176–90. Manchester: Manchester University Press, 1991.

———. "Practical Geometry and Operative Knowledge." *Configurations* 6, no. 2 (1998): 195–222.

———. "Shopping for Instruments in Paris and London." In *Merchants and Marvels: Commerce and the Representation of Nature in Early Modern Europe*, ed. Paula Findlen and Pamela H. Smith, 370–95. New York: Routledge, 2002.

———. "Mathematics, Instruments and Navigation, 1600–1800." In *Mathematics and the Historian's Craft*, ed. Glen Van Brummelen and Michael Kinyon, 43–55. New York: Springer, 2005.

———. *Navigation: A Very Short Introduction*. Oxford: Oxford University Press, 2017.

Bensaúde, Joaquim. *Histoire de la science nautique portugaise à l'époque des grandes découvertes: collection de documents publiés par ordre du Ministère de l'instruction publique de la République portugaise (décret du 29 décembre 1913)*. Munich: C. Kuhn, 1914.

Berkel, Klaas van, Albert Van Helden, and Lodewijk C. Palm. *A History of Science in the Netherlands: Survey, Themes, and Reference*. Leiden: Brill, 1999.

Bertucci, Paola. *Artisanal Enlightenment: Science and the Mechanical Arts in Old Regime France*. New Haven: Yale University Press, 2017.

Bethencourt, Francisco, and Diogo Ramada Curto, eds. *Portuguese Oceanic Expansion, 1400–1800*. Cambridge: Cambridge University Press, 2007.

Blair, Ann M. "Scientific Readers: An Early Modern Perspective." *Isis* 95, no. 33 (2004): 420–30.

———. "Student Manuscripts and the Textbook." In *Scholarly Knowledge: Textbooks in Early Modern Europe*, ed. Emidio Campi, Anthony Grafton, Simone de Angelis, and Anja-Silvia Goeing, 39–74. Genève: Librairie Droz, 2008.

———. *Too Much to Know: Managing Scholarly Information before the Modern Age*. New Haven: Yale University Press, 2010.

———. "Conrad Gesner's Paratexts." *Gesnerus* 73, no. 1 (2016): 73–122.

Blakemore, Richard. "The Ship, the River, and the Ocean Sea: Encounters with Space in the Maritime Community of Early Modern London." In *Maritime History and Identity*, ed. D. Redford, 98–119. London: I. B. Tauris, 2013.

Bleyerveld, Yvonne, and Ilja M. Veldman. *The Netherlandish Drawings of the 16th Century in the Teylers Museum*. Leiden: Primavera Pers, 2016.

Boistel, Guy. "Le problème des 'longitudes à la mer' dans les principaux manuels de navigation français autour du XVIIIe siecle." *Sciences et techniques en perspective [France]* 3, no. 2 (1999): 253–84.

———. "De quelle précision a-t-on réellement besoin en mer?" In *Mesurer le Ciel et la Terre*, ed. Evelyne Barbin and Guy Boistel. *Histoire & mesure* 21, no. 2 (Dec. 2006): 121–56.

———. "Training Seafarers in Astronomy: Methods, Naval Schools, and Naval Observatories in Eighteenth- and Nineteenth-Century France." In *The Heavens on Earth: Observatories and Astronomy in Nineteenth-Century Science and Culture*, ed. David Aubin, Charlotte Bigg, and H. Otto Sibum, 148–73. Durham, NC: Duke University Press, 2010.

Boxer, Charles Ralph. "The Dutch East-Indiamen: Their Sailors, Their Navigators, and Life on Board, 1602–1795." *Mariner's Mirror* 59 (1962): 84–86.

———. *The Dutch Seaborne Empire, 1600-1800*. London: Hutchinson, 1965.

———. *The Portuguese Seaborne Empire, 1415-1825*. London: Hutchinson, 1969.

Braudel, Fernand. *The Mediterranean and the Mediterranean World in the Age of Philip II*. Tr. Sian Reynolds. 2 vols. New York: Collins; Harper & Row, 1972.

Bréard, Charles. "Les *Memoires* de Jean Doublet, de Honfleur." *Revue historique* (1880): 58–59.

Brockliss, L. W. B. *French Higher Education in the Seventeenth and Eighteenth Centuries: A Cultural History*. Oxford: Clarendon Press, 1987.

Broecke, Steven Vanden. "The Use of Visual Media in Renaissance Cosmography: The Cosmography of Peter Apian and Gemma Frisius." *Paedagogica Historica: International Journal of the History of Education* 36, no. 1 (2000): 130–50.

Brook, Timothy. *Vermeer's Hat: The Seventeenth Century and the Dawn of the Global World*. New York: Bloomsbury Press, 2008.

Brown, Gary I. "The Evolution of the Term 'Mixed Mathematics.'" *Journal of the History of Ideas* 52, no. 1 (Mar. 1991): 81–102.

Bruijn, Jaap R. "De personeelsbehoefte van de VOC overzee en aan boord, bezien in Aziatisch en Nederlands perspectief." *BMGN* 91, no. 2 (1976): 218–48

———. *The Dutch Navy of the Seventeenth and Eighteenth Centuries*. Columbia: University of South Carolina Press, 1993.

———. *Schippers van de VOC in de achttiende eeuw aan de wal en op zee*. Amsterdam: Bataafsche Leeuw, 2008.

———. *Commanders of Dutch East India Ships in the 18th Century*. Woodbridge: Boydell & Brewer, 2011.

Bruijn, Jaap R., and F. Gaastra, eds. *Dutch-Asiatic Shipping in the 17th and 18th Centuries*. The Hague: Nijhoff, 1987.

Bruijn, Jaap R., and W. F. J. Mörzer Bruyns. *Anglo-Dutch Mercantile Marine Relations, 1700-1850: Ten Papers*. Amsterdam: Rijksmuseum Nederlands Scheepvaartmuseum, 1991.

Brummelen, Glen Van. *Heavenly Mathematics: The Forgotten Art of Spherical Trigonometry*. Princeton, NJ: Princeton University Press, 2013.

Bryant, Arthur. *Samuel Pepys: The Saviour of the Navy.* Cambridge: Cambridge University Press, 1938.

Bryden, D. J. "Evidence from Advertising for Mathematical Instrument Making in London, 1556–1714." *Annals of Science* 49 (1992): 301–36.

———. "Early Printed Ephemera of London Instrument Makers: Trade Catalogues." *Bulletin of the Scientific Instrument Society* 64 (2000): 13–16.

Burger, C. P. *Amsterdamsche Rekenmeesters en Zeevaartkundigen in de zestiende eeuw.* Amsterdam: C. L. Van Langenhuysen, 1908.

———. "Oude Hollandsche zeevaart-uitgaven. De oudste uitgaven van het 'Licht der Zeevaert.'" *Tijdschrift voor boek- en bibliotheekwezen* 6 (1908): 119–37.

Burke, Peter. "The Renaissance Dialogue." *Renaissance Studies* 3, no. 1 (1989): 1–12.

———. *Venice and Amsterdam.* Cambridge: Polity, 1994.

———. *A Social History of Knowledge: From Gutenberg to Diderot.* Cambridge: Polity, 2000.

Cabantous, Alain. *La mer et les hommes: pêcheurs et matelots dunkerquois de Louis XIV à la Révolution.* Dunkerque: Westhoek-Éditions, 1980.

———. *Dix mille marins face à l'océan: les populations maritimes de Dunkerque au Havre aux XVIIe et XVIIIe siècles (vers 1660-1794): étude sociale.* Paris: Publisud, 1991.

Cajori, Florian. "Comparison of Methods of Determining Calendar Dates by Finger Reckoning." *Archeion* 9 (1928): 31–42.

Campbell-Kelly, Martin, Eleanor Robson, Mary Croarken, and Raymond Flood, eds. *The History of Mathematical Tables from Sumer to Spreadsheets.* Oxford: Oxford University Press, 2003.

Campi, Emidio, Anthony Grafton, Simone de Angelis, and Anja-Silvia Goeing, eds. *Scholarly Knowledge: Textbooks in Early Modern Europe.* Genève: Librairie Droz, 2008.

Cañizares-Esguerra, Jorge, and Erik R. Seeman, eds. *The Atlantic in Global History, 1500-2000.* Upper Saddle River, NJ: Prentice Hall, 2007.

Canova-Green, Marie-Claude. "Dance and Ritual: The *Ballet des Nations* at the Court of Louis XIII." *Renaissance Studies* 9, no. 4 (Dec. 1995): 395–403.

Carlier, José María Blanca. "Los colegios de pilotos, la academia de guardiamarinas y oros centros docentes de la Armada." *Revista de Historia Naval* 11, no. 40 (1993): 41–57.

Caspari, Fritz. *Humanism and the Social Order in Tudor England.* New York: Teachers College Press, 1968.

Cavell, S. A. *Midshipmen and Quarterdeck Boys in the British Navy, 1771-181.* Woodbridge: Boydell Press, 2012.

Chapais, Thomas. *Jean Talon, intendant de la Nouvelle-France (1665-1672).* Québec: Impr. de S.-A. Demers, 1904.

Chapuis, Olivier. *A la mer comme au ciel: Beautemps-Beaupré & la naissance de l'hydrographie moderne, 1700-1850.* Paris: Presses de l'Université de Paris-Sorbonne, 1999.

Charlton, Kenneth. *Education in Renaissance England.* London: Routledge and Kegan Paul, 1965.

Charon, Annie, Thierry Claerr, and François Moureau, eds. *Le livre maritime au siècle des Lumières.* Paris: Presses de l'Université de Paris-Sorbonne, 2005.

Chartier, Roger. *L'éducation en France du XVIe au XVIIIe siècle.* Paris: Société d'édition d'enseignement supérieur, 1976.

Cipolla, Carlo M. *Guns, Sails and Empires: Technological Innovation and the Early Phases of European Expansion, 1400-1700.* [New York]: Minerva Press, 1965.

Clarke, A. W. "Trinity House and Its Relation to the Royal Navy." *Royal United Service Institution Journal* 72 (Feb.–Nov. 1927): 512–20.

Clifton, Gloria. "The London Instrument Makers and the British Navy, 1700–1800." In *Science and the French and British Navies, 1700-1850*, ed. Pieter van der Merwe, 24–33. London: National Maritime Museum, 2003.

Cohen, Margaret. *The Novel and the Sea.* Princeton, NJ: Princeton University, 2010.

Collins, Edward. "Portuguese pilots at the Casa de la Contratación and the Exámenes de Pilotos." *International Journal of Maritime History* 26, no. 2 (2014): 179–92.

Cook, Harold J. *Matters of Exchange: Commerce, Medicine, and Science in the Dutch Golden Age.* New Haven: Yale University Press, 2007.

Cooke, James Herbert. "The Shipwreck of Sir Cloudesley Shovell on the Scilly Islands in 1707, from Original and Contemporary Documents Hitherto Unpublished," Read at a Meeting of the Society of Antiquaries, London, Feb. 1, 1883.

Cormack, Lesley B. *Charting an Empire: Geography at the English Universities, 1580-1620.* Chicago: University of Chicago Press, 1997.

———. "Maps as Educational Tools in the Renaissance." In *The History of Cartography*, ed. J. B. Harley and David Woodward, 3.622–36. Chicago: University of Chicago Press, 2007.

Cormack, Lesley B., S. A. Walton, J. A. Schuster. *Mathematical Practitioners and the Transformation of Natural Knowledge in Early Modern Europe.* Springer, 2017.

Cotter, Charles H. *A History of Nautical Astronomy.* London: Hollis & Carter, 1968.

———. "Early Tabular, Graphical, and Instrumental Methods for Solving Problems of Plane Sailing." *Revista da Universidade de Coimbra* 26 (1978): 105–22.

Crisenoy, J. de. "Les écoles navales et les officiers du vaisseau depuis Richelieu jusqu'à nos jours." *Revue maritime et coloniale* 10, 11 (1864): 759–91; 86–127.

Crone, Ernst. "Het gissen van vaart en verheid, het loggen en de tabaksdoos van Pieter Holm." *De Zee* 50–51 (1928-29): 154–55.

———. "Pieter Holm en zijn Zeevaartschool." *De Zee: Zeevaartkundig Tijdschrift* 52 (1930): 136–44, 185–95, 270–80, 352–62, 416–24, 489–97, 560–68, 642–51, 704–16.

———. *Cornelis Douwes, 1712-1773: Zijn Leven en Zijn Werk; Met Inleidende Hoofdstukken over Navigatie en Zeevaart-Onderwijs in de 17de en 18de eeuw.* Haarlem: H. D. Tjeenk Willink & zoon, 1941.

———. "Pieter Holm and His Tobacco Box." Ed. Edwin Pugsley, tr. Dirk Brouwer. Mystic, CT: Marine Historical Association, 1953.

———. "Afgunst en Ruzie Tussen Leermeesters in de Stuurmanskunst." Offprint from *Vereeniging Nederlandsch Historisch Scheepvaart Museum Amsterdam: Jaargang 1959, 1960 en 1961.* Amsterdam, 1962.

———. "How Did the Navigator Determine the Speed of His Ship and the Distance Run?" *Nederlands Vereniging Zeegeschiedenis* 19 (1969): 5–12. Repr. *Revista da Universidade de Coimbra* 24 (1969): 1–17.

Crowther, Kathleen M., and Peter Barker. "Training the Intelligent Eye: Understanding Illustrations in Early Modern Astronomy Texts." *Isis* 104, no. 3 (Sept. 2013): 429–70.

Crowther, Kathleen, Ashley Nicole McCray, Leila McNeill, Amy Rodgers, and Blair Stein. "The Book Everybody Read: Vernacular Translations of Sacrobosco's *Sphere* in the Sixteenth Century." *Journal for the History of Astronomy* 46, no. 1 (Feb. 2015): 4–28.

Cruikshank, E. A. *The Life of Sir Henry Morgan.* Toronto: Macmillan, 1935.

Cuesta Domingo, Mariano, ed. *La obra cosmográfica y náutica de Pedro de Medina.* Madrid: BCH, 1998.

Dackerman, Susan, et al., *Prints and the Pursuit of Knowledge in Early Modern Europe.* Cambridge, MA: Harvard Art Museums, 2011.

Dainville, François de. *La géographie des humanistes.* Paris: Beauchesne et ses fils, 1940.

———. "L'Enseignement des mathématiques dans les collèges jésuites de France du XVIe au XVIIIe siècle." *Revue d'histoire des sciences et de leurs applications* 7, nos. 1–2 (1954): 6–21, 109–23.

———. *L'Éducation des jésuites (XVIe–XVIIIe siècles)*. Paris: Éditions de Minuit, 1991.

———. "L'Instruction des Gardes de la marine à Brest en 1692." *Revue d'histoire des sciences et de leurs applications* 9, no. 4 (1956): 323–38.

Darby, Madge. "John Gore's 'Young One' " *Cook's Log*, 21.4 (1998): 1550.

Darnton, Robert. "First Steps toward a History of Reading." *Australian Journal of French Studies* 23 (1986): 5–30.

Daston, Lorraine. "Nationalism and Scientific Neutrality under Napoleon." In *Solomon's House Revisited: The Organization and Institutionalization of Science*, ed. Tore Frängsmyr, 95–119. Canton, MA: Science History Publications, 1990.

———, ed. *Biographies of Scientific Objects*. Chicago: University of Chicago Press, 2000.

———, ed. *Things That Talk: Object Lessons from Art and Science*. New York: Zone Books, 2004.

Daston, Lorraine, and Elizabeth Lunbeck, eds. *Histories of Scientific Observation*. Chicago: University of Chicago Press, 2011.

Daumas, Maurice. *Les instruments scientifiques aux XVII et XVIII siècles*. Paris: Presses Universitaires de France, 1953.

———. *Scientific Instruments of the 17th & 18th Centuries*. Tr. Mary Holbrook. New York: Praeger, 1972.

Davids, C. A. "Van Anthonisz tot Lastman. Navigatieboeken in Nederland in de zestiende en het begin van de zeventiende eeuw." In *Lucas Jansz. Waghenaer van Enckhuysen. De maritieme cartografie in de Nederlanden in de zestiende en het begin van de zeventiende eeuw*, ed. C. Koeman et al., 73–88. Enkhuizen: Vrienden van het Zuiderzeemuseum, 1984.

———. "Het Zeevaartkundig Onderwijs Voor de Koopvaardij in Nederland Tussen 1795 en 1875. De Rol van het Rijk, de Lagere Overheid en het Particuliere Initiatief." *Tijdschrift voor Zeegeschiedenis* 4 (1985): 164–90.

———. *Zeewezen en Wetenschap: De Wetenschap en de Ontwikkeling van de Navigatietechniek in Nederland Tussen 1585 en 1815*. Amsterdam: De Bataafsche Leeuw, 1986.

———. "Het navigatieonderwijs aan personeel van de VOC." In *De VOC in de kaart gekeken: cartografie en navigatie van de Verenigde Oostindische Compagnie 1602-1799*, ed. Patrick van Mil and Mieke Scharloo, 65–74. Den Haag: SDU, 1988.

———. "On the Diffusion of Nautical Knowledge from the Netherlands to North-Eastern Europe, 1550–1850." In *From Dunkirk to Danzig: Shipping and Trade in the North Sea and the Baltic, 1350-1850: Essays in Honour of J. A. Faber*, ed. W. G Heeres and J. A Faber, 217–36. Amsterdamse Historische Reeks 5. Hilversum: Verloren Publishers, 1988.

———. "Finding Longitude at Sea by Magnetic Declination on Dutch East Indiamen, 1596–1795." *American Neptune* 50, no. 4 (1990): 281–90.

———. "Universiteiten, Illustre Scholen en de Verspreiding van Technische Kennis in Nederland, eind 16e-begin 19e eeuw." *Batavia Academica* 8 (1990): 3–34.

———. "Ondernemers in kennis. Het zeevaartkundig onderwijs in de Republiek gedurende de zeventiende eeuw." *De zeventiende eeuw* 7, no. 1 (1991): 37–48.

———. "The Transfer of Technology between Britain and the Netherlands, 1700–1850." In *Anglo-Dutch Mercantile Marine Relations, 1700-1850: Ten Papers*, ed. J. R. Bruijn and W. F. J. Mörzer Bruyns, 7–24. Amsterdam: Rijksmuseum Nederlands Scheepvaartmuseum, 1991.

———. "The Bookkeeper's Tale. Learning Merchant Skills in the Northern Netherlands in the Sixteenth Century." In *Education and Learning in the Netherlands, 1400-1600*, ed. Jaap Moolenbroek, Koen Goudriaan, and Ad Tervoort, 235–52. Leiden: Brill, 2004.

———. *The Rise and Decline of Dutch Technological Leadership: Technology, Economy and Culture in the Netherlands, 1350-1800*. 2 vols. Leiden: Brill, 2008.

Davids, C. A., John Everaert, and Jan Parmentier, eds. *Peper, Plancius en Porselein: De reis van het schip Swarte Leeuw naar Atjeh en Bantam, 1601-1603*. Zutphen: Walburg Pers, 2003.

Davids, C. A., J. A. van der Veen, and E. A. de Vries. "Van Lastman naar Gietermaker. Claes Hendricksz. Gietermaker (1621–1667) en zijn leerboeken voor de stuurmanskunst." *Tijdschrift voor Zeegeschiedenis* 8, no. 2 (1989): 149–77.

Davies, J. D. *Gentlemen and Tarpaulins: The Officers and Men of the Restoration Navy*. Oxford: Clarendon Press, 1991.

———. *Kings of the Sea: Charles II, James II and the Royal Navy*. Barnsley, UK: Seaforth Publishing, 2017.

Davis, Natalie Zemon. "Sixteenth-Century French Arithmetics on the Business Life." *Journal of the History of Ideas* 21, no. 1 (Mar. 1960): 18–48.

Dear, I. C. B., and Peter Kemp, eds. *The Oxford Companion to Ships and the Sea*. New York: Oxford University Press, 2005.

Dear, Peter. *Discipline and Experience: The Mathematical Way in the Scientific Revolution*. Chicago: University of Chicago Press, 1995.

———. "The Meanings of Experience." In *The Cambridge History of Science: Early Modern Science*, ed. Katharine Park and Lorraine Daston, 3.106–31. Cambridge: Cambridge University Press, 2006.

Dekker, Elly. *Globes at Greenwich: A Catalogue of the Globes and Armillary Spheres in the National Maritime Museum, Greenwich*. Oxford: Oxford University Press and the National Maritime Museum, 1999.

———. "Epact Tables on Instruments: Their Definition and Use." *Annals of Science* 50, no. 4 (July 1993): 303–24.

———. "Globes in Renaissance Europe." In *History of Cartography*, ed. J. B. Harley and David Woodward, 3.135–73. Chicago: University of Chicago Press, 2007.

Dening, Greg. *Mr Bligh's Bad Language: Passion, Power, and Theatre on the Bounty*. Cambridge: Cambridge University Press, 1992.

Depping, Georges Bernard, ed. *Correspondances administratives sous le règne de Louis XIV* 4, no. 1 (1855): 558–59.

Dickinson, Harry W. *Educating the Royal Navy: Eighteenth- and Nineteenth-Century Education for Officers*. Hove, Sussex: Psychology Press, 2007.

Dickson, Rod, ed. *HMS Guardian and the Island of Ice: The Lost Ship of the First Fleet and Lieutenant Edward Riou, 1789-1790*. Carlisle, W. Australia: Hesperian Press, 2012.

Dijksterhuis, E. J. *The Mechanization of the World Picture*. Oxford: Clarendon Press, 1961.

Dingemanse, Clazina. "Rap van Tong, Scherp van Pen: Literaire Discussiecultuur in Nederlandse Praatjespamfletten (circa 1600–1750)." Ph.D. dissertation, Universiteit van Utrecht, 2008.

Doe, Erik van der, Perry Moree, Dirk J. Tang, and Peter de Bode, eds. *Buitgemaakt en teruggevonden: Nederlandse brieven en scheepspapieren in een Engels archief*. Zutphen: Walburg Pers, 2013.

Driessen-van het Reve, Jozien J. "Wat tsaar Peter de Grote (1672–1725) leerde van Holland en de Hollanders." *Leidschrift Rusland en Europa. Westerse invloeden op Rusland*, 24.2 (2009): 15–33.

Driver, Felix, and Luciana Martins. "John Septimus Roe and the Art of Navigation, ca. 1815–1830." *History Workshop Journal*, no. 54 (Autumn 2002): 144–61.

Dubiez, F. J. " 'Int Schrijfboek' (I), De Amsterdamse boekdrukker en boekverkoper Cornelis Claeszoon." *Ons Amsterdam* 12 (1960): 206–13.

Dull, Jonathan R. *The French Navy and the Seven Years' War*. Lincoln: University of Nebraska Press, 2005.

———. *The Age of the Ship of the Line: The British & French Navies, 1650-1815*. Lincoln: University of Nebraska Press, 2009.

Dunn, Richard, and Rebekah Higgitt. *Ships, Clocks, and Stars: The Quest For Longitude*. New York: Harper Design, 2014.

———, eds. *Navigational Enterprises in Europe and Its Empires, 1730-1850*. Basingstoke: Palgrave Macmillan, 2015.

Dúo, Gonzalo. "L'enseignement de la science nautique en Labourd au XVIIIe siècle." *Zainak: Cuadernos de Antropología-Etnografía*, Arrantza eta itsasoa Euskal Herrian = La Pêche et la Mer en Euskal Herria = La Pesca y el Mar en Euskal Herria 21 (2002): 411–18.

Duzer, Chet Van. *The World for a King: Pierre Desceliers' Map of 1550*. London: British Library Publishing, 2015.

Eamon, William. *Science and the Secrets of Nature: Books of Secrets in Medieval and Early Modern Culture*. Princeton, NJ: Princeton University Press, 1994.

Eddy, Matthew D. "The Shape of Knowledge: Children and the Visual Technologies of Paper Tools in the Late Enlightenment." *Science in Context* 26, no. 2 (June 2013): 215–45.

Eisenstein, Elizabeth. *The Printing Press as an Agent of Change*. Cambridge: Cambridge University Press, 1982.

Ellerton, Nerida F., and M. A. (Ken) Clements. *Rewriting the History of School Mathematics in North America 1607-1861: The Central Role of Cyphering Books*. New York: Springer Science & Business Media, 2012.

———. *Samuel Pepys, Isaac Newton, James Hodgson, and the Beginnings of Secondary School Mathematics: A History of the Royal Mathematical School Within Christ's Hospital, London 1673-1868*. New York: Springer, 2017.

Elliott, John H. *Empires of the Atlantic World: Britain and Spain in America, 1492-1830*. New Haven: Yale University Press, 2006.

Enenkel, K. A. E., and Wolfgang Neuber. *Cognition and the Book: Typologies of Formal Organisation of Knowledge in the Printed Book of the Early Modern Period*. Leiden: Brill, 2004.

Enthoven, V., S. Murdoch, E. Williamson, and B. C. O. N. Teensma. *The Navigator: The Log of John Anderson, VOC Pilot-Major, 1640-1643*. Leiden: Brill, 2010.

Fauque, Danielle. "Les écoles d'hydrographie en Bretagne au XVIIIe siècle." *Mémoires de la Société d'histoire et d'archéologie de Bretagne* 78 (2000): 369–400.

———. "Pierre Bouguer et L'Affaire du Jaugeage.' 1721–1726." *Revue d'histoire des Sciences* 63, no. 1 (2010): 23.

Feingold, Mordechai. *The Mathematicians' Apprenticeship: Science, Universities and Society in England, 1560-1640*. Cambridge: Cambridge University Press, 1984.

Ferguson, Eugene S. *Engineering and the Mind's Eye*. Cambridge, MA: MIT Press, 1992.

Ferreiro, Larrie D. *Ships and Science: The Birth of Naval Architecture in the Scientific Revolution, 1600-1800*. Cambridge, MA: MIT Press, 2006.

———. "Spies versus Prizes: Technology Transfer between Navies in the Age of Trafalgar." *Mariner's Mirror* 93.1 (2007): 16–27.

Feyfer, Henk J. de. *Het Licht der Zeevaert: Friese Bijdragen aan het Zeevaartonderwijs*. Leeuwarden: Miedema Pers, 1974.

Finamore, Daniel, ed. *Maritime History as World History*. Salem, MA: Peabody Essex Museum, 2004.

Findlen, Paula, and Pamela H. Smith, eds. *Merchants and Marvels: Commerce and the Representation of Nature in Early Modern Europe*. New York: Routledge, 2002.

Fischer, Lewis R., ed. *The Market for Seamen in the Age of Sail*. Research in Maritime History, no. 7. St. John's, Nfld.: International Maritime Economic History Association, 1994.

Forbes, R. B. (Robert Bennet). *Seamen, Past and Present: A General Compilation of Opinions on Their Condition and the Means for Their Improvement*. Boston: James F. Cotter & Co., 1878.

Franklin, James. "Diagrammatic Reasoning and Modelling." In *1543 and All That*, ed. Guy Freeland and Anthony Corones, 53–116. Dordrecht: Kluwer, 2000.

Frasca-Spada, Marina, and Nicholas Jardine, eds. *Books and the Sciences in History*. Cambridge: Cambridge University Press, 2000.

Fury, Cheryl. *Tides in the Affairs of Men: The Social History of Elizabethan Seamen, 1580-1603*. Westport, CT: Greenwood Press, 2002.

———, ed. *The Social History of English Seamen, 1485-1649*. Woodbridge: Boydell Press, 2011.

Gaastra, Femme S. *The Dutch East India Company: Expansion and Decline*. Zutphen: Walburg Pers, 2003.

Garralón, Marta García. *Taller de Mareantes: El Real Colegio Seminario de San Telmo de Sevilla (1681-1847)*. Sevilla: Fundación Cajasol, 2008.

———. "The Education of Pilots for the Indies Trade in Spain during the Eighteenth Century." *International Journal of Maritime History* 21, no. 2 (Dec. 2009): 189–220.

———. "La formación de los pilotos de la carrera de Indias en el siglo XVIII." *Anuario de Estudios Atlánticos* 1, no. 55 (Jan. 2009): 159–228.

Geistdoerfer, Patrick. "La formation des officiers de marine: de Richelieu au XXIe siècle, des gardes aux 'bordaches' " *Techniques & Culture. Revue semestrielle d'anthropologie des techniques*, 45 (June 2005).

Gelderblom, Oscar. *Zuid-Nederlandse Kooplieden en de Opkomst van de Amsterdamse Stapelmarkt, 1578-1630*. Amsterdam: Uitgeverij Verloren, 2000.

Gent, Rob H. van. "Het Sterrenlied in het Hollandse Zeevaartonderwijs. Berijmde Instructies voor het Vinden van de Sterrenbeelden en het Uur van de Nacht." *Gewina* 28 (2005): 208–21.

Gerritsen, Johan. "Printing at Froben's: An Eye-Witness Account." *Studies in Bibliography* 44 (Jan. 1991): 144–63.

Gingerich, Owen. "Astronomical Paper Instruments with Moving Parts." In *Making Instruments Count: Essays . . . Presented to Gerard L'Estrange Turner*, ed. R. G. W. Anderson, J. A. Bennett, and W. F. Ryan, 63–74. Aldershot, UK: Variorum, 1993.

Glete, Jan. *Navies and Nations: Warships, Navies, and State Building in Europe and America, 1500-1860*. Stockholm: Almqvist & Wiksell International, 1993.

Grendler, Paul F. *Schooling in Renaissance Italy: Literacy and Learning, 1300-1600*. Baltimore: Johns Hopkins University Press, 1989.

Grier, Jason. "Navigation, Commercial Exchange and the Problem of Long-Distance Control in England and the English East India Company, 1673-1755." Ph.D. dissertation, York University, 2018.

Groebner, Valentin. *Who Are You? Identification, Dissimulation, Surveillance in Early Modern Europe*. New York: Zone Books, 2007.

Guibert, Michel Claude. *Mémoires pour servir à l'histoire de la ville de Dieppe*. Ed. Michel Hardy. 2 vols. Dieppe: A. Renaux, A. Leblanc, 1878.

Guillén Tato, Julio F. *Europa Aprendió a Navegar en Libras Españoles*. Barcelona: Instituto Historico de Marina, 1943.

Gulizia, Stefano. "Printing and Instrument Making in the Early Modern Atlantic, 1520–1600: The Origin and Reception of Pedro de Medina's Navigation Manual." *Nuncius* 31, no. 1 (Jan. 2016): 129–62.

Haan, F. de. "De 'Academie de Marine' te Batavia, 1743–1755." *Tijdschrift voor Indische Taal-, Land- en Volkenkunde* 38 (1895): 551–621.

Hanna, Mark. *Pirate Nests and the Rise of the British Empire, 1570-1740*. Chapel Hill: University of North Carolina Press, 2015.

Harkness, Deborah E. *The Jewel House: Elizabethan London and the Scientific Revolution*. New Haven: Yale University Press, 2007.

Harris, G. G. *The Trinity House of Deptford, 1514-1660*. London: Athlone Press, 1969.

Hattendorf, John B., ed. *The Oxford Encyclopedia of Maritime History*. Oxford: Oxford University Press, 2007.

Hébert, Elisabeth, ed. *Le traité de navigation de Jean-Baptiste Denoville, 1760, édition prestige*. 2 vols. Paris: Edition Point de vues, 2008.

Helden, Albert Van, and Thomas L. Hankins. "Introduction: Instruments in the History of Science." *Osiris* 2, no. 9 (1994): 1–6.

Hicks, Robert D. "Navigating the *Mary Rose*." In *The Archaeology of the Mary Rose*, vol. 2, *Mary Rose: Your Noblest Shippe. Anatomy of a Tudor Warship*, ed. P. Marsden, 346–60. Portsmouth: Mary Rose Trust, 2009.

———. *Voyage to Jamestown: Practical Navigation in the Age of Discovery*. Annapolis, MD: Naval Institute Press, 2011.

Hilster, Nicolàs de. "The Hoekboog (Double Triangle): A Reconstruction." *Bulletin of the Scientific Instrument Society* 108 (2011): 20–33.

Holtz, Grégoire. "Hakluyt in France: Pierre Bergeron and Travel Writing Collections." In *Richard Hakluyt and Travel Writing in Early Modern Europe*, ed. Daniel Carey and Claire Jowitt, ch. 5. Farnham, Surrey, UK: Ashgate, 2012.

Hoogendoorn, Klaas. *Bibliography of the Exact Sciences in the Low Countries from ca. 1470 to the Golden Age (1700)*. Leiden: Brill, 2018.

Hope, Ronald. *A New History of British Shipping*. London: J. Murray, 1990.

Howell, Colin D., and Richard J. Twomey, eds. *Jack Tar in History: Essays in the History of Maritime Life and Labour*. Fredericton, NB: Acadiensis Press, 1991.

Howse, H. Derek. "Some Early Tidal Diagrams." *Revista da Universidade de Coimbra* 33 (1985): 365–85.

Hunter, Michael Cyril William. *John Aubrey and the Realm of Learning*. New York: Science History Publications, 1975.

Iliffe, Rob. "Mathematical Characters: Flamsteed and Christ's Hospital Royal Mathematical School." In *Flamsteed's Stars: New Perspectives on the Life and Work of the First Astronomer Royal, 1646-1719*, ed. Frances Willmoth, 115–44. Woodbridge, Suffolk: Boydell Press with the National Maritime Museum, 1997.

Israel, Jonathan. *Dutch Primacy in World Trade, 1585-1740*. Oxford: Clarendon Press, 1989.

———. *Conflicts of Empires: Spain, the Low Countries and the Struggle for World Supremacy, 1585-1713*. London: Hambledon Press, 1997.

———. *The Dutch Republic: Its Rise, Greatness, and Fall, 1477-1806*. New York: Oxford University Press, 1998.

Janzen, Olaf. "A World-Embracing Sea: The Oceans as Highways, 1604–1815." In *Maritime History as World History*, ed. Daniel Finamore, 102–14. Salem, MA: Peabody Essex Museum; Gainesville: University Press of Florida, 2004.

Jardine, Lisa. *Going Dutch: How England Plundered Holland's Glory*. London: HarperPress, 2008.

Jiménez, Elisa María Jiménez. *El Real Colegio Seminario de San Telmo de Sevilla, 1681-1808: su contribución al tráfico marítimo con América y su significado en la historia de la ciudad en el siglo XVIII*. Seville: Universidad de Sevilla, 2002.

Johns, Adrian. "How to Acknowledge a Revolution." *American Historical Review* 107, no. 1 (Feb. 2002): 106–25.

Johnson, Donald, and Juha Nurminen. *The History of Seafaring: Navigating the World's Oceans*. London: Conway, 2010.

Johnston, Stephen. "Making Mathematical Practice: Gentlemen, Practitioners and Artisans in Elizabethan England." Ph.D. dissertation, Cambridge University, 1994.

Jonkers, A. R. T. "North by Northwest: Seafaring, Science, and the Earth's Magnetic Field (1600–1800)." Ph.D. dissertation, Vrije Universiteit, 2000.

———. *Earth's Magnetism in the Age of Sail*. Baltimore: Johns Hopkins University Press, 2003.

Karp, Alexander, and Gert Schubring. *Handbook on the History of Mathematics Education*. New York: Springer Science & Business Media, 2014.

Keller, Vera. "Art Lovers and Scientific Virtuosi?" *Nuncius* 31, no. 3 (Jan. 2016): 523–48.

Kennedy, Ludovic. "The Log of the Guardian, 1789–1790." *Naval miscellany* 4 (1952): 301–55.

Kennerley, Alston, and Percy Seymour. "Aids to the Teaching of Nautical Astronomy and Its History from 1600." *Paedagogica Historica: International Journal of the History of Education* 36, no. 1 (2000): 151–75.

Keuning, Johannes. "Cornelis Anthonisz." *Imago Mundi* 7 (1950): 51–65.

Kindred, Sheila Johnson. "James Cook: Cartographer in the Making 1758–1762." *Royal Nova Scotia Historical Society Journal* 12 (2009): 54–81.

Kirk, Rudolf. *Mr. Pepys upon the State of Christ-Hospital*. Philadelphia: University of Pennsylvania Press, 1935.

Knight, C. "Navigation Instruction in 1677." *Mariner's Mirror* 16, no. 4 (1930): 422–23.

Knighton, C. S. *Pepys and the Navy*. Stroud: Sutton, 2003.

———. "A Century on: Pepys and the Elizabethan Navy." *Transactions of the Royal Historical Society* 14 (2004): 141–51.

Koeman, Cornelis. "Lucas Janszoon Waghenaer: A Sixteenth Century Marine Cartographer." *Geographical Journal* 131, no. 2. (June 1965): 202–12.

———. "Flemish and Dutch Contributions to the Art of Navigation in the XVIth Century." *Revista da Universidade de Coimbra* 34 (1988): 493–504.

Kool, Marjolein. *Die conste vanden getale: Een studie over Nederlandstalige rekenboeken uit de vijftiende en zestiende eeuw, met een glossarium van rekenkundige termen*. Hilversum: Verloren, 1999.

Koyré, Alexandre. *From the Closed World to the Infinite Universe*. Baltimore: Johns Hopkins University Press, 1958.

Krogt, C. J. van der. *Advertenties voor Kaarten, Atlassen, Globes e.d. in Amsterdamse Kranten, 1621-1811*. Utrecht: HES, 1985.

Kusukawa, Sachiko, and Ian Maclean. *Transmitting Knowledge: Words, Images, and Instruments in Early Modern Europe*. Oxford: Oxford University Press, 2006.

Laan, K. ter. *Letterkundig woordenboek voor Noord en Zuid*. Den Haag and Djakarta: G. B. van Goor Zonen's Uitgeversmaatschappij, 1952.

Lamb, Ursula. "Science by Litigation: A Cosmographic Feud." *Terrae Incognitae* 1, no. 1 (Jan. 1969): 40–57.

———. *Cosmographers and Pilots of the Spanish Maritime Empire*. Aldershot, Hampshire: Ashgate Variorum, 1995.

Lane, Kris E. *Pillaging the Empire: Piracy in the Americas, 1500-1750*. Armonk, NY: M. E. Sharpe, 1998.

Laprise, Raynald. *Le Diable Volant: l'histoire, l'époque, la vie et les moeurs des flibustiers, pirates et corsaires de la Jamaïque . . . durant la seconde moitié du XVIIe siècle*. Québec, 2000–2006. http://membre.oricom.ca/yarl/Proue/H/H.html (Oct. 2018).

Larkin, Jill H., and Herbert A. Simon. "Why a Diagram Is (Sometimes) Worth Ten Thousand Words." *Cognitive Science* 11, no. 1 (1987): 65–100.

Latham, Robert, ed. *Catalogue of the Pepys Library at Magdalene College, Cambridge*. 7 vols. in 11 vols. Cambridge: D. S. Brewer, 1978.

Latour, Bruno. *Science in Action: How to Follow Scientists and Engineers through Society*. Cambridge, MA: Harvard University Press, 1987.

Law, John. "On the Methods of Long-Distance Control: Vessels, Navigation and the Portuguese Route to India." In *Power, Action and Belief: A New Sociology of Knowledge?*, ed. John Law 234–63. London: Routledge and Kegan Paul, 1986.

———. "Technology and Heterogeneous Engineering: The Case of Portuguese Expansion." In *The Social Construction of Technological Systems*, ed. Wiebe Bijker, Thomas Hughes, and Trevor Pinch, 111–34. Cambridge, MA: MIT Press, 1987.

Lefèvre, Wolfgang, ed. *Picturing Machines 1400-1700*. Cambridge, MA: MIT Press, 2004.

Le Bouëdec, Gerard. *Activités maritimes et societés littorales de l'Europe atlantique*. Paris: Armand Colin, 1997.

Le Goff, T. J. A. "The Labour Market for Sailors in France." In *"Those Emblems of Hell"? European Sailors and the Maritime Labour Market, 1570-1870*, ed. P. C. van Royen, Jaap R. Bruijn, Jan Lucassen, 287–327. St. John's, Nfld.: International Maritime Economic History Association, 1997.

Lesger, Clé. *The Rise of the Amsterdam Market and Information Exchange: Merchants, Commercial Expansion and Change in the Spatial Economy of the Low Countries, ca. 1550-1630*. London: Ashgate, 2006.

Levot, P. "Les écoles d'hydrographie de la marine à la XVIIe siècle." *Revue maritime et coloniale* 44 (1875): 165–69.

Levy-Eichel, Mordechai. " 'Into the Mathematical Ocean': Navigation, Education, and the Expansion of Numeracy in Early Modern England and the Atlantic World." Ph.D. dissertation, Yale University, 2015.

Linebaugh, Peter, and Marcus Buford Rediker. *The Many-Headed Hydra: Sailors, Slaves, Commoners, and the Hidden History of the Revolutionary Atlantic*. Boston: Beacon Press, 2000.

Long, Pamela O. "Power, Patronage, and the Authorship of *Ars*: From Mechanical Know-How to Mechanical Knowledge in the Last Scribal Age." *Isis* 88, no. 1 (Mar. 1997): 1–41.

———. *Openness, Secrecy, Authorship: Technical Arts and the Culture of Knowledge from Antiquity to the Renaissance*. Baltimore: Johns Hopkins University Press, 2001.

———. *Artisan/Practitioners and the Rise of the New Sciences, 1400-1600*. Portland: Oregon State University Press, 2011.

Long, Pamela O., David McGee, Alan M. Stahl, and Franco Rossi. *The Book of Michael of Rhodes: A Fifteenth-Century Maritime Manuscript*. Cambridge, MA: MIT Press, 2009.

López Piñero, José María. *Ciencia y técnica en la sociedad española*. Barcelona: Editorial Labor, 1979.

———. *El arte de navegar en la España del Renacimiento*. Barcelona: Editorial Labor, 1986.

Lottum, Jelle, van. "The Necessity and Consequences of Internationalisation: Maritime Work in the Dutch Republic in the 17th and 18th Centuries." In *The Sea in History: The Early Modern World / La mer dans l'histoire: La période moderne*, ed. Christian Bouchet and Gérard Le Bouëdec, 3.839–51, Woodbridge: Boydell Press, 2017.

Lunsford, Virginia West. *Piracy and Privateering in the Golden Age Netherlands.* Houndmills, Basingstoke, Hampshire: Palgrave Macmillan, 2005.

Lutun, Bernard. "Des Ecoles de marine et principalement des écoles d'hydrographie (1629–1789)." *Sciences et techniques en perspective* 34 (1995): 3–30.

MacGregor, Arthur. "The Tsar in England: Peter the Great's Visit to London in 1698." *Seventeenth Century* 19, no. 1 (Spring 2004): 116–47.

Maddison, Angus. *The World Economy: A Millennial Perspective.* Washington, DC: Brookings Institution Press, 2001.

Mahoney, Michael S. "Christiaan Huygens: The Measurement of Time and Longitude at Sea." In *Studies on Christiaan Huygens*, ed. H. J. M. Bos et al., 234–70. Lisse: Swets, 1980.

Marley, David F. *Pirates of the Americas.* Santa Barbara: ABC-CLIO, 2010.

Marquardt, Karl Heinz. *Eighteenth-Century Rigs & Rigging.* London: Conway Maritime Press, 1992.

Marr, Alexander. *Between Raphael and Galileo: Mutio Oddi and the Mathematical Culture of Late Renaissance Italy.* Chicago: University of Chicago Press, 2011.

May, William Edward. "The Last Voyage of Sir Clowdisley Shovel." *Journal of Navigation* 13 (1960): 324–32.

———. *A History of Marine Navigation.* Henley-on-Thames, Oxfordshire: G. T. Foulis, 1973.

McGee, David. "From Craftsmanship to Draftsmanship: Naval Architecture and the Three Traditions of Early Modern Design." *Technology and Culture* 40 (1999): 209–36.

McHugh, Evan. *The Drovers.* London: Penguin, 2010.

McNeill, John Robert. *Atlantic Empires of France and Spain: Louisbourg and Havana, 1700-1763.* Chapel Hill: University of North Carolina Press, 1985.

Mémain, René. *La marine de guerre sous Louis XIV. Le matériel. Rochefort, arsenal modèle de Colbert.* Paris: Librairie Hachette, 1937.

Merton, Robert K. *Science, Technology and Society in Seventeenth Century England.* Bruges, Belgium: Saint Catherine Press, 1938.

Meskens, Ad. "Michiel Coignet's Nautical Instruction." *Mariner's Mirror* 78 (1992): 257–76.

———. "Mathematics Education in Late Sixteenth-Century Antwerp." *Annals of Science* 53, no. 2 (Mar. 1996): 137.

———. *Practical Mathematics in a Commercial Metropolis: Mathematical Life in Late 16th Century Antwerp.* Dordrecht: Springer, 2013.

Mills, Allan A. "The 'Eye Error' of the Cross Staff, with a Method for Calculating the Original Dimensions of Modified or Missing Parts." *Bulletin of the Scientific Instrument Society* 48 (1996): 15–18.

Molhuysen, P. C., and P. J Blok, eds. *Nieuw Nederlandsch biografisch woordenboek.* 10 vols. Leiden: Sijthoff, 1911.

Morison, Samuel Eliot. *The Great Explorers: The European Discovery of America.* New York: Oxford University Press, 1978.

Mörzer Bruyns, Willem F. J. "Nederlands Zeevaartkundeboeken in de Periode 1800–1945." *Tijdschrift voor Zeegeschiedenis*, no. 3 (1985): 235–47.

———. *The Cross-Staff: History and Development of a Navigational Instrument.* Amsterdam: Vereeniging Nederlandsch Historisch Scheepvaart Museum; Zutphen: Walburg Pers, 1994.

———. *Schip Recht door Zee: De octant in de Republiek in de achttiende eeuw.* Amsterdam: Koninklijke Nederlandse Akademie van Wetenschappen, 2003.

Mosley, Adam. "Objects, Texts and Images in the History of Science." *Studies in History and Philosophy of Science Part A* 38, no. 2 (June 2007): 289–302.

Mukerji, Chandra. "Tacit Knowledge and Classical Technique in Seventeenth-Century France: Hydraulic Cement as a Living Practice among Masons and Military Engineers." *Technology and Culture* 47, no. 4 (2006): 713–33.

Murray, Alexander. *Reason and Society in the Middle Ages.* Oxford: Clarendon Press, 1978.

Nash, M. D., ed. *The Last Voyage of the Guardian, Lieutenant Riou, Commander, 1789-1791.* 2nd ser. 20. Cape Town: Van Riebeeck Society, 1989.

Navarro García, Luis. "Pilotos, maestres y señores de naos de la Carrera de Indias." *Archivo Hispalense* (Seville) 46–47, nos. 141–46 (1967): 241–95.

———. "La gente de mar en Sevilla en el siglo XVI." *Revista de Historia de América* (Mexico) 67–68 (1969): 1–64.

Navarro Brotóns, Víctor. *Bibliographia physico-mathematica hispanica (1475-1900).* Valencia: C.S.I.C., 1999.

Navarro-Loidi, Juan, and José Llombart. "The Introduction of Logarithms into Spain." *Historia Mathematica* 35, no. 2 (May 2008): 83–101.

Netten, D. H. (Djoeke) van. "Een boek als carrièrevehikel. De zeemansgidsen van Blaeu." *Cultuur in de Nederlanden in interdisciplinair perspectief* 27, no. 2 (Mar. 14, 2012): 214–31.

———. *Koopman in Kennis: De Uitgever Willem Jansz Blaeu.* Zutphen: Walburg Pers, 2014.

Nicolson, Marjorie Hope. *Pepys' Diary and the New Science.* Charlottesville: University Press of Virginia, 1965.

Nieborg, J. P. *Indië en de zee: De opleiding tot zeeman in Nederlands-Indie, 1743-1962.* Amsterdam: De Bataafsche Leeuw, 1989.

Neuville, Didier. *Les Établissements scientifiques de l'ancienne marine.* Paris: Berger-Levrault, 1882.

Ong, Walter J. *Orality and Literacy: The Technologizing of the Word.* Ed. princ. 1982. London: Routledge, 2002.

Oosten, F. C. van. " 'Hear Instruction and Be Wise': The History of a Naval College on Java in the Eighteenth and Nineteenth Century." *Mariner's Mirror* 55.3 (Aug. 1969): 247–62.

Oosterhoff, Richard J. *Making Mathematical Culture: University and Print in the Circle of Lefevre d'Etaples.* New York: Oxford University Press, 2018.

Parias, Louis-Henri, and René Rémond, eds. *Histoire générale de l'enseignement et de l'éducation en France.* Paris: Nouvelle Librairie de France, 1981.

Pearce, Ernest Harold. *Annals of Christ's Hospital.* London: Methuen, 1901.

[Pech de Cadel, Flavien]. *Histoire de l'École navale et des institutions qui l'ont précédée par un ancien officier.* Paris: Quantin, 1889.

Pérez-Mallaína Bueno, Pablo Emilio. "Los libros de náutica españoles del siglo XVI y su influencia en el descubrimiento y conquista de los océanos." In *Ciencia, vida y espacio en Iberoamérica*, ed. José Luis Peset, 457–84. Madrid: C.S.I.C., 1989.

———. *Spain's Men of the Sea: Daily Life on the Indies Fleets in the Sixteenth Century.* Tr. Carla Rahn Phillips. Baltimore: Johns Hopkins University Press, 1998.

Perl-Rosenthal, Nathan. *Citizen Sailors.* Cambridge, MA: Harvard University Press, 2015.

Peters, Hubert J. M. W. *The Crone Library: Books on the Art of Navigation Left by Dr. Ernst Crone to the Scheepvaart Museum in 1975 and Books on the Same Subject Acquired by the Museum Previously.* Amsterdam: De Graaf/Nieuwkoop, published for the Vereeniging Nederlandsch Historisch Scheepvaart Museum, 1989.

Phillips, Edward J. *The Founding of Russia's Navy: Peter the Great and the Azov Fleet, 1688-1714.* Westport, CT: Greenwood Press, 1995.

Plumley, N. "The Royal Mathematical School within Christ's Hospital: The Early Years. Its Aims and Achievements." *Vistas in Astronomy* 20, no. 1 (1976): 51–56.

Poelje, Otto van. "Gunter Rules in Navigation." *Journal of the Oughtred Society* 13 no. 1 (2004): 11–22.

Polak, Jean. *Bibliographie maritime française: depuis les temps les plus reculés jusqu'à 1914*. Grenoble: Éditions des ₄ seigneurs, 1976.

Polak, Jean, and Michèle Polak. *Bibliographie maritime française: depuis les temps les plus reculés jusqu'à 1914. Supplément*. Grenoble: J. P. Debbane, 1983.

Polak, Michèle. "Les livres de marine en français au XVIe siècle." In *La France et la mer au siècle des grandes découvertes*, ed. Philippe Masson and Michel Vergé-Franceschi, 41–54. Paris: Tallandier, 1993.

Polanyi, Michael. *Personal Knowledge: Towards a Post-Critical Philosophy*. New York: Harper & Row, 1964.

Porteman, Karel. "The Use of the Visual in Classical Jesuit Teaching and Education." *Paedagogica Historica: International Journal of the History of Education* 36, no. 1 (2000): 178–196.

Portuondo, María M. *Secret Science: Spanish Cosmography and the New World*. Chicago: University of Chicago Press, 2009.

Postma, Johannes, and Victor Enthoven, eds. *Riches from Atlantic Commerce: Dutch Transatlantic Trade and Shipping, 1585-1817*. Leiden: Brill 2003.

Price, Audrey. "Mathematics and Mission: Deciding the Role of Mathematics in the Jesuit Curriculum." *Jefferson Journal of Science and Culture* 1, no. 4 (2016): 29–40.

Price, Leah. "Reading: The State of the Discipline." *Book History* 7 (2004): 303–20.

Pritchard, James S. "French Developments in Hydrography with Particular Reference to the St. Lawrence River during the Reign of Louis XIV (1665–1709)." MA thesis, University of Western Ontario, 1965.

———. *Anatomy of a Naval Disaster: The 1746 French Naval Expedition to North America*. Montreal: McGill-Queen's University Press, 1995.

Pulido Rubio, José. *El Piloto Mayor de la Casa de la Contratación de Sevilla, Pilotos Mayores del Siglo XVI (datos Biográficos)*. Sevilla: Tip. Zarzuela, 1923.

———. *El Piloto mayor de la Casa de la Contratación de Sevilla: pilotos mayores, catedráticos de cosmografía y cosmógrafos*. Sevilla: Escuela de Estudios Hispano-Americanos, 1950.

Quinn, David Beers. *The Hakluyt Handbook*. Vol. 1. London: Hakluyt Society, 1974.

Randles, W. G. L. "From the Mediterranean Portulan Chart to the Marine World Chart of the Great Discoveries." *Imago Mundi* 40 (1988): 115–18.

———. "Pedro Nuñes and the Discovery of the Loxodromic Curve." *Revista da Universidade de Coimbra* 35 (1990): 119–30.

Ransom, Roger L. "British Policy and Colonial Growth: Some Implications of the Burden from the Navigation Acts." *Journal of Economic History* 28, no. 3 (Sept. 1968): 427–35.

Rasor, Eugene L. *English/British Naval History to 1815: A Guide to the Literature*. Westport, CT: Greenwood, 2004.

Raven, G. J. A., and N. A. M. Rodger, eds. *Navies and Armies: The Anglo-Dutch Relationship in War and Peace, 1688-1988*. Edinburgh: J. Donald, 1990.

Redford, Duncan, ed. *Maritime History and Identity: The Sea and Culture in the Modern World*. London: I. B. Tauris, 2014.

Rediker, Marcus Buford. *Between the Devil and the Deep Blue Sea: Merchant Seamen, Pirates, and the Anglo-American Maritime World, 1700-1750*. Cambridge: Cambridge University Press, 1989.

———. *Villains of All Nations: Atlantic Pirates in the Golden Age*. Boston: Beacon Press, 2004.

Reeves, Eileen. *Galileo's Glassworks: The Telescope and the Mirror*. Cambridge, MA: Harvard University Press, 2008.

Ridder-Symoens, Hilde de, ed. *A History of the University in Europe*. Vol. 2, *Universities in Early Modern Europe (1500-1800)*. Cambridge: Cambridge University Press, 1996.

Rivington, Charles A. *Pepys and the Booksellers*. York, England: Sessions Book Trust, 1992.

Roberts, Lissa Louise, Simon Schaffer, and Peter Robert Dear, eds., *The Mindful Hand: Inquiry and Invention from the Late Renaissance to Early Industrialisation*. Amsterdam: Koninklijke Nederlandse Akademie van Wetenschappen, 2007.

Robson, John. "Edward Riou (1762–1801)." *Cook's Log* 34, no. 2 (2011): 47.

Roche, John J. "Harriot's 'Regiment of the Sun' and Its Background in Sixteenth-Century Navigation." *British Journal for the History of Science* 14, no. 3 (Nov. 1981): 245–61.

Rodger, N. A. M. *The Wooden World: An Anatomy of the Georgian Navy*. London: Collins, 1986.

———. "Lieutenants' Sea-Time and Age." *Mariner's Mirror* 75 (1989): 269–72.

———. *The Safeguard of the Sea: A Naval History of Britain*. London: HarperCollins in association with the National Maritime Museum, 1997.

———. "Commissioned Officers' Careers in the Royal Navy, 1690–1815." *Journal for Maritime Research* 3, no. 1 (Dec. 2001): 85–129.

———. "Honour and Duty at Sea, 1660–1815." *Historical Research* 75, no. 190 (Nov. 2002): 425–47.

———. *The Command of the Ocean: A Naval History of Britain, 1649-1815*. London: Allen Lane, 2004.

———. "Queen Elizabeth and the Myth of Sea-Power in English History." *Transactions of the Royal Historical Society* 14 (2004): 153–74.

———. "From the 'Military Revolution' to the 'Fiscal-Naval State.'" *Journal for Maritime Research* 13, no. 2 (2011): 119–28.

Rodríguez, António Acosta, Adolfo Luis González Rodríguez, and Enriqueta Vila Vilar. *La Casa de la Contratación y la Navegación entre España y las Indias*. Seville: Universidad de Sevilla, 2003.

Rommelse, Gijs. *The Second Anglo-Dutch War (1665-1667): Raison d'état, Mercantilism and Maritime Strife*. Hilversum: Verloren, 2006.

Rose, Susan. "Mathematics and the Art of Navigation: The Advance of Scientific Seamanship in Elizabethan England." *Transactions of the Royal Historical Society* 14, no. 6 (Sept. 2002): 175–84.

———. "English Seamanship and the Atlantic Crossing c. 1480–1500: Was the Crossing of the Atlantic beyond the Capabilities of English Seamen in the Second Half of the Fifteenth Century?" *Journal for Maritime Research* 4, no. 1 (Sept. 2002): 127–38.

Rösler, Irmtraud. "'De Seekarte Ost VND West to Segelen...': On Northern European Nautical 'Fachliteratur' in the Late Middle Ages." *Early Science and Medicine* 3, no. 2 (1998): 103–18.

Roy, Joseph-Edmond. "La cartographie et l'arpentage sous le régime français." *Bulletin des recherches historiques* 1 (1895): 49–56.

Roy, Pierre-Georges. "Jean Deshayes, hydrographe du Roi." *Bulletin des recherches historiques* 22, no. 5 (1916): 129–38.

Royen, P. C. van. *Zeevarenden op de Koopvaardijvloot: Omstreeks 1700*. Amsterdam: Bataafsche Leeuw, 1987.

———. "Foreigners Aboard the Dutch Merchant Marine around 1700." In *From Dunkirk to Danzig: Shipping and Trade in the North Sea and the Baltic, 1350-1850: Essays in Honour of*

J. A. Faber . . . , ed. W. G. Heeres and J. A. Faber. 391–404. Hilversum: Verloren Publishers, 1988.

———. "Personnel of the Dutch and English Mercantile Marine (1700–1850) An Introductory Paper." In *Anglo-Dutch Mercantile Marine Relations, 1700-1850: Ten Papers*, ed. J. R. Bruijn and W. F. J. Mörzer Bruyns, 103–14. Amsterdam: Rijksmuseum Nederlands Scheepvaartmuseum, 1991.

———. "Mariners and Markets in the Age of Sail: The Case of the Netherlands." In *Market for Seamen in the Age of Sail,* ed. Lewis R. Fischer, 47–58. Liverpool: Liverpool University Press, 1994.

Royen, P. C. van, Jaap R. Bruijn, and Jan Lucassen, eds. *"Those Emblems of Hell"? European Sailors and the Maritime Labour Market, 1570-1870.* St. John's, Nfld.: International Maritime Economic History Association, 1997.

Ruddock, Alwyn A. "The Trinity House at Deptford in the Sixteenth Century." *English Historical Review* 65, no. 257 (Oct. 1950): 458–76.

Russo, François. "L'Enseignement des sciences de la navigation dans les écoles d'hydrographie aux XVIIe et XVIIIe siècles." In *Deuxième colloque international d'histoire maritime, le navire et l'économie du Moyen Age au XVIIIe siècle, principalement en Méditerranée,* ed. M. Mollat, 177–93. Paris: S.E.V.P.E.N, 1958.

Salmond, Anne. *The Trial of the Cannibal Dog: Captain Cook in the South Seas*. London: Allen Lane, 2003.

Sandman, Alison. "Educating Pilots: Licensing Exams, Cosmography Classes, and the Universidad de Mareantes in Sixteenth Century Spain." In *Ars Nautica: Fernando Oliveira and His Era; Humanism and the Art of Navigation in Renaissance Europe (1450-1650)*, ed. I. Guerreiro and F. Contente Domingues, 99–110. Cascais, Portugal: Patrimonia, 1999.

———. "Cosmographers vs. Pilots: Navigation, Cosmography, and the State in Early Modern Spain." Ph.D. dissertation, University of Wisconsin, Madison, 2001.

Sandman, Alison, and Eric H. Ash. "Trading Expertise: Sebastian Cabot between Spain and England." *Renaissance Quarterly* 57, no. 3 (Autumn 2004): 813–46.

Sawers, Larry. "The Navigation Acts revisited." *Economic History Review,* 45.2 (May 1992): 262–84.

Schaffer, Simon. "Astronomers Mark Time: Discipline and the Personal Equation." *Science in Context* 2, no. 1 (1988): 115–45.

Schapelhouman, Marijn. "David Vinckboons." In *The Dictionary of Art*, ed. Jane Turner, 32.586–88. New York: Grove, 1996.

Scharloo, Mieke, and Patrick van Mil, eds. *De VOC in de kaart gekeken: cartografie en navigatie van de Verenigde Oostindische Compagnie, 1602-1799.* 's-Gravenhage: SDU, 1988.

Schepper, Susanna L. B. de. " 'Foreign' Books for English Readers: Published Translations of Navigation Manuals and Their Audience in the English Renaissance, 1500–1640." Ph.D. dissertation, University of Warwick, 2012.

Schilder, Günter. "Cornelis Claesz (c. 1551–1609): Stimulator and Driving Force of Dutch Cartography." *Monumenta Cartographica Neerlandica*, vol. 7. Alphen aan den Rijn: Canaletto/Repro-Holland, 2003.

———. *Early Dutch Maritime Cartography: The North Holland School of Cartography*. Leiden: Hes & De Graaf Pub B.V., 2017.

Schilder, Günther, and Marco van Egmond, "Maritime Cartography in the Low Countries during the Renaissance." In *History of Cartography*, ed. J. B. Harley and David Woodward, 3.1384–1432. Chicago: University of Chicago Press, 2007.

Schmidt, Benjamin. *Innocence Abroad: The Dutch Imagination and the New World, 1570-1670*. Cambridge: Cambridge University Press, 2001.

Schmidt, Suzanne Karr. "Art — A User's Guide: Interactive and Sculptural Printmaking in the Renaissance." Ph.D. dissertation, Yale University, 2006.

Schmidt, Suzanne Karr, and Kimberly Nichols. *Altered and Adorned: Using Renaissance Prints in Daily Life*. Chicago: Art Institute of Chicago, 2011.

Schotte, Margaret. "A Hydrographer's Library: Jean Deshayes and Navigational Education in Early 18th-Century Québec." MA thesis, University of Toronto, 2007.

———. "Expert Records: Nautical Logbooks from Columbus to Cook." *Information & Culture: A Journal of History* 48, no. 3 (2013): 281–322.

———. "Leçons enrégimentés: l'évolution du journal maritime en France / Regimented Lessons: The Evolution of the Nautical Logbook in France." *Annuaire de Droit Maritime et Océanique* 31 (June 2013): 91–115.

———. "A Calculated Course: Creating Transoceanic Navigators, 1580–1800." Ph.D. dissertation, Princeton University, 2014.

———. "Sailors, States, and the Co-Creation of Nautical Knowledge." In *A World at Sea: Maritime Practices in Global History, 1500-1900*, ed. L. Benton and N. Perl-Rosenthal. Philadelphia: University of Pennsylvania Press, forthcoming.

———. "Ships' Instruments in the Age of Print: Making Markets and Meaning." In *Routledge Research Companion to Marine and Maritime Worlds, 1400-1800: Oceans in Global History and Culture*, ed. Claire Jowitt, Craig Lambert, and Steve Mentz. London: Routledge, forthcoming.

Selm, Bertus van. *Een menighte treffelijcke boecken. Nederlandse boekhandelscatalogi in het begin van de zeventiende eeuw*. Utrecht: HES, 1987.

Serjeantson, Richard. "Proof and Persuasion." In *The Cambridge History of Science: Early Modern Science*, ed. Katharine Park and Lorraine Daston, 3.132–75. Cambridge: Cambridge University Press, 2006.

Sherman, Claire Richter, and Peter M. Lukehart, eds. *Writing on Hands: Memory and Knowledge in Early Modern Europe*. Carlisle, PA: Trout Gallery of Dickinson College, 2000.

Shirley, John William, and F. David Hoeniger, eds. *Science and the Arts in the Renaissance*. Washington, DC: Folger Shakespeare Library, 1985.

Silverberg, Joel. "Fathers of American Geometry: Nathaniel Bowditch and Benjamin Peirce." *Proceedings of the Canadian Society for History and Philosophy of Mathematics* 25 (2012): 154–75.

Sir Francis Drake's Nautical Almanack, 1546. London: Nottingham Court Press and Magdalene College, Cambridge, 1980.

Skelton, R. A. "Captain James Cook as a Hydrographer." *Mariner's Mirror* 40, no. 2 (1954): 91–119.

Smith, Pamela H. *The Body of the Artisan: Art and Experience in the Scientific Revolution*. Chicago: University of Chicago Press, 2004.

Smith, Pamela H., and Benjamin Schmidt. *Making Knowledge in Early Modern Europe: Practices, Objects, and Texts, 1400-1800*. Chicago: University of Chicago Press, 2007.

Sobel, Dava. *Longitude: The True Story of a Lone Genius Who Solved the Greatest Scientific Problem of His Time*. New York: Walker, 1995.

Soll, Jacob. *The Information Master: Jean-Baptiste Colbert's Secret State Intelligence System*. Ann Arbor: University of Michigan Press, 2009.

Sommervogel, Carlos, and Augustin de Backer. *Bibliothèque de la Compagnie de Jésus: nouvelle édition*. 11 vols. Bruxelles: Oscar Schepens, 1890.

Sonar, Thomas. "The 'Regiments' of Sun and Pole Star: On Declination Tables in Early Modern England." *GEM—International Journal on Geomathematics* 1, no. 1 (Aug. 2010): 5–21.

Sorrenson, Richard. "The Ship as a Scientific Instrument in the Eighteenth Century." *Osiris*, ser. 2, 11 (1996): 221–36.

Spence, Jonathan D. *The Memory Palace of Matteo Ricci*. New York: Penguin Books, 1985.

Struik, D. J. "Mathematics in the Netherlands during the First Half of the XVIth Century." *Isis* 25, no. 1 (May 1936): 46–56.

Suarez, Michael F., and H. R. Woudhuysen, eds. *The Book: A Global History*. Oxford: Oxford University Press, 2014.

Sullivan, F. B. "The Naval Schoolmaster during the 18th Century and the Early 19th Century." *Mariner's Mirror* 62, no. 3 (1976): 311–26.

Sumira, Sylvia. *Globes: 400 Years of Exploration, Navigation, and Power*. Chicago: University of Chicago Press, 2014.

Taillemite, Étienne. *Dictionnaire des Marins Français*. Paris: Editions maritimes & d'outre-mer, 1982.

———. *Marins français à la découverte du monde: de Jacques Cartier à Dumont d'Urville*. Paris: Fayard, 1999.

Tanner, J. R., ed. *A Descriptive Catalogue of the Naval Manuscripts in the Pepysian Library at Magdalene College, Cambridge*. London: Printed for the Navy Records Society, 1903.

Taton, René. *Enseignement et diffusion des sciences en France au XVIII siècle*. Paris: Hermann, 1964.

Taylor, E. G. R. [Eva Germaine Rimington]. "The Early Navigator." *Geographical Journal* 113 (June 1949): 58–61.

———. *The Mathematical Practitioners of Tudor & Stuart England*. Cambridge: Institute of Navigation at the University Press, 1954.

———. "John Dee and the Nautical Triangle, 1575." *Journal of Navigation* 8, no. 4 (1955): 318–25.

———. *The Haven-Finding Art: A History of Navigation from Odysseus to Captain Cook*. London: Hollis and Carter for the Institute of Navigation, 1956.

———. *The Mathematical Practitioners of Hanoverian England, 1714-1840*. London: Cambridge University Press for the Institute of Navigation, 1966.

Teitler, G. *The Genesis of the Professional Officers' Corps*. Beverly Hills, CA: Sage, 1977.

Terrot, Noël. *Histoire de l'éducation des adultes en France: la part de l'éducation des adultes dans la formation des travailleurs, 1789-1971*. Paris: L'Harmattan, 1997.

Thomas, Keith. "Numeracy in Early Modern England. The Prothero Lecture." *Transactions of the Royal Historical Society*, 5th ser. 37, no. 1 (1987): 103–32.

Thornton, Cliff. "John Gore's 'Young One'—An Update." *Cook's Log* 30, no. 2 (2007): 34.

Thrower, Norman. "Samuel Pepys FRS (1633-1703) and the Royal Society." *Notes and Records of the Royal Society of London* 57, no. 1 (2003): 3–13.

Toulouse, Sarah. "Marine Cartography and Navigation in Renaissance France." In *The History of Cartography*, ed. David Woodward, 3.1550–68, ch. 52. Chicago: University of Chicago Press, 2007.

Tribble, Evelyn. "'The Chain of Memory': Distributed Cognition in Early Modern England." *SCAN | Journal of Media Arts Culture* 2, no. 2 (Sept. 2005): 135–55.

Trollope, William. *A History of the Royal Foundation of Christ's Hospital: With an Account of the Plan of Education, the Internal Economy of the Institution, and Memoirs of Eminent Blues*. London: William Pickering, 1834.

Turner, Anthony John. "Advancing Navigation in Eighteenth-Century France: Teaching and Instrument-Making in the Port of Rochefort." *Mariner's Mirror* 91, no. 4 (Nov. 2005): 531–47.

Turner, J. C., and P. C. van de Griend, eds. *History and Science of Knots*. Singapore: World Scientific, 1996.

Valleriani, Matteo. *Galileo Engineer*. Berlin: Springer Verlag, 2010.

Vergé-Franceschi, Michel. "Un enseignement éclaire au XVIIIe siècle: l'enseignement maritime dispensé aux gardes." *Revue Historique* [France] 276, no. 1 (1986): 29–56.

———. *Marine et éducation sous l'ancien régime*. Paris: Editions du Centre national de la recherche scientifique, 1991.

———. *La marine française au XVIIIe siècle: guerres, administration, exploration*. Paris: SEDES, 1996.

———. "Entre ciel et mer: des marins, la tête dans les étoiles. En mer, la science fait reculer la nuit." In *La nuit dans l'Angleterre des Lumières*, ed. Suzy Halimi, 49–71. Paris: Presses Sorbonne Nouvelle, 2009.

Verlinden, C. "De Nederlandse vertaling van het 'Arte de Navegar' van Pedro de Medina en de nautische 'Onderwijsinghe' van Michiel Coignet (Antwerpen, 1580)." *Collectanea Maritima* 3 (1987): 5–20.

Verwey, H. de la Fontaine. "Amsterdamse uitgeversbanden van Cornelis Claesz en Laurens Jacobsz." *Jaarboek van het Genootschap Amstelodamum* 65 (Amsterdam, 1973): 56–72.

Vries, Jan de, and Ad van der Woude. *The First Modern Economy*. Cambridge: Cambridge University Press, 1997.

Walton, Gary M. "The New Economic History and the Burdens of the Navigation Acts." *Economic History Review* 24, no. 4 (Nov. 1971): 533–42.

Ward, Robin. *The World of the Medieval Shipmaster: Law, Business and the Sea, c.1350–c.1450*. Woodbridge, UK: Boydell Press, 2009.

Warnsinck, J. C. M. *De Kweekschool voor de Zeevaart en de Stuurmanskunst 1785-1935*. 's-Gravenhage: Vaderlandsch Fonds ter Aanmoediging van 's Lands Zeedienst, 1935.

Warwick, Andrew. *Masters of Theory: Cambridge and the Rise of Mathematical Physics*. Chicago: University of Chicago Press, 2003.

Waters, David Watkin. "The Development of the English and the Dutchman's Log." *Journal of the Institute of Navigation* 9, no. 1 (1956): 70–88.

———. *The Art of Navigation in England in Elizabethan and Early Stuart Times*. New Haven: Yale University Press, 1958.

———. *The Rutters of the Sea: The Sailing Directions of Pierre Garcie; A Study of the First English and French Printed Sailing Directions, with Facsimile Reproductions*. New Haven: Yale University Press, 1967.

———. *The Iberian Bases of the English Art of Navigation in the Sixteenth Century*. Coimbra: UC Biblioteca Geral 1, 1970.

———. "Renaissance Cosmography." *History of Science* 12, no. 3 (1974): 227–230.

———. *Science and the Techniques of Navigation in the Renaissance*. London: National Maritime Museum, 1976.

———. *English Navigational Books, Charts and Globes Printed down to 1600*. Lisbon: Instituto de Investigaçao Científica Tropical, 1984.

White, C. E. "Calendar Mnemonics and Mnemonic Calendars." *Journal of the Royal Astronomical Society of Canada* 36 (Apr. 1942): 133–36.

Willmoth, Frances. *Sir Jonas Moore: Practical Mathematics and Restoration Science*. Woodbridge, Suffolk: Boydell & Brewer, 1993.

———. *Flamsteed's Stars: New Perspectives on the Life and Work of the First Astronomer Royal, 1646-1719*. Woodbridge, Suffolk: Boydell Press with the National Maritime Museum, 1997.

Wise, M. Norton. *The Values of Precision*. Princeton, NJ: Princeton University Press, 1995.

Wolfe, Charles T., and Ofer Gal, eds. *The Body as Object and Instrument of Knowledge: Embodied Empiricism in Early Modern Science*. Dordrecht: Springer, 2010.

Wright, John Middleton. *Dead Reckoning Navigation*. London: Coles, 1968.

Yates, Frances Amelia. *The Art of Memory*. Chicago: University of Chicago Press, 1966.

Yeldham, F. "An Early Method of Determining Dates by Finger Reckoning." *Archeion* 9 (1928): 325–26.

Yeo, Richard. *Notebooks, English Virtuosi, and Early Modern Science*. Chicago: University of Chicago Press, 2014.

Youngs, Tim, and Peter Hulme, eds. *The Cambridge Companion to Travel Writing*. Cambridge: Cambridge University Press, 2002.

Zilsel, Edgar. "The Origins of William Gilbert's Scientific Method." *Journal of the History of Ideas* 2, no. 1 (Jan. 1941): 1–32.

———. *The Social Origins of Modern Science*. Ed. Diederick Raven, Wolfgang Krohn, and Robert S. Cohen. Boston: Springer, 2000; Kluwer, 2003.

Index

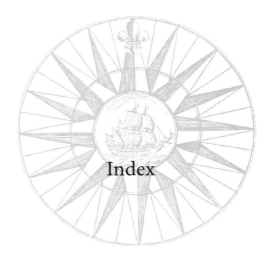

Page numbers in *italics* indicate figures. References to plates are preceded by *pl.*

accounting, 42, 53, 101, 112, 180, 217n77
accuracy: compasses, 244n88; cross-staffs, 211n11; dead reckoning, 80, 154; Denys on, 84; illustrations, 107; increasing, 80, 170; Lastman on, 216n67, 244n90; maps, 46, 50–51; Riou on, 166; tables, 46; volvelles, 16, 199n27
Adams, George, *162*, 242n56
Adams, John, 240n32
Adams, Thomas, 157
administrators' roles, 8, 196n38
algebra, 101, 103, 154, 176
almanacs: early, 35, 53, 100, 153; maritime use, 9, *60*, 161; outdated, 54, 97
altitude: calculating latitude with double altitudes, 161; calculating longitude, 153; Dutch approach, 52, 56, 57, 117, 131, 135, 136; English approach, 95, 111, *172*, 173; French approach, 72, 84; globes, 110; Riou observations, 161, 166; sextants and octants, 152
amateur as term, 46, 156
Americas, imprints, 22, 240n36
amplitude, 96, 97, 135, 155
Amsterdam Admiralty, 121
Anhaltin, Christiaen M., *122*

Anthonisz., Cornelis, 45
Apianus, Peter, 49, 56, 239n22
apprenticeships, 41–42, 95, 237n94
arithmetic, 11, 99, 101, 111, 113, 117, 157
Armada, Spanish, 41, 53
artisans, 6, 8, 11, 201n8. *See also* tacit knowledge
Asson, Assuerus van, *114*, 115, 124–27, *125*, 133, 135–36, 177, 180
astrolabes, 39, 53, 97, 215n50
astronomy: observations, 40, 123, 153; Ptolemaic, 18, 50; in schools, 111, 143; science of, 39, 87; shift to math, 99, 102. *See also* cosmography; ephemerides; regiments; stars
Atlantic Ocean, 17, 28, 59, 68, 71, 96–97, 214n43
atlases: images, 28, *44*, *123*; innovation, 50–51, 69, 204n49; reference works, 107, 179. *See also* Blaeu, Willem Janszoon; charts; Waghenaer, Lucas Janszoon
Australia, voyage to, 149, 160
Avril, Vincent, 212n21

backstaffs, 39, *39*, 111, 170, *172*, 173, 203n36, 241n54

287

balance of theory and practice: England, 23, 96, 99, 102–5, 156, 174–76; France, 84–91, 173–76; general shifts in, 5, 7–8, 12, 15, 173–77; Netherlands, 41, 44, 47, 61, 116–17, 120, 122–23, 143–44, 174–76; in Riou, 170–71; Spain, 22–23; in student workbooks, 181, 183–84. *See also* hands-on training
Banks, Joseph, 160–61, 242n62
battles, naval, 71, 169
Beeckman, Isaac, 231n13
Berthelot, F., 210n2, 211n16
Bezie, Jean, 246n27
Bezout, Étienne, 176
Bion, Nicholas, *78*, 236n80
Blaeu, Willem Janszoon: on accuracy of maps and tables, 46; as examiner, 131; globe instruction, 110; as publisher, 27, 142, 178; texts, 43–45, *44*, 47, 209n101, 228n88; on Waghenaer, 50–51
Blagrave, John, 206n65
Bligh, William, 241n47
Blondel Saint-Aubin, Guillaume, 89, 210n8, 216n66, 219n107
Blundeville, Thomas, 100
Boer, Jan de, 176
Boissaye du Bocage, Georges, 108, 210n8, 228n90
books: aboard ship, 150, *172*, 220n18, 226n72, 235n62, 243n73; absent topics in, 57–58; annotated, *36*, 234n53; critics of, 10, 23, 46, 89; interleaved, *125*, 136, 246n29; shift to learning from, 4, 11, 61, 144–47, 177–80, 184. *See also* logbooks; nautical manuals; notebooks; publishing; workbooks
bookshops, 48–50, *50*, 54, 74, 104, 177
Borough, Stephen, 24
Boteler, Nathaniel, 6–7, 217n85
Bougard, René, 71, 84
Bouguer, Pierre, 244n89, 246n28
Bourne, William, 23, 100, 202n20, 202nn22–23, 207n73, 226n66
Bowditch, Nathaniel, 197n44, 237n97
Breen, Joost van: on estimating speed, 139, 140; as examiner, 8, 120; lack of math in approach, 194n16; text, 2, 3–4, *4*, 7–8, 139; on theory vs. practice, 120
Briggs, Henry, 210n7, 211n14
Broucke, Jan van den, *36*, 54–57, *55*, *56*, 58, 106, 110

Brouscon, Guillaume, 73

calendars: Dutch approach, 11, 59, 117, 142; English approach, 100, 155, 223n35; in exams, 96, 155; on Holm's box, *141*; rhymes, 221n22; terms, 35–36
captains: career, 9; exams, 131; negligent, 218n94; role, 30; wages, 201n12
cartography. *See* atlases; charts; Mercator charts; portolans
Carvalho da Costa, António, 211n15
Casa de la Contratación, 17–25, 105. *See also* education, Spain
Cats, Jacob, 8, 120, 231n14
celestial navigation: math required, 30, 176; techniques, 18, 38–41. *See also* cosmography; latitude; longitude
Champlain, Samuel de, 73
Channel, The, 59, 71, 84, 94, 133, 160
Charles IX, 212n20
charts: and dead reckoning, 35; Dutch approach, 45, 50, *50*, 118, 120, 131, 134, 135, 136, 201n8, pl. 1; English approach, 96, 97, 155; French approach, 226n65; portolans, 38, 215n50; printing improvements, 10, 47; as source, 9; Spanish approach, 23, 199n25, 199n29, 228n91; spread of usage, 9, 40–41, 202n26. *See also* Corrections; Mercator charts
Child, Thomas, 110
Christ's Hospital, 95, 101
Claesz., Cornelis, 42, 48–54, 177, 203n32, 207n73
Clements, Thomas, 149–50, *151*, 169
clocks. *See* timekeeping
coastal profiles, 38, 51, *162*, 205n57, pl. 2
Coignet, Michiel: on accuracy of printed materials, 205n59; career, 207n80; on charts, 204n43; on large navigation, 39; on longitude, 40; on small navigation, 38; translation and appendix to Medina, 24, *48*, 49, 51–53, 56, 57
Colbert, J. B. A., Marquis de Seignelay, 86
Colbert, Jean-Baptiste, 64, 69–71, 72, 74, 84–85, 86–89, 181
Collins, John, 83, 89
Colom, Jacob Aertsz., *123*
Colson, John, 95, 111
common navigation. *See* small navigation

compass: accuracy, 244n88; "boxing," 202n18; compass rose, *32,* 38, 57–58, 117, *118, 182*; Dutch approach, 31, 45, 57–58, 117, *118,* 120, 136; English approach, 96; French approach, 57, 219n96, 226n65; on *Guardian,* 242n56; portolans, 38, 215n50; in workbooks, *182.* See also traverse boards
composed/compound courses. See traverse course diagrams
computation book (Riou), 161, *163,* 164, 166
conscription, 70
Cook, James, 150, 157–58, 160, 165, 177, 242n62
Cooper, Richard, 105
Corrections: Denys, 64, *78, 80,* 81–84, 90, 178; diagrams, *78, 80, 82*; Dulague, 170–71; Jarichs van der Ley, 81–83, 90; Lastman, 83; spread of, 89–90
corsairs. See privateers
Cortés, Martín: dialogues, 226n66; globe instruction, 110; influence of, 24; texts, *16,* 18–19, 21–22, 24, 58, 198n14, 199n27
cosmography: Dutch approach, 52, 58, 117–18, 134, 143, 209n101; English approach, 54, 100, 101–2, 103, 111, 155, 223n34, 240n42; French approach, 53, 84; vs. geography, 198n7; Spanish approach, 11, 18–20, *20, 21,* 58, 59, 179–80
Coubert, G., 108, 110, 210n2, 227n77
counting, metrical, 139–40
Crockett, Richard, 95–96, 110–11, 177
cross-staffs, 39, 45, *50,* 97, 123, 211n11, 226n65
Cuningham, Robert, 22, 37, 53, *54*
currents, in exams, 133, 135
cypher books. See workbooks

Dam, Jan Albertsz. van, 115, 116, 120, 232n27, 234n50, 245n11
Davis, John, 47, 64, 203n36, 212n25, 225n61
Davis quadrant, 95, 97, *98, 106, 172, 173*
dead reckoning: by Riou, 150, 151; techniques, 31–35. See also Corrections
Decker, Ezechiel de, 41, 210n7
De Graaf, Abraham, 115, 118–19, 234n50, 234n53
de Latre [Delastre], Capt., 71
demonstrations in exams, 23, 97
Denys, Guillaume: influence of, 64, 88–91, 178; on speed, 138; as teacher, 63–64, 68, 69–71, 72, 90, 106; texts, 69, 70, 71, 72, 74–84, *75, 77*
Deshayes, Jean, 227n83, 228n86, 232n28
d'Estrées, Jean, 85–86
diagrams: compass rose, *32,* 57–58; Corrections, *78, 80, 82*; in Dutch texts, *34,* 51, 55, *55, 56,* 59, *82,* 145; in English texts, 31, *32,* 101, 127, 221n18; in French texts, *76, 77, 78, 79,* 127; as learning tool, 12, 61, 79, 107–8, 127–30; by Riou, 157; rule of thumb, 36, *37, 55, 55,* 59, 135, 181, *182*; in Spanish texts, 18, 19–21, *20, 21,* 59; traverse diagrams, 123, *126,* 127–30, *128, 129,* pl. 5; traverse tables, *34*; trigonometry, 66, *67, 76, 77, 78,* 79; in workbooks, 115–16, 125, *126,* 135, 136, 157, *159,* 181, *182.* See also illustrations
dialogues: educational, 25, 88, 106, 178; model exams, 131, 133, 134; popular, 46
Dias, Pedro, 25
dictation, 106, 107, 227n74
Diderot, Denis, 193n10, 231n14
Dieppe as hub, 69. See also education, French
Dighton, Robert, *165*
Discovery (ship), 157–58
Doncker, Hendrick, 142, 178
Doublet, Jean-François, 71–72, 106, 177
Douwes, Cornelis, 146, 231n20
drawing: on charts, 40–41; Dutch approach, 127, 129; English approach, 107–8, 111, 112, 157, *159*; French approach, 107–8; as learning tool, 12, 107–8, 127, 179, 180
du Buc [Dubuc], Nicolas, 69, 214n44, 246n21
Dulague, V. F. N., 170–71, 177
Dutch East India Company. See VOC
Dutch West India Company. See WIC

École Royale d'hydrographie, 63–64, 69–71, 72, 212n20
economy, correlated to maritime careers, 28, 29, 68, 100, 120
education: free lectures, 70, 87, 105, 212n19; general, 6, 12, 194n22; length of programs, 23, 174, 219n97; overview of approaches, 5, 6–8; shift to learning from books, 4, 11, 61, 144–47, 177–80, 184; university pedagogy, 5, 13, 18, 25, 99, 101, 104–5, 174. *See also* balance of theory and practice; educators; exams; hands-on training; publishing; *and specific types*

education, Dutch: alignment with exam, 117, 118, 130, 135; approach, 27–28, 43–45, 117, 120–23; balance of theory and practice, 41, 44, 47, 61, 116–17, 120, 122–23, 143–44, 174–76; cram schools, 141–42; 18th c. changes, 153, 174–80; influence on other countries, 104–5, 116, 118, 181; portrayals, *26, 27–28, 43–45, 44, 122*; public lectures, 43, 45, 121. *See also* exams, Dutch; mathematics; publishing, Dutch; trigonometry; workbooks

education, English: apprenticeships, 95, 237n94; approach, 100–113, 153, 174–80; balance of theory and practice, 23, 96, 99, 102–5, 156, 174–76; competition for personnel, 6–7; drawing in, 107–8, 111, 112, 157, *159*; interest in other countries' programs, 5, 24–25, 104–5; late 18th c., 174–80; military tactics in, 241n45; portrayals, *92*; public lectures, 105; of Riou, 156–57; teachers, 45, 157, 244n5; traverse diagrams, 127; workbooks, 107, 157, *158, 159, 177, 183*, pl. 3. *See also* exams, English; mathematics; publishing, English; trigonometry

education, French: approach, 63–64, 69–72, 84–91, 153, 154, 156, 173–80; balance of theory and practice, 84–91, 173–76; competition for personnel, 6, 68–69; dictation, 227n74; influence on other countries, 104–5, 181; instruments, 72, 107, 108, 225n65; interest in other countries' programs, 5, 24, 25; late 18th c., 153, 154, 156, 173–80, 175; model ships, 228n85; notebooks, 226n70; public lectures, 87; traverse diagrams, 127. *See also* exams, French; publishing, French

education, Spain: approach, 17–25, 59, 228n91; cosmography focus, 11, 18–20, 58, 59, 179–80; influence of, 24–25, 51–53, 104–5, 181; late 18th c., 153, 177, 180

educators: as authors, 54, 87, 89, 122, 131, 142; Dutch, 45, 54, 122, 244n5; English, 45, 157, 244n5; exams for, 230n103, 244n5; French, 45, 63–64, 68, 69–71, 86–87, 89, 90, 106, 174, 212n19; Jesuits as, 86, 87; loaning of textbooks, 226n70; qualifications, 7–8, 45, 174; Spanish, 17–18, 24; wages, 70

Eisenstein, Elizabeth, 9

England. *See* education, English; exams, English; publishing, English; Riou, Edward

English East India Company, 200n37

ensigns, French exams, 154–55

epact: in exams, 96, 234n52; on Holm's box, *141*; rule of thumb, 35–37, *37*, 55, *55*; in texts, *37*, 55, *55*, 142, 209n103; in workbooks, 136, *182*

ephemerides, 40, 53, 133, 181. *See also* almanacs

estimation, 32, 80–84, 90, 111, 120, 170. *See also* speed, estimating

Euclid, 19, 72, 77, 101, 103, 106, 135, 230n12

examiners: Dutch, 8, 25, 45, 53, 117, 118, 120, 121, 131; English, 95; French, 25, 68, 154; Spanish, 17–18, 24; textbooks by, 131, 142–43; wages, 23, 234n47

exams: for educators, 230n103, 244n5; 18th c. changes, 174, 175, 176–77; Spanish, 17–18, 23–24, 25. *See also* examiners; exams, Dutch; exams, English; exams, French

exams, Dutch: approach, 130–37, 146, 153–54, 175, 176; for assistant navigators, 30; class alignment with, 117, 118, 121, 130, 135; first, 25, 131; images, *132, 134, 144, 145*; as mandatory, 131, 233n45; oral, *132*, 133; practice exams, 118, 123–27, 131–37, *132, 134*, 178, 239n26; on training ships, 222n23

exams, English: approach, 5, 93–99, 110–11; cosmography, 155; first, 25; instruments, 95, 97; for lieutenants, 93, 96–97, 99, 135, 241n49; for masters, 5, 93–95, 97–99, 135; math in, 95, 97, 103, 106, 110, 155, 156; practice exams, 155–56, 171; Riou, 241n49

exams, French: Doublet, 72; 18th c. changes, 153, 154–55, 175, 176, 177; examiners, 25, 68, 154

expertise: books as evidence of, 7, 54, 116, 178; of educators, 7–8; judgment, 47, 86–87, 90, 169–71; regional variation, 59, 86, 94, 104; replicating, 5, 28, 41, 43, 164; time to develop, 84, 96, 175. *See also* balance of theory and practice

Figueirdo, Manoel de, 212n25
fishing, 6, 68, 71, 121, 200n4, 211n16
Flamsteed, John, 103, 108, 222n26, 223n46
fothering, 150, 164–65, 171
Fournier, Georges, S.J., 73–74

France: conscription, 70; lack of sailing culture, 68–69; need for navigators, 6, 68–69; regulations, 30, 68, 86, 244n2. *See also* education, French; exams, French; publishing, French

Gale, George, 96, 97
García de Palacio, Diego, 22, 226n66
Gardner, Daniel, pl. 9
gentlemen vs. tarpaulins, 102, 105, 156
geography vs. cosmography, 198n7
geometry: Dutch approach, 45, 117, 118, 135–36, 137, 143, 233n35; English approach, 95, 99, 101, 103, 157; French approach, 154; Spanish approach, 19
Germans, educated in Netherlands, 116
Gerritsz., Adriaen, 207n77
Gerritsz., Hessel, 121
Gietermaker, Claas: *Driehoex-rekening*, 68, 211n14, 234n56; as examiner, 131; influence of, 178; on theory vs. practice, 120; on tides, 209n103; *Vermaeck der Stuerlieden*, 37, 211n14, 234n51; Veur on, 143. *See also 't Vergulde Licht*
globes: depictions of, 27, 28, 44, *44*, *50*, *109*; Dutch approach, 123, 143; English approach, 72, 108–10, *109*, 112, 178; French approach, 178; hands-on math instruction, 110
Golden Number, 35, 37, 53, 55, 142, 143, 224n49. *See also* rule of thumb
Goos, Pieter, 214n49, 236n88
Gore, John, 164, 167
great navigation. *See* celestial navigation
Grenville, Richard, 25
Gresham College, 106, 225nn62–63
Guardian (ship): images, *148*, *152*, *162*, *163*, *165*, *168*, pl. 8; Indian Ocean voyage, 149–52, 160–70
Guards (stars), 40, 52, 58
Guérard, Jean, 211n18
Gunter, Edmund, 64–65, 202n21, 215n61
Gunter's scale, *65*, 72, 157, pl. 3

Hadsell, Charles, 93–95, 99
Haeyen, Albert, 145, 205n54
Hakluyt, Richard, 25, 225n56
Halley, Edmond, 103
hands, counting on. *See* rule of thumb

hands-on training: Dutch, 41–42, 44–45, 120, 122–23; English, 105, 107, 108, 112; French, 88, 108, 173–74; small navigation, 175. *See also* balance of theory and practice; tacit knowledge
Harrison, John, 40, 164
Hartsoeker, Nicholas, 232n27
Hill, Joseph, 232n27
Hobbes, Thomas, 223n40
hoekboog (angle staff), 39, 203n36
Hoffman, H. M., *182*
Holm, Pieter, 122, 140–42
Hooke, Robert, 103
horizon, artificial, 161
Hues, Robert, 110

icebergs, watering with, 159
illustrations: accuracy of, 107; compass rose, 32, 57–58, 117, *118*, *182*; costs, 197n43; Dutch approach, 50–51, 56, 57, 59; as learning tool, 12, 61, 79, 107–8, 127–30, 178–79, 233n40; portrayals of navigators and learning, 2, 3–4, *4*, 26, 27–28, 43–45, *44*, 92, 122, *172*, *173*; printing improvements, 10, 12; ropes, 221n18; rule of thumb, *36*, *37*, *55*, 55, 59, 135, 181, *182*; Spanish approach, 18, 19–21, 57, 58, 59. *See also* diagrams; tables; volvelles
Indian Ocean voyage, 149–52, 160–70
information: crossing borders, 10, 181, 183; overload, 11, 47
instrument makers, 53, *162*, *163*, 227n79, 239n21, 241n54, 242n56
instruments: construction in texts, 19, 53; Dutch approach, 120, 122, 123, 146, 201n8; in Dutch exams, 133, 153–54; in Dutch texts, 2, 4, 45, 53, 143, 194n18; English approach, 106, 107, 108–10, 111–12; in English exams, 95, 97; in English texts, *172*, *173*, 224n49; French approach, 65, *66*, 72, 107, 108, 225n65; on *Guardian,* 150, 161; individualized, 139, 180, 226n65; inventories, 38, 111, 230n9; late 18th c. changes, 175–76, 178; in lieu of math, 65, 89, 141, 146, 157; in Québec, 228n86; in Spanish texts, 16, 18, 19, 22. *See also specific instruments*
internationalism: competition for personnel, 193n10, 213n35; Dutch labor market, 201n11; interest in other countries' pro-

internationalism (*continued*)
grams, 5, 24–25, 104–5, 181; publishing, 10–11

Jarichs van der Ley, Jan Hendrick, 47, 81–83, *82*, 90, 208n85
Java, 121, 234n57, 246n26
Jesuits, 86, 87, 228n91, 233n40
journals. *See* logbooks

Keteltas, Barent Evertsz., 208n85
Kinckel, Hendrik August van, *144*, *145*, 146
knots, speed: Holm's box, *141*; log and line, 32, *78*, 138, 202n23, 235n70
knots, tying: in classes, 175; in exams, 5, 96, 222n24; on ships, 99; in texts, 58, 221n18
Komis, P. R., *119*
Kruik, Dirk, 123–24
Kweekschool voor de Zeevaart (Amsterdam), 121, 123, *124*, 153

Lakeman, Mathijs Sijverts, 216n67
landmarks: estimating speed with, 235n73; navigating by, 31
large navigation. *See* celestial navigation
Lastman, Cornelis: on accuracy, 216n67, 244n90; on authority, 89; on Corrections, 83; on estimating speed, 139, 140; as examiner, 131; on log and line, 202n21; texts, 66, 116, 148, 230n9
latitude: calculating with double altitudes, 161; calculating with stars, 18, 40, 52, 58; calculating with Sun, 18, 39–40, 111, 146; calculating with trigonometry, 65–66; Dutch approach, 45, 120, 136, 146, 153; English approach, 155; French approach, 69, 74, 84, 215n55; latitude sailing, 40, 203n35; Riou observations, 161; Spanish approach, 18; tables, 40. *See also* Corrections; quadrants
lead and line, *78*, 192n2
Le Cordier, Samson, 66, 88–89, 107, 210n2, 228n91
lectures, public, 43, 45, 87, 105, 121
Lévêque, Pierre, 245n6
lieutenants' exam, 93, 96–97, 99, 135, 241n49
Linschoten, Jan Huyghen van, 226n66

literacy: increasing, 97, 144, 174, 177; low, 18, 41, 46, 103
Little Dipper, 203n35. *See also* Guards (stars)
log and line, 2, 3, 4, 32, *78*, 138–39, 236n80
logarithms: adoption of, 65–66; Dutch approach, 62, 66, 133, 137, 143; English approach, 64, 101, 111, 157; French approach, 79, 84; scales, 66, 146; tables, 62, 64, 79
logboards, *172*, *173*, 232n33
logbooks: dead reckoning, 35; in exams, 96; requirements for, 154, 176; Riou, 150, 152, 161–63, *162*, *163*, 166–68, *168*; as source, 9; as term, 202n26; in texts, 42, 53, 64, 87, *172*, *173*; traverse courses, 127
Long, Nathaniel, 95–96, 110–11, 177
longitude: calculating challenges, 39, 40; calculating with Mayer's method, 152–53; calculating with quadrants, 65; contests, 81, 204n41, 216n72; development of Corrections, 81–83; Dutch approach, 47, 153, 201n8; English approach, 166, 225n58, 240n42; French approach, 77, 84; interest in, 183. *See also* magnetic variation
Louis XIV, 86, 88
loxodromes. *See* rhumbs
lunar distance method, 152–53

magnetic variation: calculating longitude, 40; Dutch approach, 131, 133, 134, 136; English approach, 97, 155; French approach, 69, 73, 74; magnets, 72
Mainwaring, Henry, 221nn17–18
Makreel, Dirk, 121, 136, 234n50, 234n53
manuscripts. *See* workbooks
maps: accuracy, 46, 50–51; in Dutch texts, 50–51, pl. 2; as learning tool, 12; master, 199n25. *See also* charts
marine timekeepers, 40, 164, 176
Maskelyne, Nevil, 153, 161, 164
masters: exams, 5, 93–95, 97–99, 135; role, 30; wages, 30
mathematics: avoidance techniques, 61, 65, 89, 141, 146, 157; in celestial navigation, 30, 176; Dutch approach, 112, 117, 120, 133, 137, 144, 146, 175–76; in Dutch texts, 62, 66–68, 118, 142, 143, 211n14, 233n35, 234n56; English approach, 100–104, 111–13, 157, 171; in English exams, 95, 97, 103, 106, 110, 155, 156; in English texts,

64, 66, *67,* 68, 100–103, 108, 155, 223n35; French approach, 71, 74–84, 113, 154, 176; in French texts, 64, 68, 69, 71, 74–84, *75,* 113, 178; increase in usage, 7, 101; mixed, 22, 101, 112; in small navigation, 176; in Spanish texts, 19, 68, 113; speed of calculations, 175–76. *See also* algebra; arithmetic; geometry; trigonometry

Mayer, Tobias, 152–53

McCulloch, Kenneth, *162,* 242n56, 242n56

Medina, Pedro de: career, 198n10, 207n81; influence of, 24, 51–53, 73; texts, 19–22, *20, 21,* 24, 49, 51–53, 56, 57, 58, 73, 100, 199n18

Mediterranean navigation, 10, 59, 68, 88, 94, 101

memorization: Dutch approach, 30–31, 42, 50, 55–56, 57, 59–61, 106; English approach, 106–7, 112; French approach, 226n67, 246n28; importance of, 181; metrical counting, 139–40; mnemonic devices, 12, 57, 58, 117; small navigation, 30–31, 42; Spanish approach, 59; texts as adjuncts to, 11, 12, 47, 106–7; trigonometry, 77

Mercator, Gerhard, 103, 198n13

Mercator charts, 41, 65, 118, 131, 134, 136, 155, 198n13

Méricourt, Nicholas Lefèvre de, 85–86

metaphors: body, 20–21, 58; chariot, 63, 74; clock, 40, 56; garden, 77; lantern, 43

Metius, Adrian, 83, 138

metrical counting, 139–40

military tactics, 154, 155, 241n45

Milliet de Chales, Claude-François, S.J., 79, 87, 211n16, 230n12

mnemonic devices, 12, 57, 58, 117. *See also* memorization; rule of thumb

model exams. *See* practice exams

model ships, 97, 108, 123, *124*

Monson, William, 107, 203n31

Moon: age of, 35, 36, 53, *60,* 95, 96, 136, *141;* calculating longitude, 152–53; observation by Riou, *163, 167;* volvelles, *48,* 53, *60. See also* epact; tides

Moore, John Hamilton, 155–56, 197n44, 245n12

Moore, Jonas, 67, 97, 101, 108

Morgan, Henry, 94

Moxon, Joseph, 232n27

music, *148,* 152. *See also* zodiac songs

nautical manuals: audience, 10, 19, 21, 57–58, 117, 142, 145–47; growing market, 28, 77, 100; number of titles, 10

navigation: importance of, 4–5; inland, 196n33, 200n4, 220n2, 227n83; by landmarks, 31; in ports, 35–37, 38; scholarship on, 5, 8; as term, 200n3. *See also* balance of theory and practice; celestial navigation; education; educators; exams; hands-on training; publishing; small navigation

Navigation Acts, 100

navigators: assistant, 30; career path, 9; need for, 6–7, 17, 41, 68–69, 222n31; scholarship on, 5–6; secrecy of, 43; status of, 3–4, 6, 9, 86, 90, 178; terms for, 197n1, 201n13, 212n21; training by, 42; wages, 3, 8–9, 30

Netherlands: interest in other countries' programs, 5, 24, 25; maritime culture, 28, *29,* 121; need for navigators and sailors, 41, 193n10; respect for navigators, 90; small navigation, 30, 31–38. *See also* education, Dutch; exams, Dutch; publishing, Dutch

New France, 6, 71, 89. *See also* Québec

Newhouse, Daniel, 224n49, vi

Newton, Isaac, 103, 110, 154

Newton, John, *109*

Newton, Samuel, 104, 107, 228n86

Nierop, Dirck Rembrantsz. van, 68, 123

nocturnals, 40, 53, 95, 97, *98,* 111, 227n80

North Star. *See* Pole Star

Norwood, Richard, 66, 202n23

notebooks: blank, 46, 49, 106; captain's, 38, 51; Riou, 152, 160, *160,* 243n71; teachers', 226n70. *See also* workbooks

numeracy: high in Netherlands, 99, 147; low, 23; required, 30, 42, 100, 176

Nuñes, Pedro, 24, 198n13, 214n48

"nurseries" for navigators, 6–7, 105, 121

observations: skilled, 4, 7, 11, 30; taught, 42, 108, 139–40

octants, 142, 145–46, 152, 153, *172,* 173, 242n56

Oostwoud, Govert M., 120

ordonnances, French, 68, 86, 244n2

Oughtred, William, 103, 229n98

Paget, Edward, 95, 104, 110–11, 174
paper tools: blank globes, 110; chart paper, 81; fill-in exams, *134*, *145*, 146; popularity, 178, 210n8. *See also* tables; volvelles
Paterson, James, 157
Pepys, Samuel: on exams, 95, 96; globe purchase, 228n86; interest in other countries' programs, 104–5, 181, 193n14, 232n27; on models, 228n85; on need for navigators, 222n31; RMS role, 102–3, 104–5, 110–11; on training, 193n14
Perkins, Peter, *92*, 101–2, 107, 138, 221n22
Peter I of Russia, 122
Phillippes, Henry, 210n8, 227n78
Phillips, Thomas, 103
physics, 87, 154
pilot term, 197n1. *See also* navigators
pirates, 23, 86, 94
Pitiscus, Bartholomaeus, 64
Plancius, Peter, 27, 205n54
poetry: authors, 3, 231n13; Riou, 160; in textbooks, 62, 101, 142, 208n93
Pole Star (Polaris): calculating latitude, 18, 40, 58; in Dutch texts, 58; in Spanish texts, *16*, 18, 19, 20–21, *21*, 59; timekeeping, 40; volvelles, *16*, 19, *34*, 51, *52*, 58, 199n27
portolans, 38, 215n50
ports, navigating, 35–37, 38
Portugal: education, 24, 211n15, 231n18; latitude sailing, 40
practice exams: Dutch, 118, 123–27, *125*, 131–37, *132*, *134*, 178, 239n26; English, 155–56, 171
practice questions, 107, 115, 118, 131, 142, 178, 180
privateers, 6, 71–72, 94, 107
problem solving: in Dutch texts, 47–48, 118; in French texts, 79–80, 89; multiple methods, 47, 80, 84–85, 108, 157, 181, 216n65; in Spanish exams, 23–24
protractors, 81, 228n91
publishing: audience, 10, 22, 177; 18th c. changes, 156, 177–80; impact of, 9–11, 47, 178; internationalism, 10–11; print culture, 5; printing improvements, 10, 12, 47; regional differences, 11–12; scholarship on, 9–10; types of materials, 9–10. *See also* illustrations; *and specific types*
publishing, Dutch: approach, 10, 11, 28, 41–43, 45–61, 66–68, 116–19, 142–47; audience, 22, 28, 46–47, 101, 112; bookshops, 48–50, 54; cosmography, 52, 58, 117–18, 143, 209n101; illustrations, 50–51, 56, 57, 59; influence, 232n27; late 18th c. changes, 177–80; math in, *62*, 66–68, 118, 142, 143, 211n14, 233n35, 234n56; practice exams, 118, 123–27, 131–37, *132*, *134*, 178, 239n26; practice questions, 115, 118, 142; printing process, 49; rule of thumb, 35–37, *36*, *37*, 117, 135, 233n35; rutters, 203n32, 207n73; support materials, 46, 49; translations, 24, 73, 181, 214n49. *See also* Claesz., Cornelis; workbooks
publishing, English: approach, 10, 11, 100–103; audience, 100–101, 112; basic seamanship, 96, 156; cosmography in, 54, 100, 101–2, 223n34, 240n42; instruments in, *172*, *173*, 224n49; later 18th c., *172*, *173*, 177–80; math in, 64, 66, *67*, 68, 100–103, 108, 155, 223n35; practice exams, 155–56; RMS textbooks, 101–2, 107
publishing, French: approach, 10, 11, 73–84, 87, 88, 89–91, 177–80; copying texts, 226n73; cosmography in, 53, 84; Denys texts, 63, 69, 70, 71, 72, 74–84, *75*, *77*; early, 73–74; instruments, 65, *66*; late 18th c. changes, 177–80; translations, 69; weather, 209n96
publishing, Spanish: approach, 10, 11, 18–22; audience, 22; illustrations, 18, 19–21, 57, 58, 59; late 18th c. changes, 177; translations, 49, 51–53, 73, 100

quadrants: Davis, 95, 97, *98*, 106, *172*, *173*; Dutch approach, 45; English approach, 95, 97, 106, 111, *172*, *173*, 227n80; French approach, 65, *66*, 72, 226n65, 227n82; Gunter's, 65; sinical, *66*, 72, 210n8, 226n65, 227n82; universal, 146
Québec, 210n2, 227n83, 228n86

Ramsden, Jesse, 242n56
ratio studiorum, 218n91, 228n91
Reenan, Jacob van, 243n81
regiments: Pole Star, 19–20, 23, 58; Sun, 19–20, 23, 64, 100. *See also* volvelles
Reyersz., Heyndrick, 46–47, 58
rhumbs: charts, 41; on compass rose, *32*; Cor-

rection rules, *80*; Dutch approach, 51, 133; in Spanish texts, 19, 21; tables, *34, 35,* 79, 142; traverse boards, 32, *33,* 35

rhymes, 32, 35, 221n22

Richelieu, Cardinal, 212n19

Riou, Edward: on accuracy, *163,* 164, 166; death, 168; education, 156–57; *Guardian,* 149–52, 160–70; logbooks, 150, 152, *160,* 161–63, *162, 163,* 166–68, *168*; portrayals, *148,* 152, 169, pl. 9; promotions, 159–60, 168; sailing with Cook, 150, 157–58, 165, 177; workbooks and notebooks, 157, *158, 159,* 160, 177

Robbertsz., Robbert "Le Canu," 205n54, 211n11

Robertson, John, 157, 171, 245n20, 246n24

Robyn, Jacobus, 134–35

ropes: English approach, 96, 156; French approach, 175; in texts, 156, 221n18

Ross, James, 166

Royal Mathematical School (RMS): apprenticeships, 95, 237n94; badge, 101, *102*; careers of graduates, 112; curriculum and methods, 101, 102–12, 175, 178; exams, 93, 95–96, 99, 110–11; images, *92, 102*; instruments, 108, 111; teachers, 157; textbooks, 101–2, 107

Royal Naval College (Portsmouth), 153

Royal Navy: calculating longitude, 153; exams, 5, 93–99, 135, 155–56. *See also* Royal Mathematical School

Royal Observatory, 108

Royal Society, 93, 102, 104, 161, 225n58

rudder on *Guardian,* 149, 166–68

Ruelle, Pieter, 122

rule of three, 34, 35, 223n35

rule of thumb: Dutch texts, 35–37, *37,* 55, *55,* 59, 117, 135, 181, *182,* 233n35; in English exams, 97; technique, 35–37

rutters, 38, 50, 207n73

Sacrobosco, Johannes de, 18, 214n48

sailors: and alcohol, 3, 41, 71, 167, 205n60, 241n50; competition for, 6–7; demographics, 29, 68, 100, 120; math tutoring, 112–13; negative stereotypes, 3, 11, 41, 43, 46, 105, 146–47, 205n60; positive stereotypes, 22, 41, 42, 46; scholarship on, 5–6, 8; wages, 29, 229n103

Saltonstall, Charles, 100–101, 194n19, 202n18

Sampson, Murray, 166, 170, 243n68

sandglasses, 32, 139

Scandinavia, 116, 232n27

Scargill, G., *183*

Schaffer, Elizabeth, 238n2

Schaffer, Philip, 238n2

schatkamer. See workbooks

schools: centralized vs. independent, 14, 25, 45, 121; finances, 70, 72, 106, 111, 121, 122; shipboard, 88, 173, 244n5. *See also specific regions*

Scientific Revolution, 5–6, 180, 184

seamanship, 5, 96, 154–56, 175

seasickness, 85, 105, 173

sectors, *65,* 146, 202n21, 216n61

Seville, Jean de, 73

sextants, 123, 146, 152, 153, 161, 163, 176

shipwrecks, 85–86, 121, 149–52, 164–69

shortcuts, 7, 56–57, 61, 99, 141

Siberg, Johannes, 231n22

sinical quadrant, *66,* 72, 210n8, 226n65, 227n82

Skay, John, 210n10

Sleutel, Jan, *126, 127, 128*

small navigation: hands-on training, 175; math in, 176; Netherlands, 28, 30, 31–38; skills needed, 38, 42, 203n35; techniques, 31–38

Smith, John, 23, 196n40, 221n18

Snellius, Willebrord, 55, 83, 138

social class: Dutch navigators, 29, 43; Dutch teachers, 45; English schools, 99, 103, 105; and level of education, 174; of navigators, 9, 13, 156, 169, 179

sounding, depth, 35, 167, 192n2

South Africa, 149–50, 167

Spain: interest in Dutch texts, 232n27; secrecy, 196n35, 199n25, 199n29. *See also* education, Spain; publishing, Spanish

speed, estimating, 31–32, 117, 137–41

sphere, the, 18–20, 72–73, 103, 118. *See also* Apianus, Peter; Sacrobosco, Johannes de

Spink, William, 229n99, 233n39, pl. 3, pl. 5

stars: calculating latitude, 18, 40, 52, 58; calculating time, 40, 56; volvelles, *16,* 19, *34,* 51, *52,* 58, 199n27; zodiac songs, 57, 58, 117. *See also* Pole Star

Stavorinus, Jan, 175–76

Steenstra, Pybo, 143
steering machines, 166–68, 170
Stevin, Simon, 229n101
students: best, 72, 89, 90, 103, 106; duties of French, 244n2; portrayals, *92*, *122*; quality of, 70, 87, 102–3, 174, 175
Sun: calculating latitude, 18, 39–40, 111, 146; declination, 69, 233n35; ecliptic, 111, 203n37; right ascension, 56
Sunday letter, 35–37, *36*, 55
sundials, 56, 111
surveying, land, 41, 108, 111, 231n23
swimming lessons, 175
system of classes, 70

tables: accuracy of, 46; astronomical, 20, 40, 46, 49, 53, 58, 118, 133, 181; Dutch preference for, 146; latitude, 40; logarithm, 62, 64, 79; longitude, 152–53; printing improvements, 12, 47; rhumb/traverse, *34*, *35*, 79, 142; solar declination, 69; tides, 11, 35, 36, 46, 49, 53–54, *54*, 59, 143; trigonometry, 62, 79, 142, 211n14
tacit knowledge: new skills, 180–81; as scientific, 6, 11, 28; traditional skills, 24, 30, 96, 166. *See also* hands-on training
Tapp, John, 100, 127
teachers. *See* educators
Tegg, T., pl. 8
telescopes, 111, 201n8
textbooks. *See* nautical manuals
theory. *See* balance of theory and practice
Thoubeau, Claude-Joachim, S.J., 87, 88, 225n65, 226n73, 227n82, 228nn85–86
thumb, rule of. *See* rule of thumb
tides: calculating, 35–37; Dutch approach, 11, 31, 59, 117, 120, 131, 134, 137, 143; English approach, 96, 97, 155; French approach, 72; Holm's box, *141*; tables, 11, 35, 36, 46, 49, 53–54, *54*, 59, 143; volvelles, 57, 59, 117
timekeeping: Cook voyages, 241n47; in exams, 95; on *Guardian*, 161, 163–64; marine timekeepers, 40, 164, 176; Pepys's interest, 225n58; by stars, 40, 56
tobacco box, 140–42, *141*
Tourville, Anne-Hilarion de Cotentin, 88
training ships, 222n23, 245n11
translations, 24, 59–61, 69, 73, 104–5, 181

travel narratives, 46, 73
traverse boards, 32, *33*, 35, 64, 127
traverse course diagrams, 123, *126*, 127–30, *128*, *129*, pl. 5
traverse tables. *See* rhumbs, tables
trigonometry: adoption of, 64, 65–68, 89; diagrams, *67*, *76*, *77*, *78*, *79*; Dutch approach, 117, 146; in Dutch exams, 115, 133, 135–36, 137; in Dutch texts, 62, 66–68, 118, 142, 143, 211n14, 234n56; in Dutch workbooks, 124–30, *125*, *126*, *128*, *129*; English approach, 99, 104, 108, 156, 157, pl. 3; in English exams, 97, 155; in English texts, 64, 66, *67*, 68, 108, 155, 223n35; French approach, 71, 74–84; in French exams, 154; in French texts, 64, 68, 69, 71, 74–84, *75*, 113, 178; Gunter's scale, 64–65; in Spanish texts, 68, 113; tables, 62, 79, 142, 211n14. *See also* quadrants
trigonometry, spherical: calculating latitude, 161; in Dutch texts, 66, 118, 143, 233n35; English approach, 157; French approach, 72, 77
Trinity House, 25, 94–96, 97–99, 110–11, 135
Troughton, John, *163*, 242n56
'*t Vergulde Licht* (Gietermaker): approach, 117–18; compass rose, *118*; later edition, 137, *138*, 146; practice exams, 131, *132*, 133; star list, 209n97; trigonometry, 211n14; Veur on, 143; volvelles, 209n103; workbooks, 115–16, 124–27, *125*, *128*, *129*, 135–36

Veen, Adriaen, 53
vernier scales, 153, 176
Vespucci, Amerigo, 17–18
Veur, Adriaan Teunisz., 66–68, 143
Vinckboons, David, *26*, 27–28, 43–45, *44*, 122
Visscher, C., 115–16, 124–27, *125*, *132*, 135, 136
visualization, 19, 108, 140. *See also* diagrams, as learning tool
Viviers, Guillaume Clément de, 87–88, 204n50, 214n49
Vlacq, Adrian, 210n7
VOC: calculating longitude, 153; competition for personnel, 193n10; equipment, 230n9, 237n96; examiners, 25, 117; founding, 200n5; influence on classes, 117, 118, 121; logbook requirement, 154; and Peter I,

232n27; publishing, 236n88; speed of calculations, 175–76; wages, 201n12. *See also* exams, Dutch

volvelles: age of Moon, *48,* 53, *60*; as learning tool, 12, 47, 178; Pole Star, *16,* 19, *34,* 51, *52,* 58, 199n27; tides, 57, 59, 117; trigonometry, *183*; in Van den Broucke, 55, *56*; in workbooks, 181, *182, 183*

Vooght, Claas, *62, 82,* 139–40

Vries, Klaas de, 137, 234n50, pl. 6

wages: captains, 201n12; educators, 70; examiners, 23, 234n47; navigators, 3, 8–9, 30; sailors, 29, 229n103

Waghenaer, Lucas Janszoon: on accuracy, 205n59; astronomy, 50, 51; dialogues, 226n66; experts, 42–43; *schatkamer* term, 116; texts, 41–42, 46, 49–51, *52,* 56, 57, 58, 206n66, pl. 1, pl. 2, pl. 4

Warius, Peter, 133, *134,* 137

watches. *See* timekeeping

weather instruction, 58

Werner, Johann, 16, 239n22

whaling, 196n33, 211n16

WIC, 117, 121, 131, 200n5

Wijkman, Anders, 115, 133, 135, 136, 177

Willaumez, J. B. P., 6, 173–74

Wilson, Edmund, 245n20, 246n24

winds: in Spanish texts, 20; term, 202n18

Witsen, Nicholas, 209n96, 225n61, 232n26

Wood, Robert, 108, 110, 111

workbooks: balance of theory and practice, 181, 183–84; dictation, 107; English, 107, 157, *158, 159,* 177, *183*, pl. 3, pl. 5; as learning tool, 9, 12, 180–81; Riou, 157, *158, 159,* 177; as source, 9, 117; volvelles in, 181, *182, 183*. *See also* workbooks, Dutch

workbooks, Dutch: for Gietermaker text, 115–16, 124–27, *125, 128, 129,* 135–36; as historical source, 9, 117; images, *114, 119, 125, 126, 128, 129, 132, 182*, pl. 6, pl. 7; as learning tool, 9, 115–16, 119–20, 123, 177; practice exams, 123–27, *125, 132, 133,* 135–37; traverse course diagrams, 123, *126,* 127–30, *128, 129,* pl. 5; trigonometry, 124–30, *125, 126, 128, 129*

Wright, Edward, 100, 200n37, 204n43

Yvounet, Paul, 214n49

Zaragoza, José, 211n15

zodiac songs, 57, 58, 117

Const der Stierliede

1 Exempel

Laet van een Scheefhoekige triangel ABC bekent zijn de zijde BC 131°
2 m̅t. en ∠A 50 gr̅. met den hoek BAC 100 gr̅. 2 m̅t. vrage na den hoek ABC Antw.

```
  131-2   BC          100-m̅t ∠BAC
  180-0               180-0
  ─────                ─────
   48-58  Comp: BC      79-58  Comp: BAC
```

Om de hoek ABC te vinden
Sinus BC tot sinus ∠BAC alsoo Sinus AC tot Sinus

ABC 48-58   79-58
9.87756 ────────── 9.99330 ──────────────── 50
                    9.88425   9.88425
                    ───────
                    19.87755
                     9.87756
                    ───────
                     9.99999 Comt Sinus van 90 gr̅
                              de ∠ABC

---

2 Exempel

Laet van dese nevenstaende Scheefhoekige triangel ABC bekent zijn de zij.
AC 50 gr̅. BC 32 gr̅. met den hoek BAC 40 gr̅. vrage na de hoek ACB Antw. 86 gr̅ 17

om den perpendiculaer DC te vinden
Radius ∠ADC tot Sinus AC alsoo Sinus CAD tot Sinus

10.00000 ──────────── 50   40
                     9.88425 ──────────── 9.80806
                                          9.80806
                                          ───────
                                          19.69231 Comt Sinus van 29 gr̅: 30 m̅t
                                                   de zijde DC

om den hoek ECB te vinden
Sinus BC tot Radius BEC alsoo Sinus BE tot di
                 ECB
                 60-30
9.93969 ───────────── 10.00000 ──────────── 50
                      9.88425   9.8884
                      ───────
                      19.88425
                       9.93969
                       ───────
                       9.94456 Comt Sinus van
                               59 gr̅ 40 m̅t voor de hoek ECB

Om de zijde AE te vinden
Tang: BAE tot tang: EBA alsoo Radius AEB tot Sinus AE
     60-30   58
10.24735 ─────────────── 10.20421 ──────────── 10.00000
                         10.00000
                         ────────
                         20.20421
                         10.24735
                         ────────
                          9.95686 Comt Sinus van 64
                                  52 m̅t voor de AE vienb Com: ib
                                  gr̅. m̅t.
                                  25-7 voor de hoek BCD
                                  61-10 voor de hoek ECB
                                  ─────
                                  86-17 voor de hoek ACB
```
```